CAMBRIDGE LIBRARY COLLECTION

Books of enduring scholarly value

History

The books reissued in this series include accounts of historical events and movements by eye-witnesses and contemporaries, as well as landmark studies that assembled significant source materials or developed new historiographical methods. The series includes work in social, political and military history on a wide range of periods and regions, giving modern scholars ready access to influential publications of the past.

William Whewell, Master of Trinity College, Cambridge

William Whewell (1794–1866) was born the son of a Lancaster carpenter, but his precocious intellect soon delivered him into a different social sphere. Educated at a local grammar school, he won a scholarship to Cambridge, and began his career at Trinity College in 1812; he went on to be elected a fellow of Trinity in 1817 and Master in 1841. An acquaintance of William Wordsworth and a friend of Adam Sedgwick, his professional interests reflected a typically nineteenth-century fusion of religion and science, ethics and empiricism. Published in 1876, and written by the mathematician and fellow of St John's College, Isaac Todhunter (1820–84), this biography combines a narrative account of Whewell's life and achievements with extracts taken from his personal correspondence. Volume 1 covers his sermons and early poetry, as well as his work on tides, moral philosophy and mechanics, and his celebrated study of the inductive sciences.

T0188049

Cambridge University Press has long been a pioneer in the reissuing of out-of-print titles from its own backlist, producing digital reprints of books that are still sought after by scholars and students but could not be reprinted economically using traditional technology. The Cambridge Library Collection extends this activity to a wider range of books which are still of importance to researchers and professionals, either for the source material they contain, or as landmarks in the history of their academic discipline.

Drawing from the world-renowned collections in the Cambridge University Library, and guided by the advice of experts in each subject area, Cambridge University Press is using state-of-the-art scanning machines in its own Printing House to capture the content of each book selected for inclusion. The files are processed to give a consistently clear, crisp image, and the books finished to the high quality standard for which the Press is recognised around the world. The latest print-on-demand technology ensures that the books will remain available indefinitely, and that orders for single or multiple copies can quickly be supplied.

The Cambridge Library Collection will bring back to life books of enduring scholarly value (including out-of-copyright works originally issued by other publishers) across a wide range of disciplines in the humanities and social sciences and in science and technology.

William Whewell, Master of Trinity College, Cambridge

An Account of his Writings

VOLUME 1

EDITED BY ISAAC TODHUNTER

CAMBRIDGE
UNIVERSITY PRESS

CAMBRIDGE UNIVERSITY PRESS

Cambridge, New York, Melbourne, Madrid, Cape Town,
Singapore, São Paolo, Delhi, Tokyo, Mexico City

Published in the United States of America by Cambridge University Press, New York

www.cambridge.org
Information on this title: www.cambridge.org/9781108038539

© in this compilation Cambridge University Press 2011

This edition first published 1876
This digitally printed version 2011

ISBN 978-1-108-03853-9 Paperback

WILLIAM WHEWELL, D.D.

MASTER OF TRINITY COLLEGE, CAMBRIDGE.

AN ACCOUNT OF HIS WRITINGS.

WILLIAM WHEWELL, D.D.

MASTER OF TRINITY COLLEGE, CAMBRIDGE.

AN ACCOUNT OF HIS WRITINGS

WITH SELECTIONS FROM HIS LITERARY AND SCIENTIFIC CORRESPONDENCE.

By I. TODHUNTER, M.A. F.R.S.

HONORARY FELLOW OF ST JOHN'S COLLEGE.

VOL. I.

London:

MACMILLAN AND CO.

1876.

𝕮𝖆𝖒𝖇𝖗𝖎𝖉𝖌𝖊:

PRINTED BY C. J. CLAY M.A.
AT THE UNIVERSITY PRESS

PREFACE.

THE late Dr Whewell died on March 6th, 1866. He had appointed as executors of his will two Fellows of Trinity College, Mr Mathison, then Senior Tutor, and Mr J. L. Hammond, then Senior Bursar of the College; but no person was specially charged with the duty of attending to the literary and scientific correspondence and manuscripts. It was at one time supposed that Mr Mathison would have examined these papers with a view to the publication of some of them; but he left college for the living of Dickleburgh in Norfolk, in 1868, and died on October 11th, 1870. Some attempts were made to obtain an editor, who would arrange the materials and write a biography of Dr Whewell, but without success. In the course of these attempts application was made to two distinguished members of Trinity College; but want of sufficient leisure in one case, and failing health in the other, constituted insuperable obstacles.

In October, 1872, I was requested to undertake the work by Dr Whewell's nearest surviving relative, through the Bishop of Carlisle, and I accepted the honourable though difficult task. The correspondence and literary remains were to be placed at my disposal, and a short memoir of Dr Whewell's life was to be drawn up by a member of his family. It was afterwards suggested that a part of the work should be devoted to the College and University career of Dr Whewell. Two members of Trinity College, to whom this was successively entrusted, were unable to carry it on; but it has been recently arranged that this part shall find a place in the forthcoming memoir of Dr Whewell's life.

I need scarcely say that I engaged in the work without any presumptuous hope that I was competent to appreciate the wide extent of learning for which Dr Whewell was so justly famous; I simply resolved to do the best which my ability and application would allow. On some of the subjects which would necessarily come before me I had expended considerable time; with others I was moderately acquainted; and where I was almost entirely unprepared I enjoyed the great advantage which accompanies a long residence at Cambridge, namely the facility of appeal to those who possess the requisite knowledge, and are most liberal in communicating it.

It was obviously impossible for me to undertake the duty of editor without the concurrence of all who had the right and the privilege to feel interested in the great reputation of the late Master of Trinity College; and my first care was to ascertain that my engaging in it would not be unwelcome to the members of that great foundation. I received from the present Master, and from Fellows of the College, such kind assurance of interest as effectually relieved me from all anxiety with respect to the apparent intrusion of a stranger into a field which might have been supposed closed against him. Mr Hammond, the surviving executor of Dr Whewell's will, entered most cordially into the design; and during the whole time which has been occupied in the work his services have been promptly and skilfully given. The assistance of his successor in the office of Senior Bursar of Trinity College, Mr Aldis Wright, has always been immediately rendered to my application. Nor must I fail to record my obligations to Sir G. B. Airy for the benefit I have obtained from correspondence and long conversations with him, and for access to his large and well-arranged collection of papers. To Mr Bushby of St John's College, a contemporary of Dr Whewell, I am deeply indebted; the treasures of information amassed during a long and observant career, and preserved in a singularly tenacious memory, have always been accessible to me, and have immediately solved many questions which it would have required much time and labour, perhaps after all spent in vain, in order to investigate.

The materials submitted to my attention consisted of two sets, namely those which had been left by Dr Whewell himself, and those which were lent to me by his friends and correspondents. I will speak first of the former set. These constituted a mass formidable on account of its extent and the confusion into which it was thrown; and if the whole of it had come into my hands at once, I should scarcely have felt sufficient energy to undertake the labour of disentangling and studying the manuscripts and letters. But the papers reached me by instalments, and at intervals, and thus I never quite lost my resolution in dealing with them. It would be difficult to convey an idea of the hopeless disorder in which the papers were involved; for though Dr Whewell was fond of Bacon's remark, that truth emerges more easily from error than from confusion, he did not conform to it in his own practice. It is scarcely too much to say that the letters and manuscripts of a long life of incessant literary activity were thrown into one promiscuous heap.

It has been my main business to reduce this chaos into a condition of intelligible arrangement, and I venture to hope that this has been accomplished with as much success as the efforts of one person were likely to secure. The manuscripts are now carefully sorted and catalogued; so that it will be easy henceforward for any specialist, if necessary, to consult all those belonging to the matter in which he may be interested. It had been a constant habit with Dr Whewell to fasten together rudely by pins leaves of manuscripts relating to the same subject; this insecure process in the course of time had generally failed, for the pins had sadly mangled the leaves, and had allowed many to escape: a neater and safer mode of connexion has been substituted for this rough treatment. With respect to the nature of the principal manuscripts I may refer to pages 147, 277, 339, and 345 of the first volume of the present work.

A large collection of letters written to Dr Whewell exists; but there is something capricious as to the rule of preservation. It certainly was not entirely a case of the survival of the fittest; for many communications remain of almost imperceptible interest, while on the other hand it becomes evident on close examination that some of considerable importance have disappeared. But on the whole the collection is remarkable both for extent and value. The number is about 3,500; without naming writers who are still living, many of them are from such men as Brewster, De Morgan, Faraday, Forbes, Hallam, Hare, Herschel, Jones, Lubbock, Lyell, Macaulay, Malthus, Murchison, Quetelet, Sedgwick, Sheepshanks, Stephen, Thirlwall. These letters are now disposed in alphabetical order and catalogued. The *envelopes* had not been preserved by Dr Whewell, so that the destruction of them is not to be attributed to those who have had the care of his papers since his decease. Dr Whewell occasionally fastened letters into printed books with which they were in some measure connected; his large library was sold at his death, except about 2,000 volumes, which by his will, and by the gift of his sister, Trinity College was allowed to select. It is possible that among the volumes thus dispersed some letters of importance may exist, and hereafter be printed.

Among Dr Whewell's own papers were two sets of letters written by himself, namely those to Mr Jones and those to Mr Hare. The former were returned by Mr Jones as may be seen from page 370 of the second volume of this work; the latter were, I believe, returned by the widow of Mr Hare, but I have seen no record of the fact.

The second set of materials submitted to my attention consisted of letters written by Dr Whewell and lent to me by his correspondents, or by their representatives. For these application was made in some cases by myself personally, but in most cases through a letter written by Mr Hammond and printed in various newspapers. My most sincere thanks are due for the kindness with which these requests for the loan of letters were received and answered, though they must frequently have caused much trouble by requiring the examination of old collections of papers. I have returned to their owners all the letters lent to me. On the whole I have examined more than 1000 letters written by Dr Whewell, including those to Mr Jones and to Mr Hare.

When all the materials were brought into an accessible state, it remained to discover the best mode of dealing with them; and this was a matter of great difficulty. A life devoted to literature and science, and unconnected with public events, cannot be made so widely interesting as to justify considerable bulk and consequent expense in printing; thus economy of space became necessary. Of letters written *to* Dr Whewell it seemed on various grounds that only an extremely sparing use could be made. The *right* to publish such letters could be obtained only from the correspondents themselves, or their legal representatives; and it became evident on enquiry that in some cases these letters would appear, if published at all, in biographies of the writers. Moreover such letters, however interesting,

illustrate the character and attainments of the writers themselves rather than of the person to whom they may be addressed;
and Dr Whewell was to be the prominent object of the present
work. After much consideration it was finally arranged that
two volumes should be published, one containing a full account
of the literary and scientific career of Dr Whewell, with a few
extracts from his unpublished manuscripts, and the other
consisting of a selection from his own letters; and this design
has been carried out in the work now produced.

In the first volume it has been my intention to record
every edition of every book, and also to notice, at least briefly,
every article or memoir, of which Dr Whewell was the
author. I trust it will not be considered that I have gone
into too much detail; if so, I must solicit indulgence on the
ground that a taste for the history of science not unfrequently
involves some fondness for bibliographical minuteness and accuracy. I hope that in any future publication connected with
the subject, even if additions are necessary to supply accidental
omissions, there will be but few errors to correct. I have been
solicitous to guard against the obtrusion of my own opinions
and judgment, being desirous of making known Dr Whewell
himself, his pursuits and attainments, and not of obscuring
them by any interpretation of mine. Accordingly I have confined myself to a few remarks in connexion with those subjects
which the course of my studies had brought specially under
my attention; and as to these I have not hesitated to record
occasionally my dissent from the conclusions at which Dr Whewell
arrived: but I have in general merely indicated this without
attempting to enforce my own opinion. The range of subjects
which had to be included is so wide, that I fear I shall seem
in many cases to touch too lightly, and dismiss too hastily, what
required more elaborate discussion, if I had been capable of

supplying it. In fact the variety of studies to which Dr Whewell turned his inexhaustible energy forms the obstacle which prevents the general reader from retaining his interest throughout volumes like the present.

It will be observed that chronological order has been followed, with certain slight deviations, which are however rather nominal than real. I have usually, when giving an account of a work, anticipated so far as to include the subsequent editions in the Chapter which discusses the first; I have grouped together all the publications which relate to a single well-defined subject, as Mechanics, the Tides, Moral Philosophy, and Poetry; and I have collected in one Chapter notices of all the scientific memoirs except those on the Tides. Thus if a reader wishes to determine the exact employment of any specified year of Dr Whewell's life, it will be necessary to consider, besides what may be explicitly assigned to that year, reprints of works already in circulation, scientific memoirs, and publications included in the Chapters which are devoted to single subjects.

Sketches of Dr Whewell's life have already appeared in various publications, which were useful to me in the early stages of my engagement on the present work: these I will now mention. An article in *Macmillan's Magazine* for April, 1866, by W. G. Clark, Public Orator in the University of Cambridge. An account in the *Proceedings of the Royal Society*, Vol. XVI. by Sir J. Herschel. An account in the *Monthly Notices of the Royal Astronomical Society*, Vol. XXVII. The article in Allibone's *Dictionary of British and American Authors* is valuable, as it records the periodicals in which reviews are given of various books written by Dr Whewell: I have however in adverting to reviews confined myself to those with which I became acquainted by independent study, and by the aid of a large collection preserved among Dr Whewell's papers. A correspondent, who was well acquainted with Dr Whewell, informs me that a very good account of him appeared soon after his death in a German periodical entitled *Unsere Zeit.* Sir H. Holland in his *Recollections of Past Life,* and Mr De Morgan in his *Budget of Paradoxes* give interesting notices of Dr Whewell. I did not see, until after my first volume was printed, a brief memoir drawn up by Sir D. Brewster, as President of the Royal Society of Edinburgh, and published in the *Proceedings* of that body, Vol. VI. It is pleasing to observe that the aged philosopher speaks in gentle and respectful language over the tomb of one with whom he had been so often in controversy. Twelve years had elapsed since the last contest, which related to the *Plurality of Worlds,* and Sir D. Brewster devotes a third of his space to this book. He again records his protest against it; but he allows that, notwithstanding all the paradoxes of Dr Whewell's *Essay,* posterity " will forgive its author on account of the noble sentiments, the lofty aspirations, and the suggestions almost divine, which mark his closing chapter on the future of the universe."

Two small errors as to time run through most of the accounts which I have seen of Dr Whewell's life: it is generally stated that he was ordained *soon* after taking his M.A. degree, whereas some years intervened between the two dates; and that a few *months,* instead of only a few *days,* intervened between his marriage and his appointment to the Mastership of his College.

In the strictly chronological part of my first volume, which reaches to the end of the thirteenth Chapter, I have in general abstained from the use of the words *Professor* or *Doctor,* except when they were strictly applicable; but in the remaining Chapters I have found it convenient to employ throughout the term

Dr Whewell, which is usually correct, and is the most familiar to us at the present time. The matter is not of great importance, but writers of biographical works seem frequently unduly lax. In a volume recently published I find under the date 1827 a conversation recorded which begins thus: "Lord Brougham, you must be aware."

I pass to my second volume; this consists of letters written *by* Dr Whewell. From the large collection entrusted to me I have selected those which throw most light on the character and pursuits of the writer; always however with the limitation that the names of living persons should be rarely introduced. There are, I think, grave objections to the practice, common in recent times, of publishing indiscriminately remarks written in the confidence of familiar friendship, which were never intended to travel beyond the person immediately addressed. The letters which constitute my second volume belong rather to the earlier than to the later period of Dr Whewell's life; and this would naturally be the case from the circumstance that the more modern correspondence contains frequent allusions to persons still living, or recently deceased, and so cannot with propriety be introduced. There is however another point to be noticed. In Dr Whewell's earlier days the expense of postage prevented all correspondence on mere trifles, and consequently letter writing was a more serious and responsible occupation than it is at present; moreover the opportunities of personal intercourse were rarer than they are now. Hence, to use one of Dr Whewell's own words, correspondents strove to make their letters *postworthy*. But in our time the facility of communication diminishes the importance of it, so that brief and hasty notes are found sufficient for the purposes of society; and thus letter writing is likely to become a lost art, or to survive only in the case of families in which some members reside in

India or the colonies, where the interchange of opinion and information may still retain the gravity and importance of older days. Dr Whewell's letters in his younger years are remarkable for their neat and accurate execution; whole pages occur written in the most regular manner without any erasure or correction, and composed with an obvious regard to literary effect. The notes of his later years, though still free from marks of correction, are short and unstudied.

The letters are arranged chronologically, and the whole series contains a good account of the many subjects which at various times gained the attention of the busy writer: indeed, they may be said to amount almost to an autobiography. Dr Whewell often made a mistake as to the year in dating a letter, retaining the *old* year after it had passed away; the wrong date is naturally most frequent at the commencement of a *new* year, but it is occasionally to be seen when the year was some weeks or even months old. These mistakes have been silently corrected; it is possible, though not very likely, that some have escaped detection.

After the Preface will be found brief notices of the persons to whom the letters are addressed.

The names of the persons to whom the letters are addressed are placed at the top of the pages; I have given to the name the *last* form which it assumed, which is generally that which is now most familiar to us, though very often it is not chronologically exact.

Some passages have been omitted from the letters; these in general relate to matters of no public interest. Any remarks made by myself in the text of the second volume are included between square brackets. A few notes have also been added. It would have been easy to extend these greatly, but economy of space was an important consideration. Often a reference to the part of the first volume which corresponds in date will sufficiently explain the object of some brief allusion; take for example that on page 132 to a paper of an amusing kind. The date shews that this must be one of three, namely, those on *English Adjectives*, on *Clipt Words*, and on the *Use of Definitions*: see Vol. I. pages 61, 62, 65.

The letters contain the names of many persons: these may be considered to form three classes.

In the first class we may put the names of persons eminent in literature or science, which must be familiar to all readers of this work; for example, Arago, Brewster, Coleridge, Laplace, Mackintosh, Malthus, Mill, Ricardo, Thirlwall, Wordsworth, Young.

In the second class we may put the names of persons of University distinction or position; which, though well known at Cambridge, are less familiar to the general reader. If the date at which a person took his B.A. degree at the University is approximately known, it is usually easy by the aid of the *Cambridge Calendar* to ascertain his subsequent career. For example, take the names mentioned on page 34 of the candidates for fellowships at Trinity College. The first four names occur in the mathematical tripos list for 1818, with notes of information against the first three; the last name occurs in the mathematical tripos list for 1817. If the date at which the degree of B.A. was taken is not known, it may be found by consulting the *Graduati Cantabrigienses*, and then as before the *Cambridge Calendar* will supply information.

The third class of names consists of such as for various reasons there may be some difficulty in identifying; the context will however often sufficiently explain the reference. On pages xxvii and xxviii will be found a list of some of these names with brief indications respecting them. Some of the

allusions however are now not intelligible; nor can it be surprising that in a series of letters extending over the duration
of a long life names and circumstances should be mentioned
which seem at the present time quite forgotten.

Of the letters written *to* Dr Whewell I have made no explicit use, except to the extent of a few extracts, almost all in
my first volume. It will be easy at any time to publish a
selection from these and from the remainder of those written
by Dr Whewell; but every year of delay will naturally augment
the number available for use. In saying this I must guard
myself from conveying the impression that the correspondence
contains what it is necessary or desirable to conceal; but still
for obvious reasons delay is advisable. For example, opinions
as to the relative claims of various candidates for an office may
well remain unpublished while some of these candidates are
still among us.

The volumes now submitted to the public will, I fear, appear
to many to be but an inadequate treatment of the subject
committed to my care, and a meagre result of the continued
labour of nearly three years. There were however many difficulties in the way. I have already hinted at the dissipation
of energy caused by the trouble of arranging the entangled mass
of material. Moreover the limits which were prescribed to my
task have in some measure rendered it more onerous. For by
the exclusion of all that belongs to the personal and domestic
life the interest of the work is seriously impaired for the general
reader; and the omission of the College and University career
deprives it of matter which might have attracted attention at
Cambridge. My own experience convinces me that the separation of a biographical work into distinct portions under different

editors, notwithstanding any apparent advantages, is really a mistake; and I find my conclusion supported by some who have taken a share in similar engagements, and by others who, wise in time, declined to accept anything short of the whole responsibility. It is the fault of the plan of the work, though it seemed at the time the best which could be devised, that my subject may appear rather as a scientific abstraction than as a living person invested with the circumstances which engage the sympathy of readers.

I do not think that adequate justice can be rendered to Dr Whewell's vast knowledge and power by any person who did not know him intimately, except by the examination of his extensive correspondence; such an examination cannot fail to raise the opinion formed of him by the study of his published works, however high that opinion may be. The evidence of his attainments and abilities which is furnished by the fact that he was consulted and honoured by the acknowledged chiefs of many distinct sciences is most ample and impressive. United with this intellectual eminence we find an attractive simplicity and generosity of nature, an entire absence of self-seeking and self-assertion, and a warm concern in the fortunes of his friends, even when they might be considered in some degree as his rivals.

I must record my gratitude for the numerous kind expressions of interest in my work which have reached me since I undertook it. Some who had thus encouraged me have passed away during the last three years, as M. Guizot, M. Quetelet, Sir C. Lyell, Professor Selwyn, and Professor Phillips. Sir C. Wheatstone also, who told me the names of the persons present on an occasion mentioned in Vol. I. page 410, no longer survives.

W. *b*

In conclusion I have only to hope that the volumes will not be quite unsatisfactory to those who sympathize with the subject to which they relate—to Dr Whewell's relatives and personal friends, now a rapidly decreasing number—to the members of that great College with which for more than fifty years he was intimately connected—and to the University which must ever regard him as a conspicuous ornament and a liberal benefactor.

I. TODHUNTER.

March 6, 1876.

ALPHABETICAL LIST OF THE PERSONS TO WHOM THE LETTERS IN THE SECOND VOLUME ARE ADDRESSED.

In the following list references are occasionally given to the *Proceedings of the Royal Society*, and to the *Monthly Notices of the Royal Astronomical Society*, for accounts of the lives of the persons named; the former work is denoted, for brevity, by *Proceedings*, and the latter by *Monthly Notices*. These accounts are in general drawn up with great care by competent writers, and are the sources from which the memoirs in ordinary biographical dictionaries are derived.

Sir George Biddell Airy, of Trinity College, the present Astronomer Royal: see Vol. I. page 28.

Dr D. Brown: see Vol. II. page 433.

Baron Bunsen was born in 1791, and died in 1860; he was eminent as a scholar, a theologian, and a diplomatist. A biography of him has been published by his widow.

John Inglis Cochrane: see Vol. I. page 294.

Augustus De Morgan, of Trinity College, was born in 1806, took his B.A. degree as fourth wrangler in 1827, and died in 1871. He was for many years Professor of Mathematics in University College, London. See Vol. I. page 60, and *Monthly Notices*, Vol. 32.

Michael Faraday was born in 1791, and died in 1867. See Vol. I. page 89, and *Proceedings*, Vol. 17. Separate biographies have also been published of him.

James David Forbes was born in 1809, and died in 1868. He was for many years Professor of Natural Philosophy at Edinburgh, and afterwards Principal of St Andrew's. See Vol. I. page 47, and *Proceedings*, Vol. 19. A volume has been published entitled *Life and Letters of James David Forbes*, 1873.

Richard Gwatkin, of St John's College, took his B.A. degree as senior wrangler in 1814. He held the college living of Barrow-on-Soar from 1832 to 1854, and died in 1870.

Sir William Rowan Hamilton, of Dublin, was born in 1805, and died in 1865: see the *Monthly Notices*, Vol. 26. There is reason to hope that a memoir of this remarkable man will eventually be published by his representatives.

William Venables Vernon Harcourt, of Christ Church, Oxford, was born in 1789, and died in 1871. See Vol. I. page 47, and *Proceedings*, Vol. 20.

Julius Charles Hare, of Trinity College, was born in 1795, took his B.A. degree in 1816, and died in 1855. A memoir of him is prefixed to the edition of the *Guesses at Truth*, published in 1866; and interesting notices of him occur in the book entitled *Memorials of a Quiet Life*. More than a hundred letters written to him by Dr Whewell are preserved.

Sir John Frederick William Herschel, of St John's College, was born in 1792, took his degree of B.A. as senior wrangler in 1813, and died in 1871. See *Proceedings*, Vol. 20, and *Monthly Notices*, Vol. 32.

Sir Henry Holland, an eminent physician, was born in 1788, and died in 1873. His *Recollections of Past Life* appeared in 1872.

Richard Jones, of Caius College, was born in 1790, took his B.A. degree in 1816, and died in 1855; a short account of him was drawn up by Dr Whewell: see Vol. I. page 227. Mr Jones was a man of eminent practical sagacity, and had probably more influence than any other person over Dr Whewell. The volume on *The Distribution of Wealth*, published by Mr Jones, and the Literary Remains which appeared after his death, seem but inadequate evidence of the great speculative ability which he must have possessed according to the belief of Dr Whewell and Sir John Herschel, and of keen judges still living. But public business and the fascinations of society absorbed his time; and thus he never effected what his friends had anticipated, and what he might have accomplished by a greater concentration of his powers, and by a more ascetic discipline. About two hundred and seventy letters written to him by Dr Whewell are preserved.

Sir George Cornewall Lewis, eminent as a statesman and scholar, was born in 1806 and died in 1863. Three short letters written to him by Dr Whewell are printed on pages 405 and 424.

Sir John William Lubbock, of Trinity College, was born in 1803, took his B.A. degree in 1825, and died in 1865. See *Proceedings*, Vol. 15, and *Monthly Notices*, Vol. 26. He was a pupil of Dr Whewell in 1823, when the latter gave up private teaching on being appointed official tutor to the college.

Sir Charles Lyell, the eminent geologist, of Exeter College, Oxford, was born in 1797 and died in 1875.

James Henry Monk, of Trinity College, took his B.A. degree as seventh Wrangler and second Medallist in 1804, and died in 1856. He became successively Professor of Greek, Dean of Peterborough, and Bishop of Gloucester. In reply to the congratulations of Bishop Monk on the appointment to the Mastership of Trinity, Dr Whewell wrote, "I look back with great pleasure to the time when you selected me as your fellow-labourer in the tuition, and consider that as the turning-point of my life, which decided my course to be what it has since been." See Vol. I. page 11.

George Morland was a master in the Lancaster Grammar School during the early part of Dr Whewell's course at Cambridge. Mr Morland afterwards took orders and became a member of St Peter's College as a ten-year man.

Sir Roderick Impey Murchison was born in 1792, and died in 1871. See *Proceedings,* Vol. 20, and *Monthly Notices,* Vol. 32. A separate memoir has been published by Professor Geikie. Some of Dr Whewell's letters are addressed to Lady Murchison.

The Marquis of Northampton was born in 1790 and died in 1850. See Vol. I. page 182.

John Phillips was born in 1800 and died in 1874. He was in early life connected with the Museum at York, and was finally Professor of Geology, at Oxford.

George Peacock, of Trinity College, was born in 1791, took his B.A. degree as second Wrangler in 1813, was appointed Dean of Ely in 1839, and died in 1858. See *Proceedings,* Vol. 9, and *Monthly Notices,* Vol. 19; the former account is probably by Sir John Herschel, and the latter by Mr De Morgan.

Lambert Adolphe Jacques Quetelet was born in 1796 and died in 1874. He was director of the Observatory of Brussels, and perpetual Secretary of the Academy. See *Proceedings,* Vol. 23, and *Monthly Notices,* Vol. 35. The extracts from Dr Whewell's letters were copied under the direction of M. Quetelet, and nearly all that he sent has been printed, though some of it is of small importance.

Thomas Rickman, well known as a writer on Gothic Architecture, was born in 1776 and died in 1835.

Hugh James Rose, of Trinity College, was born in 1795, took his B.A. degree as fourteenth Wrangler and Senior Medallist in 1817, and died in 1838. He was distinguished as a scholar, a theologian, and a preacher. He suffered much from

ill health, and to this cause was mainly due his frequent change
of residence and occupation. He finally became Principal of
King's College, London. He published in 1834 a lecture entitled
An Apology for the Study of Divinity; in this he rather depreciates some of the human sciences. The following sentence
occurs : "The most, as far as I know, which has ever been
said for these sciences, as they can affect the human mind, has
been said by one whom I can never name without the strongest
emotions of respect and regard. Mr Whewell has declared his
conviction, in one of his minor works, that habits of inductive
reasoning are best learned from wide acquaintance with natural
philosophy." In a note which Mr Rose shortly before his death
wrote to Dr Whewell, he said, "My many months of miserable
weakness have prevented me even from thanking you for all
your valuable books—thanks richly due from me, though of
little worth to you. But I cannot persuade myself (sentenced
as I am to leave home for a milder climate), to go without a
Hail and Farewell, without offering you the warmest best wishes
of an old friend for your welfare and happiness, for the increase
(if that may be) of your usefulness, and with it of your already
great name." Mr Rose had a brother, younger than himself,
named Henry John, who was a fellow of St John's College, and
in 1837 took the college living of Houghton Conquest. He died
in 1873.

Adam Sedgwick, of Trinity College, was born in 1785, and
took his B.A. degree as fifth Wrangler in 1805. In 1818 he was
appointed Professor of Geology in the University of Cambridge,
and held that office until his death in 1873. See *Monthly Notices*,
Vol. 33.

Sir James Stephen, of Trinity Hall, took his LL.B. degree
in 1812. In 1849 he was appointed Professor of History in the
University of Cambridge, and died in 1859. A memoir of him is

prefixed to the edition of his *Essays on Ecclesiastical Biography* which was published in 1860.

Henry Wilkinson of St John's College, took his B.A. degree as second Wrangler in 1814. He left college in 1820 to become Master of Sedbergh School, and died in 1838. Mr Wilkinson was a man of considerable attainments, and long after his death Dr Whewell spoke of him with strong expressions of esteem and regard.

The following is the list of Scientific Societies with which Dr Whewell was connected, and the date of his election. H. M. denotes Honorary Member.

1819 Cambridge Philosophical Society.
1820 Royal Society.
1821 Royal Astronomical Society.
1827 Geological Society.
1829 Gesellschaft für Naturwissenschaft und Heilkunde Heidelberg.
1831 Royal Geological Society of Cornwall. H. M.
1832 Yorkshire Philosophical Society. H. M.
1832 Naval and Military Library and Institution. H. M.
1832 Society of Antiquaries.
1835 Shropshire and North Wales Natural History Society. H. M.
1835 Bristol Institute for the Advancement of Science. H. M.
1836 Institute of British Architects. H. M.
1837 Institution of Civil Engineers. H. M.
1837 Royal Geographical Society.
1839 Société Française de Statistique Universelle. H. M.
1840 Boston Society of Natural History. H. M.
1840 Institut d'Afrique. Vice President.
1843 Literary and Philosophical Society of Manchester. H. M.
1845 Royal Society of Edinburgh. H. M.
1849 Académie Royale de Belgique. Associate.
1851 Societas Naturæ Scrutatorum Helvetorum.
1854 Historic Society of Lancashire and Cheshire. H. M.
1857 Kaiserlich-königliche Geographische Gesellschaft, Vienna. H. M.
1857 Académie des Sciences Morales et Politiques. Correspondent.

He was also elected an Honorary Member of the Royal Irish Academy not later than 1837.

ADDITIONS.

I have omitted to mention in its proper place a College paper belonging to the year 1854. Lord Palmerston, on behalf of the Government, requested information from the Colleges of Cambridge, with a view to some proposed reforms. A reply was returned from Trinity College, in the form of a letter addressed to the Vice-Chancellor of the University, dated January 10th, 1854, and signed by Dr Whewell. Copies were printed on five quarto pages, and the letter is also published in the Blue Book, entitled *Correspondence respecting the Proposed Measures of Improvement in the Universities and Colleges of Oxford and Cambridge*. The letter is a statement and defence of the existing college practice. Perhaps the letter was drawn up rather by the governing body of Trinity College than by Dr Whewell himself; for it is known that when his attention was first directed to Lord Palmerston's application, he was favourable to some reform, especially with respect to sinecure and non-resident fellowships.

A volume, entitled *Fugitive Poems connected with Natural History and Physical Science*, collected by the late Professor Daubeny, was published in 1869. It contains two pieces by Dr Whewell, both of which however had previously appeared in the *Sunday Thoughts and other Verses*: these pieces are the *Seal and the Sea Mew*, and the whole of the verses relating to the excavations at Bartlow. See Vol. I. page 169.

NAMES of some of the persons mentioned in the letters ; see Preface, page xv. The figures give the pages of the second volume where the names occur ; if anything is known respecting the persons it is added.

There are in the letters allusions to some circumstances now unknown; for example, *Kenty's pamphlet* on page 28, the *Romeo and Juliet* friend on page 44, the *war cry* on page 106, and the *answer* by Chevallier on page 239.

CONTENTS OF THE FIRST VOLUME.

CHAPTER XVII.

CHAPTER XVIII.

CHAPTER XIX.

CHAPTER XX.

CHAPTER XXI.

The second Volume consists of letters written by Dr Whewell, arranged in chronological order.

CHAPTER I.

1794...1819.

IT has been said of Dr Whewell that, "a more wonderful variety and amount of knowledge in almost every department of human inquiry was perhaps never in the same interval of time accumulated by any man." Such a statement, made by a philosopher so eminent as Sir John Herschel, will naturally excite an interest in the career of the subject of his commendation. That career, as manifested in printed works and in correspondence, I now propose to trace; and I trust that no imperfection in the mode of treatment will prevent the reader from recognising the lineaments of a great and good man.

A remark respecting the name may deserve some attention on account of the person to whom it is due. The following sentence occurs in a letter addressed to Professor Whewell on Jan. 17, 1839, by the late James Bailey, the well-known editor of the English edition of Forcellini's Latin Lexicon: "I have often thought of communicating to you the origin of your name, which is a very rare one, and mispronounced by all South-country men. It is a corruption of Wheelfell, a place between the rivers North Tyne and Read." Dr Whewell himself believed his name to be identical with "Wyvill."

William Whewell was born at Lancaster on May 24, 1794. He was sent first to the grammar school of his native town, but afterwards to that of Heversham in order to be qualified for holding an exhibition at Trinity College, Cambridge, connected with

w

it. In a speech at the opening of a new school-house at Lancaster in 1852, Dr Whewell referred with approbation to the training which he had received in Arithmetic, Practical Geometry, and Mensuration; and he acknowledged in appropriate terms the kindness of the master, Mr Rowley. He spoke of the new interest, and the new dignity, which had been given in recent times to the schoolmaster's position, and of the improvement in the conduct of the scholars: it would seem however from the tone of his speech, and from some expressions in his correspondence, that his recollections of the Lancaster school of his youth were not altogether favourable. On Sept. 28, 1854, Dr Whewell preached at Heversham, and his sermon contains allusions to his former familiarity with the place.—"...I say that many such persons do constantly, in their thoughtful moods, turn away their minds from the scenes of movement and struggle, of pomp and splendour it may be, of excitement and expectation—turn them away to some early remembered spot, in which they can imagine such peace and quiet to reside, as they cannot conceive are to be found in the regions to which their subsequent career has introduced them. The mild sunshine and the dew of the morning seem to them still to linger in the places to which their memory thus turns; while all the way that they have since travelled is scorched by the glare of too bright a day, and filled with the stifling dust of crowded paths." "Some of us who are now here assembled can remember the aspect of these walls, and the sound of these services, for more than one generation of men; some, it may be, for more than two generations, as generations are commonly reckoned. To the eyes of such, how entirely is the aspect of the congregation changed! How few are there of the faces, which were once so well known to us, if in no other way, at least by their regular appearance in their accustomed place on the stated occasions of worship! The countenances of those to whom we were wont to look with respect and affection as our teachers, our guides and directors, our friends and advisers, are here no longer. The silver hairs of those times no longer grace their well-known seat of worship: and heads which then wore the unchanged hue of the spring of life are now hoary with the advance of winter. Eyes then mild with the wise kindliness which long years had brought, are now closed for ever: and cheeks which even in age wore the healthy hue which these rural scenes foster, have ere this been blanched by the last paleness."

A few of Dr Whewell's early exercise books have been preserved, and they shew that at school he paid great attention to classical studies, including versification in both Greek and Latin. He entered Trinity College in October, 1812. Some notion of his college life may be gathered from his letters, especially those addressed to Mr Morland, a master in the Grammar School at Lancaster; it is obvious that he entered heartily into the studies and the rational amusements of the place, manifesting the vigorous earnestness which belonged to all he undertook throughout his life. He was fortunate in the society he cultivated, especially in his intercourse with two eminent persons of somewhat maturer age; namely, Herschel, his academic senior by three years, and Richard Jones, who, though his contemporary in the University, was about four years older than himself. He was also intimate with Mr Gwatkin and Mr Wilkinson, both of St John's College, who were two years senior to himself. His familiar associates of his own college seem to have been men of his own year, or junior to himself; among the former we find Julius Charles Hare, Thomas Paynter, and Richard Sheepshanks; and, among the latter, Hugh James Rose. It was, probably, through the medium of Mr C. Bromhead of his own year and college, that Mr Whewell became acquainted with an elder brother, Sir E. Bromhead of Caius College, who took his B.A. degree in 1812. Some years later Sir E. Bromhead introduced to Mr Whewell, and through him to the Cambridge Philosophical Society, the famous mathematical memoirs of George Green.

Some of Mr Whewell's friends, however, were of a less studious turn than those who have been mentioned, and tradition still retains the name of one in whose company not a little of his time was supposed to have been wasted.

In November, 1813, Mr Whewell delivered a Latin declamation in the college chapel, according to a custom which lingered until recent years; his subject was Cæsar and Brutus.

In 1814, Mr Whewell gained the Chancellor's Medal for the

best English prize poem on the subject of Boadicea. The poem was printed in the *Classical Journal*, number xix., and has been reproduced in various collections of University Prize Poems; it consists of 340 lines. Sir John Herschel, himself a cultivator of versification, speaks of this as "a spirited production, which may be read with pleasure as something beyond a college exercise, and evidencing that strong vein of poetical talent which showed itself on many subsequent occasions."

In the poem, speaking of some wonders which were reported to have occurred, Mr Whewell says:

> Yes, they have mark'd; and speak in portents dread
> The wrath that trembles o'er th' oppressor's head.
> Push'd from its base his idol Victory falls,
> Unbodied furies howl along the walls,
> ....

I quote these lines for the sake of the word *unbodied*. In the well-known poem by Shelley, entitled *To a Skylark*, which is dated 1820, we have the line

> Like an unbodied joy whose race is just begun.

The word *unbodied* here always puzzled me, and I remember some years since seeing a paper in which a critic pointed out many apparent inaccuracies or misprints in Shelley's poems, and conjectured that here for *unbodied* we ought to read *embodied*, which is used more than once by the poet elsewhere. I do not know whether the occurrence of the word in *Boadicea* lends any support to the received text of Shelley's line.

A candidate for a poetical prize in the University would naturally avail himself of an opportunity to celebrate the praises of beauty.

> O Beauty! heaven-born Queen! thy snowy hands
> Hold the round earth in viewless magic bands;
> From burning climes where riper graces flame
> To shores where cliffs of ice resound thy name,
> From savage times ere social life began
> To fairer days of polish'd, soften'd man,
> To thee, from age to age, from pole to pole,
> All pay the unclaim'd homage of the soul.

But *snowy* seems scarcely a satisfactory epithet when we consider the *burning climes;* we must at least restrict it to the *candidior nive,* and omit the *frigidiorque manus* of a famous line.

A letter dated Lowick-bridge, July 21, 1814, from Joshua King, who was senior wrangler in 1819, to Mr Whewell, conveys the writer's warmest thanks for the valuable and much esteemed present of the prize poem,—the perusal of which he does not doubt will give him infinite pleasure.

I have already alluded to the instruction in the elements of mathematics which Mr Whewell received at the Lancaster Grammar School; he also had the benefit of some training from Mr John Gough, the famous blind mathematician of Kendal: see the *Biographical Notices of some Liverpool Mathematicians,* by Mr T. T. Wilkinson, in Vol. XIV. of the *Transactions of the Historic Society of Lancashire and Cheshire.* A letter is preserved addressed to Mr William Whewell, Friarage, Lancaster, dated Kendal, Sep. 26, 1814, and signed J. Gough; it relates to the collision of hard bodies, but is not in accordance with the theory on that subject adopted by writers on mechanics.

At the beginning of 1815, Mr Whewell was looking forward to that part of his academical exercises which was called *Keeping an Act.* He says in a letter to his friend Mr Morland on January 3, 1815, "I have not time to explain the whole ceremony to you at present, but it consists in a person getting up into a box to defend certain mathematical and moral questions, from the bad arguments and worse Latin of three men who are turned loose into an opposite box to bait him with syllogisms. In the mean time I am seized with an inconceivable desire to read all manner of books at once, and have at this present writing no less than two folios and six quartos of different works upon my table, which would prove to any one who is in the habit of reading that I am very idle."

In the early part of 1815 a fever broke out at Cambridge, and Mr Whewell left the place for a short time, and visited London in company with Mr Gwatkin.

An interesting fact with respect to Mr Whewell's associates at this time is recorded in a letter to him from T. Forster, dated Tunbridge Wells, Dec. 24, 1841. "We have all made some advances in mere *physical* science, but in *metaphysics*, as far at least as I am concerned, I am not conscious of having advanced one single step, since the period when you and I and Herschel and Babbage used to meet at our Sunday morning's philosophical breakfasts in 1815." This Mr Forster seems not to have taken any degree at the University; his name occurs as the author of 35 papers in the Royal Society's Catalogue of Scientific Papers. He was about seven years older than Mr Whewell, who was thus the youngest of the philosophical party.

Mr Whewell spent the Long Vacation of 1815 in Cambridge, preparing himself for the approaching mathematical examination; his friend Mr Gwatkin urged upon him the necessity of vigorous study, but it appears from Mr Whewell's reply that he by no means neglected amusement and relaxation.

On taking his B.A. degree in 1816, Mr Whewell was second wrangler; it had been anticipated that he would have been senior, but that distinction was gained by Edward Jacob of Caius College, then under twenty years of age. At the end of a week's examination the first and second wranglers stood out clearly from all their competitors; the next eight wranglers were provisionally bracketted, as being nearly equal in merit, and according to the custom then prevalent they had to undergo a further scrutiny in order to settle their relative places. The order of the names was preserved in the award of the Smith's prizes, Mr Jacob obtaining the first, and Mr Whewell the second.

Edward Jacob possessed great ability, which had been carefully directed by his college so as to bear with concentrated force on the mathematical competition. Mr Whewell's own taste would naturally lead him to the more varied study which has always been encouraged at Trinity College; that great foundation can, without danger, pay less exclusive regard to mere academic triumphs than its smaller rivals. Mr Jacob went to the bar and justified the reputation he had gained in the University; but he died at an early age, as years are reckoned in that laborious profession, and thus he did not reach the conspicuous eminence which had been anticipated for him; so that his fame mainly rests on the fact that he outstripped so formidable a competitor in the Cambridge race.

The father of Edward Jacob was William Jacob, known for his writings connected with Political Economy, and especially for his work on the *Production and Consumption of the Precious Metals.* Mr Jones, the friend of Mr Whewell, was intimate with the Jacob family—attracted to them probably by his devotion to the same studies as the father, and by his connexion with the same college as the son—and the name of Jacob is occasionally mentioned in the letters of the two correspondents, but the allusions are generally to the father and not to the son.

The list of graduates for the year 1816, though exhibiting a good supply of names afterwards distinguished, has only one besides that of Mr Whewell which became known in science, namely that of his friend and fellow-collegian Mr Sheepshanks.

Mr Whewell, in accordance with a common practice, engaged in private tuition after taking his B.A. degree, and continued in this occupation for about seven years. Among his pupils were Mr Thorp, now Archdeacon of Bristol, Mr Mansel, probably son of the former Master of Trinity College, Mr Dodsworth, Mr Heneage, Mr Kenelm Digby, the late Sir J. W. Lubbock, and the late Mr Horace Waddington, afterwards Under Secretary for the Home Department. The tutor does not seem, however, to have taken very kindly to his employment, since it absorbed the time which he was eager to spend in incessant study.

In the summer of 1816 Mr Whewell passed some weeks at Burlington with a party of pupils. In a letter to his friend Mr Wilkinson he records a fall from his horse, which made him blind and deaf for five minutes, but left no permanent injury.

In the latter part of the year we find him complaining of one of the inevitable vexations of prolonged residence in college—the departure of friends from Cambridge. He says in a letter to Mr Morland, dated Nov. 10, 1816, with reference to Mr Herschel and Mr Jones : "Two of my most intimate acquaintances, and I will add two men of the greatest intellectual powers and attainments that I ever saw or ever expect to see, have left the university ; and their departure has made an irrecoverable gap in my enjoyments."

In March, 1817, Mr Whewell says in a letter to Mr Herschel

that he has six pupils, and that he is or ought to be reading for a fellowship. Towards the end of the month he was a prominent actor in a scene which he himself describes in a letter to his friend Mr Rose, and which is still well remembered by one who was present. The *Union* is the name given to a society at Cambridge, composed principally of the younger residents in the University, which furnishes the advantages of a library and reading rooms, and encourages debates on set subjects among its members. It now possesses an ample and well-arranged building, but in those early days of its history it was in a less flourishing condition, and hired a room at the Red Lion Inn. Mr Whewell was President in the month of March 1817, when Dr Wood, at that time Vice-Chancellor, took with him the Proctors, Mr Okes of Caius College, and Mr French then of Pembroke College and afterwards Master of Jesus College, together with a tutor from Trinity and from St John's, and proceeded to the place of meeting. The Proctors were sent into the room to desire the members to disperse, and to meet no more. The President, mindful of the dignity of his position, requested the messengers to withdraw that the Society might consult on the matter. This could not be granted, but a deputation, consisting of the President, Mr Thirlwall, and Mr Sheridan, was permitted to have an interview with the Vice-Chancellor. The deputation was urgent, but the Vice-Chancellor was obdurate; and all that was conceded was permission to finish the current debate, and to retain the room merely for the purpose of reading.

Succeeding Vice-Chancellors, however, did not exhibit much interest in the matter, and so the debates were after no long interruption renewed, but in order to evade the formal prohibition some obvious artifices were employed; a member for instance would move that a certain newspaper should be discontinued, and then criticise at length the political principles which that newspaper advocated.

It is natural to conjecture that Mr Whewell hesitated for some time as to the choice of his future profession; he does not seem to have been fond of lecturing or of private tuition, and he did not take orders until about 1825. On this subject we may give an extract of a letter to him; the writer was Richard Whit-

combe, who by the evidence of his letters, and by the testimony
of those who still remember him, was a man of considerable
ability; he was an intimate friend of Mr Hugh James Rose, went
to the bar, and probably died early. He is, I presume, the author
of a short article published in Vol. I. of the division of History
and Biography in the *Encyclopædia Metropolitana*, entitled *Me-
nander, Middle and New Comedy.* Mr Whitcombe says in a
letter dated April 29, 1817: "But how stands the case with you?
What is your beau ideal of life? To attain to eminence in, and
efficiently serviceable to your country—or to gain a name, in an
age better able to appreciate merit than this, by pushing on the
land-marks of knowledge? I think you know me too well to
accuse me of flattery—and therefore I will say that your object
ought to be of one of these kinds. If you are outrageously modest,
remember that your *model* must be of perfection. Seriously, you
cannot but feel that you have talents to succeed to any extent
you wish—flattery from me to you is out of the question—it
would be absurd, but it would also be what is far worse. Your
talking, then, of not having prospects of success in the law, I over-
rule at once. I ensure your succeeding there—but, if I must give
you a decided opinion, I do not wish to see you enter on it. I
think that your mind is calculated to do more, and to receive more
enjoyment in the active and unfettered pursuit of knowledge of its
own selection than it would in the shackles of any profession.
The knowledge of law, as a science, studied philosophically, I have
no doubt is delightful—even the detail of it I should like—and
yet—Special Pleading!—but will your thirst of knowledge be
slaked from this one stream how pure so ever? and if, in the
midst of your professional labours, you ever cease to sigh for more
frequent and extended opportunities of pursuing genuine Science,
will it not be when you have torpified the vigour of a Mind which
was not endowed with strong powers for inactivity or prostitution.
Do not dread the necessity of an eternity of Cambridge. Live in
London if you like—your fellowship will support you for the present,
without drudging your intellects at £40 per annum for stupid
pupils, and I shall be vexed if you do any such thing after October.
And, when you see how very little talent has raised very many

men do not doubt that you may raise yourself by your mind, without sticking a wig on your head. Literature, I honestly think, is your natural bent. If so, *ruat cœlum*, follow it! But you must judge for yourself."

We see the impression which Mr Whewell's ability had already made on a clever friend, and we may infer that the friend had been consulted as to the choice of a profession: moreover the advice given was substantially, if not literally, followed. A passage in a sermon which Dr Whewell preached in his college chapel on Oct. 23, 1864, seems to recall the dread of a continued residence at Cambridge which he had himself felt nearly half a century before. "But some of you may continue to be resident members of this community for many years, or it may be for the whole of your lives. Such a lot may, at the present moment, appear to you dreary and desolate. Yet it need not be so. Such a lot is not at all inconsistent with the constant culture of all kindly affections, with an enduring enjoyment of literary pleasures, and with a truly Christian spirit, enduring all things and waiting calmly for the end in faith and hope."

In the year 1817 Mr Rose proposed to translate from the French the work of Lacroix on the *Application of Algebra to Geometry*, and Mr Whewell half promised to assist him, and to furnish some notes; but the scheme was never carried out, for Mr Whewell was not very earnest, and Mr Rose himself seems to have been much more devoted to literature and theology than to mathematics.

Mr Whewell spent the Long Vacation of 1817 in Cambridge, preparing for the ensuing Fellowship Examination at his college. A long letter addressed to his friend Mr Rose in September discourages the latter from proceeding with an ambitious scheme he had formed of starting a Cambridge Review: Mr Whewell points out with great judgment the difficulties which would beset such a young staff of writers as it was proposed to employ.

Mr Whewell was elected Fellow of Trinity College in October, 1817; five others obtained Fellowships at the same time; namely, G. Waddington, J. Wigram, and Moody, who were a year senior, and Sheepshanks and E. B. Elliott, who were of the same year.

Mr Whewell seems to have been disappointed that a Fellowship was not awarded to Mr Higman who was just below himself in the Mathematical Tripos of 1816; he says in a letter that he is "very angry with the seniors for condemning Higman to another year of academic trifling."

Immediately after his election to a Fellowship, Mr Whewell set out on a visit to Mr Jones, but he failed to see his friend in consequence of a succession of errors as to time and place.

In March, 1818, we find Mr Whewell engaged with private pupils during the evenings—perhaps an indication that he had increased his number of them.

In a letter to Mr Herschel, dated June 19, 1818, Mr Whewell expressed that high opinion of the merits of Simon Stevinus with respect to the early history of Mechanics, which he zealously maintained in his later writings: see the *History of the Inductive Sciences*, Book VI. Chapter 1. This letter also alludes to the work on which the writer was then engaged: " I have nearly finished my Statics, and feel tempted to publish it by itself; for Dynamics, to treat it as I am going on, working out all kinds of problems, will take up much time. I shall say nothing about the metaphysics of the subject as yet; though my conscience pricks me on sharply to take up my testimony against the abomination of motion, and many other abominations, with which the faith has been polluted for two hundred years."

In the Long Vacation of 1818, Mr Whewell took a reading party into Wales ; he had intended to settle at Barmouth, but finding that place unsuitable he proceeded to Caernarvon. One of the pupils could not be persuaded to read steadily, and Mr Whewell declined to accept the usual fee ; the long correspondence between them during subsequent years shews how completely the tutor gained the affection of his somewhat unsatisfactory pupil.

While in Wales, Mr Whewell accepted the post of Lecturer on Mathematics in Trinity College, which was offered to him by Mr Monk, then tutor, and afterwards Bishop of Gloucester.

In the year 1819, Mr Whewell took his M.A. degree; he says in a letter to his friend Mr Morland, dated March 23, 1819:

"Next week I and my contemporaries take the first step to becoming Masters of Arts, which is attended with divers advantages, among which free admission to the public library is one which does not least excite my cupidity."

In the autumn, Mr Whewell was proceeding to France in company with Mr Sheepshanks, but the packet "Nancy" in which they sailed came into collision with another vessel and was sunk. The two friends lost everything they had with them; but they were fortunately landed next morning at Brighton, where Mr Jones was staying, and he helped to refit them.

In this year the Cambridge Philosophical Society was founded; and Mr Whewell was one of the original members.

In November, 1819, Mr Whewell began his long course of authorship by the publication of a volume on *Mechanics;* and we shall find it convenient to trace the history of his writings on the subject. The next Chapter however must necessarily be of little interest except to those who make a special study of that part of Mixed Mathematics.

CHAPTER II.

In the year 1819, Mr Whewell first appeared as an author by the publication of his work entitled *An Elementary Treatise on Mechanics;* this forms an octavo volume; the Title, Preface, and Contents occupying xxii pages, and the text 346 pages, besides two pages of Errata.

The treatise may be considered as one which, in conjunction with the publications of Peacock and Herschel, introduced the continental mathematics, in order to replace the system of fluxions which had so long prevailed at Cambridge. A copious account both of the elementary and the higher parts of Statics is given, and the Differential and Integral Calculus are very freely used. In Dynamics only the elementary parts are treated.

In the notice of Dr Whewell published by the Royal Astronomical Society, the following opinion is pronounced on the treatise: "It was a work of great value, strikingly logical and accurate. It is considered by one of our most eminent living mathematicians to have been very far in advance of any then existing text-book in the clearness and correctness of the treatment of bodies in contact, and in the precision with which the assumptions involved in the laws of motion and the composition of forces are stated and illustrated." Sir John Herschel quotes nearly the whole of this opinion of the work, which he assigns to "one excellently qualified to judge of its merits."

The treatise on Mechanics passed through many editions, but changed its character considerably in its course ; it will be convenient to trace its history, and also that of the kindred publications, so as to collect in one place all that I have to say on the subject. The octavo form of the elementary treatise on Mechanics was preserved throughout.

In 1821 appeared a *Syllabus of an Elementary Treatise of Mechanics. With Corrections and Additions;* this occupies fifty octavo pages, besides a leaf which contains the title and preliminary notice : there is one plate. The Syllabus consists of the enunciations of those propositions in the Elementary Treatise which are most simple and important; there are references to that treatise for the demonstrations, but in some cases additions or alterations are given.

The second edition of the Elementary Treatise on Mechanics appeared in 1824; the title-page says " with numerous improvements and additions." The Title, Preface, and Contents now occupy xxviii pages, and the text 342 pages, besides a page of Errata. The Preface begins thus: " The present Volume is offered as a republication of one which has already appeared ; but the changes which it has undergone in substance and arrangement are such, that it may be doubtful whether it should be considered as a new edition or a new work." The second edition differs widely from the first, especially in the Statics, so that, as the author suggests, it may be regarded as a new work.   On the title-page of the first edition these words appeared: " Vol. i. containing Statics and part of Dynamics." They are not reproduced on the title-page to the second edition; and at the close of the Preface it is stated that it has been thought advisable to publish the Treatise on Dynamics as a separate work.   Accordingly this appeared in 1823, and I shall speak of it hereafter.

Mr Jones when he heard of this plan wrote, " Your first literary Babe will think you a most unnatural Papa for not letting your youngest darling call it Brother."

A troublesome characteristic of Dr Whewell's writings appears in the Preface, namely the phrase, " as I have stated elsewhere," without any indication of the place intended by " elsewhere:" in this case it is I presume the Preface to the *Dynamics* of 1823. And on page 288 we have an example of a habit to which he was always prone, which seems to me a grave fault in scientific works, namely, an unnecessary change of language where there should be no change in the sense. A section is headed, "Planes of Quickest and Slowest Descent," and almost immediately afterwards we have, "It is required to find the plane of shortest descent." - Here instead of *shortest* we ought to have had *quickest*, both as more suitable and as having been already employed.

The third edition of the Elementary Treatise on Mechanics appeared in 1828; the title-page says "with improvements and additions." The Title, Preface, and Contents occupy xxiv pages, and the text 369 pages, besides a page of Errata. This edition is substantially the same as the second with the addition of two chapters, one on *Friction*, and the other on the *Connexion of Pressure and Impact*. This may be considered as the best edition of the work in one volume; as we shall see it was subsequently separated into parts. In a bookseller's catalogue I have found a *Supplement* to the third edition recorded; but I have not seen such a Supplement.

The fourth edition of the Elementary Treatise on Mechanics appeared in 1833; the title-page says "with improvements and additions." The Title, Preface, and Contents now occupy xviii pages, and the text 280 pages. The principal part of the preface to the third edition is embodied in the preface to the fourth. The portions of the work which assumed the student to possess a knowledge of Analytical Geometry and the Differential Calculus, except the last chapter, were now withdrawn, and issued in a separate volume under the title: *Analytical Statics. A Supplement to the fourth edition of an Elementary Treatise on Mechanics*, 1833. This is an octavo volume; the Title, Preface and Contents occupy viii pages, and the text 152 pages. The investigations concerning the Forms of Bridges on various hypotheses, and the discussion of the Species of the Elastic Curve, which constituted part of the former editions of the *Mechanics*, were now omitted. On the other hand additions are made respecting Suspension Bridges and the Strength of Materials. We may also notice a demonstration of the proposition called the *Parallelogram of Forces*, which is substantially the same as Poisson published in 1833 in

the second edition of his *Traité de Mécanique:* Mr Whewell
had independently invented this demonstration, as we learn from
a letter of his to Professor J. D. Forbes. " There is a proof of
the composition of forces which I am obliged to assure you was
my own invention. I excogitated it during a ride; wrote it out
and sent it to the press;—and then, and not sooner, began to
read Poisson's new edition, and found it given there with the
same steps almost without a variation."

Pages 229...269 of the fourth edition of the Elementary
Treatise on Mechanics are new: these treat on the *Friction of
Bodies in Motion,* and on the *Principle of Work.*

There is an octavo sheet called *Additions in the fourth edition
of an Elementary Treatise on Mechanics,* 1833. It relates to
Friction and to Work, and was, I suppose, intended for the convenience of purchasers of the earlier editions; it seems to end
abruptly.

The fifth edition of the Elementary Treatise on Mechanics
appeared in 1836; the title-page says " with considerable improvements and additions." The Title, Preface, and Contents
occupy xvi pages, and the text 326 pages. The preface is rather
polemical, referring to the famous attack on Mathematics which
had been made by Sir W. Hamilton, in No. 126 and No. 127
of the Edinburgh Review, and to Mr Whewell's *Thoughts on
the Study of Mathematics.*

The fifth edition, like the fourth, presents itself as strictly
an *elementary* work, by being confined to those parts of the
subject which do not require the higher mathematics; this is the
most comprehensive edition of the treatise in this restricted form,
as it was much reduced in bulk in the next edition.

We read in the Preface " I have added several Articles...upon
the Theory of Arches. The theory of the equilibrated arch, which
I had introduced in the earliest editions of the work, I rejected in
the subsequent editions, since, in the way in which it was then
treated, it was quite inapplicable in practice." The statement is
rather vague; but I think the fact is that the matter about
arches appeared in the first three editions, and was rejected in
the fourth.

Again, we read in the Preface, "I have also introduced into the present volume several of the simpler propositions which respect the doctrine of the rotatory motion of a rigid body, transferring them from my Treatise on Dynamics." Accordingly, pages 237...271 relate to this and kindred subjects.

The sixth edition was published in 1841; the title-page says "with extensive corrections and additions." The Title, Contents, and Preface occupy viii pages, and the text 124 pages. Thus although the title-page speaks of *additions,* it is obvious that these are more than counterbalanced by extensive *omissions.* The Preface begins thus: "This Edition of the Elementary Treatise of Mechanics has been entirely rewritten, and the plan of the former editions has been so much modified, that this might perhaps more properly be considered as a new work."

The portions which occupied pages 237...315 of the fifth edition are now withdrawn. The account of the *Mechanical Powers* is also omitted; but it was published again in 1845 as a *Supplement* to the sixth edition: this Supplement consists of 29 octavo pages with two plates. Moreover, in the sixth edition, speaking generally, examples are not introduced to illustrate general propositions and general formulæ. The Preface says, "Such examples, proposed and solved by the living instructor, or under his eye, are of far more service to the learner, than those which he finds ready solved in the pages of his book." This indicates a considerable change in opinion, for in the original form of the work the number of problems solved was very great; and we know from the correspondence of the author that he purposely adopted this plan.

Towards the conclusion of the Preface Mr Whewell refers to a work on Engineering, which he was about to publish as a sequel to the present Elementary Treatise on Mechanics. Accordingly, this appeared in 1841, under the title of *Mechanics of Engineering;* the Title, Dedication, Preface, and Contents occupy xii pages, and the text 216 pages; it is dedicated to Professor Willis, whose *Treatise on Mechanism* was just about to appear. The *Mechanics of Engineering* is not an independent work, as the title might perhaps suggest; the student is supposed to have already acquired the principles of Mechanics from some treatise on the subject, and here the principles "are traced to their exemplification in the practical business of the Engineer." The subject of *oblique arches* may be noticed as a new investigation in the theory; to this Mr Whewell attached great importance, as we see from his Preface, and from the *History of the Inductive Sciences,* 3rd edition, Vol. II, page 447.

The subject of *Work* or *Labouring Force,* which had been introduced into the fifth edition of the *Mechanics,* occupies much space in the volume. Mr Whewell acknowledges in his Preface his obligations to Navier, Poncelet and others, but says, "I have attempted, however, to present this matter in a more systematic shape than my predecessors."

The publication of the *Mechanics of Engineering* led to some correspondence between Mr Whewell and Professor Mosely. The latter claimed as his own a certain principle of the *limiting angle of resistance* which he found used in Arts. 106...108 and 111...113 of the work; and also the two conditions in which the general theory of the arch is summed up in Art. 113 : he added that he had subsequently corrected this theory of the arch. I have not seen Mr Whewell's replies, but he seems to have said that he had not read the writings of Professor Mosely, and that the matters in question were implicitly common to many authors. Professor Mosely's last letter concludes with acknowledging the very candid and frank manner in which Mr Whewell's part of the controversy had been carried on.

The seventh edition of the *Elementary Treatise on Mechanics* was published in 1847; the title-page says " with extensive corrections and additions." The Title, Preface, and Contents occupy viii pages, and the text 191 pages. We read in the Preface, "In the Sixth Edition, the innovations were, perhaps, inconveniently large, including omissions of portions of the previous editions. In the present edition, I have restored some of these omitted portions, especially the *Mechanical Powers....*" Mr Whewell refers to the *Mechanics of Engineering* as intended to be a *Supplement* to the *Elementary Treatise on Mechanics,* and says, "The *Elementary Treatise* and the *Mechanics of Engineering* taken together, made up a much more complete Treatise than any of the former editions of my work." He should have mentioned also the *Analytical Statics*, so that the *three* works taken together might be considered to represent the original publication of 1819.

The Elementary Treatise on Mechanics appears to have been translated into German; I have seen in a bookseller's catalogue a record of the first edition of the translation in 1841 and of the second in 1849.

The Elementary Treatise on Mechanics, as it thus passed through seven editions, must have exercised a powerful influence on Mathematical education for a long period. I cannot speak of the work from the experience either of a student or a teacher; for it had practically disappeared from general use before I entered the University, though it lingered officially in Trinity College until a later date. But it is easy to see the characteristic merits and defects of the work. Among the former I should place prominently the distinction between Statics and Dynamics, to which attention was forcibly drawn in the Preface, and which was well maintained in the work; as the two subjects differ in origin and in the nature of the principles on which they rest, Mr Whewell acted judiciously in keeping them separate. It is I conceive a matter of regret that some modern writers shew a desire to obliterate the distinction, and to treat Statics as a particular case of Dynamics; this course seems to me to impair the educational value of the study of Mechanics, without any compensating advantage to science. Another merit in Mr Whewell's treatise is the soundness of principle which pervades it; the student may safely trust himself to a guide who has such a wide and accurate knowledge of the subject he professes to teach: it must be allowed, however, that the inaccuracies of the printer were numerous and troublesome.

On the other hand the Treatise cannot be considered attractive in form. The matter does not seem well arranged; problems are thrown in promiscuously, and the methods of solution too often take a cumbrous geometrical form, quite devoid of elegance. In the earlier editions the explanatory matter is far too diffuse, and is wisely retrenched in the later editions.

But the great objection to the work is the perpetual alteration which deprived it of all stability and permanence. No sooner had teachers become familiar with one edition than another would appear in which the subject had been rather revolutionized than modified; and in each edition the preface expounded, with characteristic energy, the paramount merits of the last constitution which had been framed. It is scarcely possible for an author to retain the unwavering confidence of his readers when his own opinions are in constant fluctuation.

It has always appeared to me that Mr Whewell would have been of great benefit to students if he had undertaken a critical revision of the technical language of Mechanics. This language was formed to a great extent by the early writers, at an epoch when the subject was imperfectly understood, and many terms were used without well-defined meanings. Gradually the language has been improved, but it is still open to objection. For example, our books on Dynamics retain the awkward phrases *accelerating force* and *moving force*, as if there were two kinds of force, instead of two different ways of estimating the effect of force. Mr Whewell was well aware of the objection to the phrases, as we learn from his Essay in the *Cambridge Philosophical Transactions* Vol. v., reprinted in his *Philosophy of the Inductive Sciences*, 2nd edition, Vol. ii. It would have been easy for him to avoid the use of the objectionable phrases, and then from the influence of his authority they would probably by this time have vanished from our books.

As a simple example of the same kind I may notice the word *pressure*, which is so unnecessarily prominent in the *Mechanics*, and used without any fixed speciality of meaning. Thus, for example, in the fifth edition of the *Mechanics* we have in the first seven pages the word *force* used simply and naturally, in various illustrations mostly of a statical kind; then suddenly, on page 8, we read that statical forces are called *pressures:* this would suggest that dynamical forces are not so called, but when we reach the Dynamics we find the word *pressure* introduced there also. Again, in the sixth edition, after nine pages in which the word *force* has been used, we read that forces exerted by means of any solid bodies are called *pressures*. But the fact is that *pressure* has always been specially appropriated to the case of force exerted by *fluid* bodies. It is not too much to say that a great improvement would be effected in the Treatise on Mechanics by replacing the word *pressure* almost invariably by *force*.

I proceed now to speak of Mr Whewell's works on Dynamics. In the year 1823, he published a volume entitled *A Treatise on Dynamics. Containing a considerable collection of Mechanical Problems*. This is in octavo; the Title, Preface, and Contents occupy xvi pages, and the text 403 pages, besides a page of Errata.

This work contains a copious account of the dynamics both of a particle and of a rigid body. Some *Corrections* were afterwards issued in 7 pages, without a date; these consist of a corrected reprint of part of Article 36, and of nearly the whole of Arts. 40 and 104. The Articles are 148, 152 and 223 respectively of the work of 1834 entitled *On the Motion of Points constrained and resisted*.

This treatise on Dynamics may be considered as substantially the second volume corresponding to the *Mechanics* of 1819, as a first volume; but, as we saw, the plan was changed in the preface to the edition of the Mechanics of 1824. The treatise on Dynamics was not reproduced in the compass of a single volume, but replaced by three, all in octavo, namely, the Introduction to Dynamics, the Dynamics, Part I, and the Dynamics, Part II: I shall now notice these three works.

In 1832 appeared *An Introduction to Dynamics, containing the Laws of Motion and the first three Sections of the Principia*. The Title, Contents, and Preface, occupy xvi pages, and the text 64 pages. This *Introduction* formed no part of the Dynamics of 1823. The design predominant in this work, and that to which we shall next proceed, is to enable a student of slender mathematical attainments to gain a respectable knowledge of the main results obtained by Newton in Physical Astronomy.

The Preface to the Introduction is written with great animation; a few passages may be quoted:

"The first section of the Principia is eminently instructive with reference to the fundamental principles of the Differential Calculus." "And it is very desirable that the mathematical student, before he rushes forward to differentiate and integrate upon the slightest provocation, should employ some thought in understanding the construction and trustworthiness of the instrument which he is so familiarly to use."

"The Differential Calculus is not necessary for a person who would merely understand the mechanics of the skies, though that instrument is indispensable for one who would himself examine or extend the calculated results."

"...the spirit of commentatorship has generally been sufficiently ready to fasten upon the objects of the intellectual admiration of mankind. Its flowers and weeds seem to spring up luxuriantly about the wheels of the car of genius, the moment there is a pause in the career."

We have on page x of the Preface the words "I have explained elsewhere;" I think the reference must be to a paper published in 1828, in the *Edinburgh Journal of Science*, Vol. VIII.

The second of the three works which replaced the original *Dynamics* bears the following title: *On the free motion of points, and on Universal Gravitation, including the principal propositions of Books I and III of the Principia; the first part of a Treatise on Dynamics.*

This was published in 1832, and was then called the first part of a new edition of the Dynamics, the work of 1823 being considered I presume as the *first* edition. The issue of 1836 is called *third* edition; the preface, however, is called *Preface to the second edition*, and is dated April 27, 1832. Thus it appears that the third edition was not furnished with a preface of its own; this is an unusual circumstance in Dr Whewell's publications, for he displayed great ability and vigour in the composition of prefaces, and rarely neglected the opportunity for such exercise which his successive editions supplied.

In the issue of 1836 the Title, Preface, and Contents, occupy xxviii pages, and the text 238 pages. The text includes the pages 1...78 and 364...372 of the Dynamics of 1823; the additional matter consists mainly of propositions taken from Newton's Principia with illustrations. The preface common to the second and third editions is long and interesting. One sentence expresses very happily what many writers must have felt, though they may not have formally made the confession: "A few years experience has a great tendency to diminish the confidence of producing what shall satisfy himself and others, with which a young author sets out: and he learns that the vivid impression of the fancied deficiencies and imperfections of preceding works which at first induced him to write, is a very insufficient warrant of his own skill and judgement."

An elaborate compliment is paid to Mrs Somerville for her *Mechanism of the Heavens.* "Our willingness to adopt a more extended study of the mechanism of the heavens into our academic system must needs increase, when these severer studies, thus shewn to be reconcilable with all the gentler train of feminine graces and accomplishments, can no longer, with any shew of reason, be represented as inconsistent with a polished taste and a familiar acquaintance with ancient and modern literature."

The merits of Newton, and the claims of the Principia on the attention of students, are also very strongly enforced.

Towards the end of the Preface we read : " The present volume appears in the character of a portion of a new edition, although much the greater part of it is entirely new matter." The volume may still be recommended to persons who are engaged in the study of Newton's Principia, though it treats mainly on subjects which now rarely appear in Examinations. It is to be regretted that in this volume, and also in that which I proceed to notice, Mr Whewell allowed himself to adopt a very objectionable notation for integrals which had a temporary existence in Cambridge, but has now fortunately vanished. As he must have known well the evils which had been produced by the adherence of English mathematicians to the old fluxional notation and language, we may wonder at his giving any countenance to the attempt thus made again to separate his countrymen from the current of European science.

The last of the three works which replaced the original *Dynamics* bears the following title : *On the motion of Points constrained and resisted, and on the motion of a Rigid Body. The second part of a new edition of a Treatise on Dynamics.* 1834. The Title, Preface, Elementary Principles and Contents occupy xxi pages, and the text 338 pages. This work is in the main a reproduction of pages 79...346, 378...388 of the original *Dynamics*, with some difference in arrangement. The other changes are not important: pages 154...158 are enlarged from the original 203 and 204; pages 177...183 are altered from the original 221...226; pages 201, 202 and 212 are new; pages 244 and 245 are enlarged from the original 257; and pages 276...287 are altered from the original 304...311, with which they correspond as to subject. There are certain portions of the original *Dynamics* which were not reproduced in the three works which replaced it; namely, pages 346...364, which are *On the Motion of a Body about two Centres of Force;* pages 372...377, which are *On some particular Cases of the Motions of three Bodies;* and pages 388...394, which are *On the Descent of Small Bodies in Fluids. On the Ascent of an Air-Bubble.* There was also on pages 394...403 of the original *Dynamics* a section on *General Mechanical Principles* which treated of the Conservation of the Motion of the Centre of Gravity, of the Conservation of Areas, of the Conservation of Vis Viva, and of Least Action. But in the later works only the third and fourth of these general principles are noticed; and they are placed, somewhat inappropriately, in the Constrained Motion of Several Points.

Here we finish with the Dynamics, a work which seems never to have been so much used as the Elementary Treatise on Mechanics, and sooner passed into that obscurity which is the fate of all academical text-books. We have next to notice a slight publication issued in 1832, and then to pass to the *Mechanical Euclid,* which closes this series of our author's works.

In the year 1832 Mr Whewell published *The first principles of Mechanics, with historical and practical illustrations.* This is in octavo; the Title, Preface, and Contents occupy xii pages, and the text 118 pages. The design was to introduce a reader who possessed the bare rudiments of Algebra and Geometry to the Elements of Mechanics. The style is attractive, and the illustrations are interesting; but it may be doubted whether the work is well adapted as a whole for any very definite class of students. Mr Whewell seems not to have been quite satisfied himself; in the preface to the *Mechanical Euclid,* 1837, he refers to his *History of the Inductive Sciences* as giving much more completely all that the present historical illustrations contained, and adds: 'I shall therefore consider these "First Principles" as now superseded, and shall not republish the work. The Practical Illustrations may be perhaps incorporated in some future publication in an improved form.'

The *Mechanics of Engineering* is perhaps the work which corresponds to this *future publication*.

In the year 1837 Mr Whewell published a work under the following title: *The Mechanical Euclid, containing the Elements of Mechanics and Hydrostatics demonstrated after the manner of the Elements of Geometry; and including the Propositions fixed upon by the University of Cambridge as requisite for the degree of B.A.*

This is a duodecimo volume; the Title, Contents, and Preface occupy viii pages and the text 182 pages. In the text we have a brief introduction to Algebra, taken from Dr Wood's treatise with the author's permission, a Book on Statics, a Book on Hydrostatics, a Book on the Laws of Motion, and finally Remarks on Mathematical Reasoning and on the Logic of Induction. The *Remarks* are important; the principles here maintained were afterwards enforced in the Philosophy of the Inductive Sciences, and the whole Essay reprinted in the *Appendix* at the end of the second edition of that work.

The propositions in the *Mechanical Euclid* seem treated with as much success as the nature of the undertaking allows. Throughout the Statics the word *force* is retained, thus effecting a great improvement over the author's other works where the word *pressure* is frequently substituted for *force*. But the experience of teachers is not favourable to the method of this book for introducing their pupils to the elements of Mechanics. The students who are really able to understand the Mechanical Euclid would in general find it advantageous to take a wider course of reading, involving more mathematics; while the majority of those for whom the book was designed would probably derive more profit from simple experimental lectures. In particular the subject of demonstrative Hydrostatics seems to be imperfectly appreciated by the untrained minds for whom it is prescribed by the University Regulations.

The second edition of the *Mechanical Euclid* appeared in the same year as the first; the title-page says "the second edition corrected." The Preface to the first edition is reprinted; and it is followed by some brief remarks stating that only a few slight alterations appeared necessary on revising the work for a second edition, and specifying these alterations.

The third edition of the *Mechanical Euclid* appeared in 1838; the title-page says "the third edition corrected." The Preface and the Remarks following it are reproduced as they stood in the second edition; and then it is stated "In the present (Third) Edition no alterations of importance have been introduced."

The fourth edition of the *Mechanical Euclid* appeared in 1843. Some questions from the Examination Papers of the current year are inserted, and also the Regulations of the University respecting the Examinations. The Book on the Laws of Motion, and the Remarks on Mathematical Reasoning, were now omitted; but the latter reappeared in 1845, as *A supplement to the fourth edition :* the supplement consisted of 64 pages. The prefatory notice to the supplement begins with a curious mistake; "In the Third Edition of the Mechanical Euclid, I omitted...:" it should be "In the Fourth Edition..."

The fifth edition of the *Mechanical Euclid* appeared in 1849; the title-page says " carefully adapted to the ordinary Examinations for the Degree of B.A." The Title, Preface, and Contents occupy xii pages, and the text 200 pages. The Preface to the fourth edition is reproduced, followed by a Preface to the fifth edition. The parts omitted in the fourth edition are now restored; some questions from Examination Papers for the current year, instead of those for the year 1843, are inserted, and also the Regulations of the University respecting the Examinations. Various small changes are to be found in this edition as compared with the third; these are made for the purpose of adapting the work more closely to the University Regulations: but on the whole the ultimate form is substantially coincident with that originally published, thus presenting a striking contrast to the author's other treatises relating to Mechanics.

An important remark from the preface to the fourth edition

may be reproduced; it is as true now as it was when it first appeared thirty years since: "...I believe every one practically acquainted with University and College Examinations and their effects, will agree with me that Euclid's Geometry is the most effective and the most valuable portion of our mathematical education."

The *Mechanical Euclid* forms the text of an article in the *Edinburgh Review*, No. 135, April, 1838, which was written by the late Thomas Flower Ellis. This article is devoted to the discussion of one point in the *Remarks on Mathematical Reasoning*. Dugald Stewart considered that Geometry was founded solely on *definitions;* Mr Whewell maintained that it was founded on *axioms and definitions;* the Reviewer supports the opinion of Dugald Stewart. Mr Whewell replied to the review in his *Philosophy of the Inductive Sciences;* see the second edition of that work, Vol. I, pages 101...111. The question is discussed by J. S. Mill in his *Logic*, Book II, Chapter V; and I will extract a few of his sentences: "The opinion of Dugald Stewart respecting the foundations of geometry, is, I conceive, substantially correct; that it is built on hypotheses; that it owes to this alone the peculiar certainty supposed to distinguish it; and that in any science whatever, by reasoning from a set of hypotheses, we may obtain a body of conclusions as certain as those of geometry, that is, as strictly in accordance with the hypotheses, and as irresistibly compelling assent, *on condition* that those hypotheses are true.... The important doctrine of Dugald Stewart, which I have endeavoured to enforce, has been contested by Dr Whewell, both in the dissertation appended to his excellent *Mechanical Euclid*, and in his more recent elaborate work on the *Philosophy of the Inductive Sciences;* in which last he also replies to an article in the Edinburgh Review, (ascribed to a writer of great scientific eminence), in which Stewart's opinion was defended against his former strictures.......But though Dr Whewell has not shaken Stewart's doctrine as to the ·hypothetical character of that portion of the first principles of geometry which are involved in the so-called definitions, he has, I conceive, greatly the advantage of Stewart on another important point in the theory of geometrical reasoning;

the necessity of admitting, among those first principles, axioms as well as definitions."

In dismissing Dr Whewell's works on Mechanics it will be convenient to recapitulate the forms they finally assumed:

1. Analytical Statics.   1833.
2. Mechanics of Engineering.   1841.
3. Elementary Treatise on Mechanics.   1847.
4. Mechanical Euclid.   1849.
5. Introduction to Dynamics.   1832.
6. Dynamics.   Part I.   1836.
7. Dynamics.   Part II.   1834.

CHAPTER III.

1820...1830.

WE resume the chronological order after the digression of the preceding Chapter.

In the year 1820 Mr Whewell was one of the Moderators at Cambridge, his colleague being his friend Mr Wilkinson of St John's College; the printed examination papers shew that the duties of the office were discharged with ability and discretion.

In the early part of this year Mr Whewell first met the present Astronomer Royal, then an undergraduate of Trinity College, but not on the *side* to which he himself belonged; this was the commencement of one of the most intimate of his scientific and social friendships.

In April, 1820, Mr Whewell was elected a fellow of the Royal Society. An eminent philosopher, through whom his admission fee was paid, described the society to him as in a state of unstable equilibrium. Sir Joseph Banks, the President, was about to resign his office, and had recommended Davies Gilbert as his successor: "But all sorts of plans, speculations, and schemes are afloat, and all sorts of people, proper and improper, are penetrated with the desire of wielding the sceptre of Science."

In the autumn Mr Whewell travelled in Switzerland in company with Mr Sheepshanks. The Astronomer Royal writing about the eclipse of the Sun in 1842 to Dr Whewell says, "I think that you saw the eclipse of 1820 from the Gemmi."

In 1821 Mr Whewell passed a few days at Oxford, and was rejoiced to find that geology was cultivated there even more vigorously than at Cambridge. He also visited the Isle of Wight, and then went on a tour among the English lakes, of which he had as yet seen only one, although his home was so near them. On this

occasion he made the acquaintance of Wordsworth, having a letter
of introduction to the poet from his brother the Master of Trinity
College.

In this year Mr Whewell addressed a letter to the Editor
of the *Museum Criticum*, which appeared on pages 514...519 of
the second volume of that publication. The letter is dated Trin.
Coll., Oct. 25, 1821, and, although it has not any name attached,
it is known to have been written by Mr Whewell: it criticises a
statement made by Professor Playfair respecting the University of
Cambridge. Professor Playfair asserted in his Dissertation on
the History of the Mathematical and Physical Sciences, that the
Cartesian system kept its ground in the University of Cambridge
for more than thirty years after the publication of Newton's
discoveries in 1687; and credit is implicitly claimed for the Scotch
Universities as being at that epoch more enlightened than the
English. Mr Whewell adduces evidence which completely redeems
the University of Cambridge from any suspicion of undervaluing
the teaching of the greatest of her sons. The substance of the
letter is reproduced in the *History of the Inductive Sciences:* see
the third edition, Vol. II, pages 147 and 453; see also *Barrow and
his Academical Times,* page vi.

In 1822 Mr Whewell seems to have had some intention of
writing on Physical Astronomy, as appears from the following passage in a letter addressed to him by Mr Jones: "I am at No. 9,
Downing Street, Westminster, in Herschel's lodgings; he is gone
to Slough for a week—direct to me here. He desired me before
he set out to beg you to let him know distinctly if you mean to
write the article Physical Astronomy for the Encyclopædia Metropolitana—because he has undertaken to do it if you do not, and
wants as long a time as possible to do it in." Mr Whewell must
have declined the task, as it was performed by Mr Herschel.

A severe fever broke out this year at Cambridge during the
summer, as in 1815. Five members of St John's College who
were attacked died after leaving the place. Mr Whewell and
several others remained during the whole time the disorder prevailed. Mr Whewell was at this time occupied with his work on
Dynamics, which he had originally proposed to call the second

volume of his Mechanics. He says, in a letter to Mr Jones, dated Sept. 23, 1822 : " My book is swelling out larger than I expected. As I do not think it will gain much by coming under the protection of the former volume, I intend to print it as a separate work, instead of calling it Volume II. I still meditate doing something about the History of the Metaphysics of Mechanics, though as yet it is only intention. Something like Smith's History of Astronomy, but with more historical facts. But if you can help me to get rid of that word of abomination, *Metaphysical,* I shall be exceedingly obliged to you. I hope you have not exhaled your wrath and brought yourself into anything like charity with it, for you may depend on it that no convention or management will ever make it permanently faithful to inductive Philosophy."

Mr Whewell had already begun to take a strong interest in Ecclesiastical Architecture, as we learn from the letters addressed to him, by Mr Sheepshanks, from various places on the continent.

Towards the end of the year the tranquillity of the University was disturbed by the contest for the seat in Parliament rendered vacant by the death of Mr Smyth. The successful candidate was Mr William Bankes, who was a strenuous advocate for Protestant ascendancy; his opponents were Mr Scarlett and Lord Hervey, of whom Mr Whewell supported the former.

In the beginning of the year 1823 Mr Whewell visited Paris. A letter from Arago regrets his absence from the Observatory when Mr Whewell called, and pays a high compliment to Mr Herschel, who had doubtless furnished a letter of introduction. Mr Whewell made the acquaintance of Constant Prevost, who had just been elected an honorary member of the Cambridge Philosophical Society. An amusing supply of Cambridge news was forwarded to Paris by Mr Romilly, Fellow of Trinity College, including the Tripos list of the year, in which Mr Airy's name stood first.

In this year Mr Whewell became the official Tutor of one of the *sides* of Trinity College, sharing the appointment for a year with Mr Brown, and afterwards taking sole possession. In the early part of the summer he made an architectural tour among the abbeys and churches of Normandy. The *Westminster Review* was started about this time, and Mr Whewell was asked to join the staff. It may be safely assumed that he declined. He seems to have taken some interest in the struggles of the Greeks with the Turks, for a letter from Mr John Bowring to Mr Douglas Kinnaird is preserved which gives "replies to the queries contained in Mr Whewell's letter." More than thirty years afterwards Sir John Bowring wrote to solicit Dr Whewell's attention to some efforts which were being made for the scientific instruction of the Chinese.

About the year 1824 some dissensions arose in the University as to the proper mode of appointing a Professor of Mineralogy. The Heads of Colleges wished to nominate two persons, of whom the Members of the Senate were to take one; but the Senate required to have an unfettered choice. Mr Whewell with Mr Sedgwick and Mr Peacock were active in opposition to the Heads. For an account of the matter we may refer to a pamphlet published by Mr Gunning, entitled *The King v. the Vice-Chancellor of Cambridge*, 1824.

A notice of some of Mr Whewell's movements during the year is supplied by an extract of a letter to him from his friend Mr Wilkinson, dated Nov. 21, 1824. Mr Wilkinson was then the Head Master of the school at Sedbergh, from which place he wrote: "When Professor Sedgwick was here last month he mentioned that you and he had projected a journey to Edinburgh this next Christmas, to hear lectures, converse with the philosophers, and for sundry other laudable objects. Do you not think you could save a few days for me out of this tour, particularly I may hope this, as Sedgwick purposes to visit Dent before his return to Cambridge. I suppose your new edition of the first volume of *Mechanics* will then be out, so you will not be in such breathless haste as when you were at the lakes with Gwatkin and Sedgwick in the summer."

Mr Whewell appears to have been ordained Deacon in 1825, though it is curious that the letters which he received even in 1823, from so intimate a friend as Mr Herschel, bore the title of *Reverend* on the address. In June of this year Mr Henslow, who then held the Professorship of Mineralogy, was appointed Professor of Botany by the king; it was supposed that this would render the Professorship of Mineralogy vacant in the October term, and Mr Whewell resolved to seek for the office. He issued the following circular to the Members of the Senate:

" Sir,

I beg leave to state to you that I intend to offer myself as a Candidate for the office of Professor of Mineralogy, on the vacancy which will be occasioned by Mr Henslow's resignation in consequence of his being appointed Regius Professor of Botany. To shew that I have paid some attention to this department of science, I may take the liberty of referring to communications on the subject of Crystallography, which I have made both to the Royal Society of London and to the Philosophical Society of Cambridge, some of which have been printed. I may venture to add that a principal reason which makes me wish to devote my exertions to the subject hereafter, is the opinion, founded on the views which my past enquiries have suggested, that the science may be most successfully and properly pursued by cultivating and extending the applications which have been made of Mathematical Principles and Laws to this branch of Natural History.

Under these circumstances, I beg to solicit your good will and support in the election to this office, and am,

Your faithful and obedient servant,

W. WHEWELL.

TRIN. COLL. CAMBRIDGE,
June 24th, 1825."

He made a tour in Germany from July to October, principally devoting himself to the study of the science. Professor Mohs was then one of its most eminent teachers, and in order to profit by his instructions Mr Whewell went to Freiberg; he found him just about to leave in order to spend his vacation at Vienna, and at the professor's invitation followed him to that place. On returning home Mr Whewell was detained nine days on the sea between Hamburgh and Harwich; and his delay in reaching Cambridge caused some anxiety to his friends there.

Mr Whewell's application for the Professorship of Mineralogy was warmly supported by many persons. Among others, Mr Brewster expressed his satisfaction: "I am glad to hear that you are a candidate for the Mineralogical chair in your University, and I shall be delighted if you succeed." Other friendly letters from the same writer, down to the year 1833, are preserved: in later times Mr Whewell and Sir David Brewster were more than once engaged in sharp cònflict. Mr Coddington, of Trinity College, had thought of offering himself as a candidate for the professorship, but only on the condition that neither Mr Peacock nor Mr Whewell would take it; and at once abandoned his intentions on learning the wishes of the latter.

Mr Whewell was ordained Priest on Trinity Sunday, 1826. In June he was engaged with Mr Airy in some laborious experiments at Dolcoath mine in Cornwall; he printed an account of these operations in 1828, and this I shall notice under that date. In the mean time a general election of members of Parliament took place; the candidates for the University of Cambridge were Sir J. S. Copley, Lord Palmerston, Mr W. Bankes, and Mr Goulburn; the first two of these were successful. Protestant ascendancy was the main principle at issue, and at this time Sir J. S. Copley was very strenuous for it. Mr Whewell regretted that his occupations at the mine prevented him from voting for Lord Palmerston, whose claims he supported.

In the latter part of the year, Mr Turton, who was the Lucasian Professor of Mathematics, afterwards Bishop of Ely, accepted a college living; the professorship was not tenable with this, and in consequence became vacant. Mr Whewell thought of trying for it if Mr Herschel would not; but he relinquished his intention on finding that by the statutes the professorship could not be held by a college tutor. Mr Airy was appointed to the office. In December Mr Whewell visited Paris in company with Professor Sedgwick.

Mr Whewell contributed to the *Encyclopædia Metropolitana* two articles; the exact date of publication is not known, but it seems to have been about 1826; and so they may be noticed here. One article refers to *Electricity*, and the other to *Archimedes*.

The treatise on Electricity in the *Encyclopædia Metropolitana*, which forms a portion of the second of the volumes devoted to *Mixed Sciences*, is ascribed in the list of contents entirely to the Rev. Francis Lunn: it is known however that some of it was written by Mr Whewell, namely Part II, called the *Theory of Electricity*, which occupies pages 140...170 of the volume. This consists essentially of a reproduction of the first of two memoirs published on the subject by Poisson, in the *Mémoires de l'Institut* for 1811. Poisson's investigations constitute one of the most fascinating specimens of mathematical physics ever produced, and well deserved to be presented to English readers; but as Mr Whewell's article was entombed in the recesses of a ponderous encyclopædia it was not very accessible to students, for the various treatises composing the work could not be procured separately until about a quarter of a century after the date at which we have now arrived. The fact that Mr Whewell contributed this article to the Encyclopædia is known from what he says himself in the preface to his work *On the Free motion of Points*, and also from a letter written by Mr Lunn preserved among the papers of Dr Whewell. In this article Mr Whewell gave the very convenient name of *Laplace's Coefficients* to certain important mathematical expressions with which Laplace was much occupied. See *Monthly Notices of the Royal Astronomical Society*, Vol. XXVII, page 211.

The other contribution by Mr Whewell to the *Encyclopædia Metropolitana* is entitled *Archimedes— Greek Mathematics*: it occupies pages 686...694 of the first volume of the Division *History and Biography* of the Encyclopædia. This contains a slight sketch of the progress made by the Greeks generally, and by Archimedes especially, in pure and mixed mathematics. A few words relating to Eratosthenes deserve notice: "Eratosthenes was a cotemporary of the Sicilian mathematicians, and was a remarkable instance of great acquirements in very different branches of knowledge. He is generally called by the ancients Eratosthenes the grammarian or philologer; and though he comes under our notice as a great geometer and astronomer, he was also a poet and an antiquary. It is seldom that one person attempts to master so many subjects, without incurring the charge and perhaps the danger of being superficial." Some portions of the *Encyclopædia Metropolitana* have been republished in crown octavo size. Mr Whewell's article on Archimedes and Greek Mathematics forms part of a volume thus issued in 1853, entitled *Greek and Roman Philosophy and Science*, in which it occupies pages 305...325.

In the year 1827 Mr Whewell was elected a Fellow of the Geological Society. In the month of February he preached a course of sermons before the University, which seems to have attracted much attention; the manuscripts are preserved, and I will notice them in a future chapter. At the request of the Master of his College, Dr Wordsworth, who was the Vice-Chancellor of the University for the year, he preached the Commencement Sermon in July; and in November he preached the Commemoration Sermon. It naturally occurred to his friends that he might well devote himself to theology; Mr Sheepshanks considered him to be at the top of the University preachers, and suggested the Divinity Professorship as a proper object for his aim; but Mr Whewell shewed no disposition to accept this advice.

In the year 1828 Mr Whewell was one of the Moderators for the Mathematical Tripos; his colleague was Mr Joshua King, afterwards President of Queens' College: the two Moderators discharged the duty of Examiners in the next year according to the usual custom.

In the early part of the year Mr Airy obtained the Plumian Professorship, and in consequence resigned the Lucasian. Mr Whewell tried to induce Mr Herschel to be a candidate for the latter, and on his declining exerted himself strenuously in favour of Mr Babbage, who received the appointment.

In March Mr Whewell was made Professor of Mineralogy. There had been a great dispute as to the mode of election, which was referred to Sir John Richardson as arbitrator: according to Mr Whewell the decision was adverse to the opinion he had held.

The Professor of Mineralogy soon shewed that he did not intend to treat his office as a sinecure. A work was published under the following title: *An Essay on Mineralogical Classification and Nomenclature; with Tables of the Orders and Species of Minerals. By W. Whewell, M.A., F.R.S., M.G.S., Fellow of Trinity College, and Professor of Mineralogy in the University of Cambridge.* Cambridge, 1828. This is in octavo. The title-page, introductory notice, and contents occupy six pages; the Essay on Mineralogical Classification and Nomenclature occupies pages i...xxxii; a leaf follows with the title *Tables of the orders and species of minerals;* and then the tables occupy pages 1...71. Two modes of classification of the mineral kingdom had already been proposed in Europe. "Mohs and Breithaupt have published treatises in which physical or external mineralogical characters determine the classes of minerals. Berzelius, Gmelin, Beudant and Leonhard, have given arrangements of the same substances according to their chemical relations." Mr Whewell lays down in his Essay the principle which he maintained in his *History and Philosophy of the Inductive Sciences,* that if the classes obtained by these two modes of arrangement agree to any extent, so far we may admit the classes to be really natural.

In the summer of 1828, Professors Airy and Whewell repeated their operations in the mine at Cornwall; and in the course of the year the latter published an *Account of Experiments made at Dolcoath Mine, in Cornwall, in 1826 and 1828, for the purpose of determining the Density of the Earth.* This is an octavo pamphlet of 16 pages, printed at the University Press. On the back of the title-page is a notice: "It is requested that this Account be not published." The pamphlet was apparently printed for private circulation; it seems to have been expected that a fuller account would afterwards be published by Professor Airy; but this was never done, so that the present pamphlet remains as the only printed record of a very arduous experiment.

Speaking generally we may say that the object was to determine the density of the Earth, and the essential part of the process was to compare the time of vibration of a pendulum at the surface of the Earth with the time of vibration of the same pendulum at a considerable depth below the surface. The experiment failed to lead to a satisfactory result, because it appeared that the pendulum could not be trusted; but besides this difficulty a serious accident occurred on each occasion. More success was obtained in 1854 by the Astronomer Royal at Harton Colliery: see the *History of the Theories of Attraction*...Vol. I, page 470.

An extract may be given from Professor Whewell's pamphlet:

"The first attempt of this kind was made in the summer of 1826, by Mr Airy and Mr Whewell, Fellows of Trinity College, Cambridge. The scene of their operations was the copper mine of Dolcoath, near Camborne in Cornwall. Their lowest station was a chamber in the rock at the depth of 1200 feet below the surface, and a small hut nearly perpendicularly over this was the higher point. At each of the stations were placed a detached pendulum and a clock with which it was to be compared by means of the method of coincidences above referred to. The two clocks were compared with one another by means of seven chronometers, which were all, during each day of the experiment, compared first with one clock, and then with another, both at the beginning and at the end of the observation. The reason for having so many of these intermediate instruments of comparison was, that any irregularity in going, or inaccuracy in comparing in one case, might be remedied in the average of them.

"The most obvious difficulties in this undertaking seemed to be overcome. The clock and pendulum were lowered to the underground chamber without injury, fixed in their places, and set in motion. In the course of the observation, the daily fatigue of the observers in descending and ascending between the two stations was not slight, as may easily be imagined when it is considered that it was the same process as clambering by ladders down and up a well 3½ times the height of the pinnacle of St Paul's; and this was accompanied by a stay of 6 or 8 hours at the bottom of the mine amid damp and dirt, and in damp and dirty clothing..."

Two detached pendulums were used, which for the sake of distinction were called by the names of their former possessors, two distinguished naval officers, Captain Basil Hall and Captain Foster. At first *Hall* was below and *Foster* above; then the pendulums were interchanged, and the conclusion is related thus:

"The next step attempted was to raise the pendulum *Foster* to the surface, in order to ascertain that it had undergone no change in its rate during the preceding operations, and this attempt led to the final catastrophe of this first experiment. The pendulum-box was dispatched from the lower station, packed in dry reeds, and drawn up a shaft nearly vertical, 1200 feet deep, by the strength of men and horses. When, however, the bucket which had contained it reached the surface, it offered to the expectant eyes of Mr Airy nothing but a few smoking embers. By some unexplained accident, perhaps by a candle-snuff falling upon the light packing materials, this pendulum, which had escaped the dangers of the North West passage, and swung successfully at Port Bowen, was burnt from the ropes which supported it, and fell very probably 1000 feet perpendicular."

In the year 1828 the experiment at Dolcoath Mine was repeated. Mr Airy and Mr Whewell were joined by Mr Sheepshanks, Fellow of Trinity College, and also by two students of the University, namely Mr W. Airy of Trinity College, and Mr Jackson of Caius College. The pendulums were kept swinging, with only such stoppages as were inevitable, from Monday morning, July 21, till the Saturday afternoon following. By calculations and observations which extended during the first fortnight of August, it appeared, as in the first experiment, that the pendulums could not be trusted: the nature of the defect in the most troublesome pendulum seemed to be discovered, and it was resolved to send the pendulum again to the lower station. In the mean time, owing to the subsidence of a vast mass of the stratum which contained the workings of the mine, the pumps were stopped which ran from the top of the mine to the bottom; so that the water immediately began to rise through the lower workings. The pendulum was sent down on the 16th of August, and brought up on the 19th; the water had now risen above 100 feet from the time of the accident, and had arrived within 14 feet of the lower station: in the course of a few days the place was beneath the surface of the water.

I may observe that Captain Basil Hall heard in 1826 that an accident had happened to one of the pendulums, and he wrote a very interesting letter to Mr Whewell asking for information;

he was anxious to know whether the sufferer was his friend or the other, as he still cherished a regard for the old pendulum which was his companion in South America.

Professor Whewell made a tour in Wales in the early part of the summer with his friend Mr Jones.

Towards the end of the year Professor Whewell was very earnest in seeking support for a proposal to admit Bachelors of Arts to the use of the University Library; but the proposal was stopped in the *Caput*, which was a part of the Academical constitution of those days.

A publication of this date on six octavo pages may, I believe, be ascribed to Professor Whewell. Its object is to shew the deplorable state of the University, which had neither lecture-rooms for its Professors, nor museums in which to place its collections. The pamphlet is introduced by the notice: "The following Statement is respectfully offered to the Members of the Senate." I extract one sentence from it, which was probably intended to carry much meaning: "The Lucasian Professor of Mathematics is directed by the foundation of his office to deliver lectures, and it is understood to be the intention of the present Professor to do so." Many years elapsed before the wants of the University, exposed in this pamphlet, were adequately supplied, and Dr Whewell was always one of the most strenuous advocates of the necessary reform.

Towards the end of the year Mr J. C. Hare preached before the University, and as he disregarded the usual limits of time, he exhausted the patience of the undergraduates; however, more than 200 of them petitioned him to print his sermon, and he complied with the request. Professor Whewell speaks of the sermon to a friend as "very much after the manner of a *guess*, but fearfully long." It may be stated on Mr Whewell's own authority, that there are no contributions of his to the *Guesses at Truth*, published by J. C. Hare and A. Hare.

In 1829 Sir N. C. Tindal, who represented the University in Parliament, having succeeded Sir J. S. Copley in that honour, was made Chief Justice, and thus an election became necessary. Professor Whewell strongly supported his pupil and friend Mr

Cavendish, the present Duke of Devonshire, who was successful; the defeated candidate was Mr George Bankes.

In the summer Professor Whewell made a tour through parts of Switzerland, Germany, and Holland. A great congress of scientific men was held at Heidelberg, which he attended, and where he made the acquaintance of M. Studer of Berne and of M. Quetelet of Brussels: twenty-one years later he renewed his intercourse with the former at Berne; with the latter he maintained an active correspondence throughout his life.

Professor Whewell suffered much from that inconvenience to which all persons are exposed who become known as the authors of works on science, namely, importunate applications from ill-informed enthusiasts, who announce as discoveries what has long been familiar, or pretend to have achieved what is known to be impossible. For a specimen we may advert to a communication about this date on *perpetual motion*, as there seems some novelty in one of the writer's remarks : " There is one consideration which has damped my ardour in all my researches after perpetual motion, and that is its encouragement of idleness, and its apparent bad effects on the lower orders of society. If these should be the result of such a discovery, may the time I have devoted to it be lost for ever, and may my present projected plan be as abortive as the former."

At the beginning of the year 1830 we find Professor Whewell again discouraging one of the schemes of his energetic friend Mr Rose; on this occasion the proposal seems to have been to join in establishing a Cambridge newspaper, which should advocate sound principles both in Church and State. In the summer Professor Whewell geologized for about a week in the neighbourhood of Bath, and then made an architectural tour among the churches of Devon and Cornwall in company with Mr Rickman. During the year Professor Whewell took an active interest in the questions which disturbed the Royal Society; he supported Mr Herschel as a candidate for the office of President against the Duke of Sussex: the latter was elected.

In September he was appointed one of the writers of the proposed *Bridgewater Treatises* by Mr Davies Gilbert, acting under the advice of Dr Blomfield, the Bishop of London. The final arrangements with respect to the publication of these works were made at a meeting in December of the majority of the selected writers, namely Sir C. Bell, Dr Roget, Dr Buckland, Mr Kirby, and Dr Prout, and were sent for their approval to Dr Chalmers, Dr Kidd, and Professor Whewell, who had not been present.

I shall now advert to the Notes on German Churches, of which the first edition belongs to this year.

In 1830 a work was published anonymously entitled *Architectural Notes on German Churches, with remarks on the Origin of Gothic Architecture.* It is in octavo, and contains altogether 118 pages, besides the title-leaf. The author had proposed to add to the title the words, " by a summer tourist:" but his friends seem to have objected to this as undignified.

The work was enlarged and republished in 1835 under the title : *Architectural Notes on German Churches. A new edition. To which is now added notes written during an Architectural Tour in Picardy and Normandy. By the Rev. W. Whewell, M.A., Fellow and Tutor of Trinity College, Cambridge.* The original matter is reproduced with the addition of a new Preface, and of the French tour ; there are now 261 pages including the title-leaf. With respect to the title Mr Hare wrote, " ...my eyes have hardly wandered beyond the title-page, and are puzzled to make out why you have chosen to startle them with a false concord." Professor Whewell replied, " I was amused at your finding bad grammar in my title-page. You see I am quite given over to a reprobate mind in the matter of concord since you left us. However, I will try to defend it when we meet."

The work was further enlarged and republished in 1842 under the title : *Architectural Notes on German Churches; with notes written during an Architectural Tour in Picardy and Normandy. By the Rev. W. Whewell, B.D., Master of Trinity College, Cambridge. The third edition, to which are added, Notes on the Churches of the Rhine, by M. F. de Lassaulx, Architectural Inspector to the King of Prussia.* The second edition is reproduced with the addition of a new Preface, and of the notes due to M. F. de Lassaulx. There are now 348 pages including the title-leaf.

There are four plates in all the editions. Thus the work grew into its final shape by the accretion of new matter, without other alteration. In the third edition the pages ix, x, 17...143 constitute the original work of 1830; the pages 1...16, and 227...333, consist of the additions which first appeared in 1835; and the rest of the volume pertains to 1842.

Professor Whewell was accompanied by Mr Rickman in his French tour; on one occasion their antiquarian zeal led to the temporary inconvenience of an arrest by a serjeant-major of the national guard: see page 294 of the third edition.

The notes of M. de Lassaulx are translated from the original, published in Klein's *Rhine Journey from Strasburg to Rotterdam.* Probably in subsequent editions of that work the notes have been revised. There is a letter from M. de Lassaulx to Professor Whewell containing some additions to them; this letter is dated May 2, 1842, and apparently it did not reach England soon enough to be of use to Professor Whewell, as his preface is dated June 27, 1842. M. de Lassaulx seems to have been much attached to his eminent English friend; in one letter he sends a plan of a house which he was building for himself, with a guest-chamber marked, which Professor Whewell was earnestly invited to occupy.

In the Preface to the third edition Professor Whewell says that his book " is copiously quoted in a striking article which has recently appeared in the *Quarterly Review."* The article is one on the *Principles of Gothic Architecture,* in No. 137 of the *Quarterly Review.*

It will be seen that in the original title these words occur: *with remarks on the origin of Gothic Architecture.* It is singular that in the following editions this part of the title disappears, though it is really the most important part. The reader from the later titles might be led to expect that various great German churches are successively described; but the plan of the work is very different: certain theoretical views of the origin and progress of Gothic Architecture are expounded, and references are given to the buildings which justify the points of the theory. But the French tour is treated otherwise, so as to be more interesting to the general reader; thus on pages 239...245 of the third edition we have a consecutive notice of the cathedral of Amiens; then on pages 245...251 a consecutive notice of the cathedral of Beauvais; and so on.

Professor Whewell deals with Gothic Architecture in a manner which is quite characteristic of his mind; he seeks for what he himself would have called the fundamental idea: see pages 20 and 306 of the third edition. If a single sentence may be adopted as a specimen of his theory we will take one from page 313 of the third edition: "A leading circumstance then, in the formation of the Gothic style, is the introduction of vertical arrangements and lines of reference, in the place of the horizontal members, which predominate in the Grecian and Roman architecture." It would be interesting to know what are the corresponding circumstances in other styles, as the Italian or the revived Classical of page 92.

Professor Whewell must have been much gratified by one of the results of his study of French churches, which was communicated to him by M. Arago in a letter dated Jan. 10, 1845: "...Vous avez sans doute oublié un entretien que nous eûmes ensemble à Edinburgh, touchant les monuments gothiques de la ville de Rouen. J'y puisai le desir d'aller visiter St Ouen. Dans ce voyage je m'associai à votre enthousiasme et, depuis lors, j'ai sans cesse harcelé le gouvernement pour le décider à achever le monument; mes vœux viennent d'être accomplis; les chambres ont voté une somme considérable pour achever le portail de St Ouen."

An interesting testimony to the vigour with which Professor Whewell pursued his architectural studies exists in the form of an interleaved copy of Rickman's work on the *Styles of English Architecture;* in this he has recorded, partly by notes and partly by drawings, many particulars respecting churches and cathedrals which came under his own observation. It has been well said by Sir John Herschel with respect to the many objects of Professor Whewell's attention that his knowledge was "not merely a general and superficial acquaintance, but one which an exact and conscientious application, such as most men give to some favourite branch of study, alone can give." In fact, in all he undertook, his motto may be said to have been the word—once rendered famous in a far less worthy application—*thorough*. A curious illustration is remembered by one of his oldest friends. In those pre-Tennysonian days every person of taste in the University delighted in the works of Henry Taylor; and Professor Whewell was found engaged in the perusal of *Philip Van Artevelde*, with Froissart by his side—carefully comparing the modern drama with the ancient chronicle.

CHAPTER IV.

1831...1833.

THE year 1831 gives a striking proof of the variety of Professor Whewell's attainments, in reviews which he wrote of three works in different departments of science, each by a master in his own subject. Scarcely any person but himself could have ventured at once on an estimate of the labours of Herschel, and Jones, and Lyell. Before however we consider these reviews, we will advert to other events of the year.

Some of the most important of Professor Whewell's scientific intimacies date from about this year, namely, those with Professor Faraday, Professor J. D. Forbes, the Rev. W. Vernon Harcourt, and the Marquis of Northampton.

The first letter from Professor Faraday which has been preserved is dated Feb. 21, 1831. It refers to some article which had been sent to him for publication : "When a friend from Cambridge (who had received it from a friend of yours) gave it to me to put into the Journal if thought fit for that purpose, you may suppose I did not hesitate a moment in my opinion." The article was probably that *On the employment of notation in Chemistry,* which was published in the *Journal of the Royal Institution of England,* 1831. The letter has the following postscript: "Mr Daniell wished me to ask you where we could pay to your account the sum (I think seven guineas) for the paper on Arches, &c." This apparently refers to an article by M. de Lassaulx, on light vaults over churches, communicated by Professor Whewell, which was published in the same volume. The payment for articles inserted in the official publications of a scientific society seems to be unfortunately discontinued in the present day.

Professor Whewell formed the acquaintance of Mr. Forbes in the month of May. The youthful Scotchman visited Cambridge with some letters of introduction, particularly one from Mr Babbage, and one from Sir D. Brewster, both addressed to Professor Whewell; and he seems to have won all hearts by his attractive manners, his striking ability, and his enthusiastic devotion to science. His diary records his own delight in the society of the eminent persons he met: see his *Life*, pages 70...73. The acquaintance thus commenced with Professor Whewell soon grew into a close intimacy, and the correspondence between the two forms one of the most interesting portions of the papers on which the present work is founded. It is remarkable that a comparative stranger like Mr Forbes gained a place in Dr Whewell's regard at least as high as that occupied by the most cherished friends even in the much-loved Trinity College itself; perhaps it might be owing in some degree rather to contrast than to identity of taste. Mr Forbes was more of an experimentalist than his friend or most of his Cambridge contemporaries; while, as a mathematician, he was not conspicuous, and for metaphysical subjects he had little of that fondness which is almost indigenous in Scotland.

The correspondence between Mr Harcourt and Professor Whewell commenced in the autumn by a letter from the former, which announced the approaching meeting at York with a view to the formation of the *British Association for the Advancement of Science*, and asked for advice and assistance. Professor Whewell was prevented from attending the meeting, because the time was that appropriated to the election of Fellows in his College, at which he had to assist; but he approved of the design, and became a very active member of the new society. To him we are indebted for the most valuable among the many results produced by the Association, namely, the *Reports* which it has published on the current state of various sciences: see page 36 of the first volume of the Reports.

Professor Whewell became acquainted with the late Marquis of Northampton through Dr Buckland, who wished to procure for him a copy of the work on German churches, which was then out of print. The Marquis was a member of Trinity College, and giving great attention to Architecture, Geology, Mineralogy, and other scientific subjects, was naturally led to cultivate the society of Professor Whewell.

Professor Whewell seems, during this year, to have taken somewhat more interest than usual in political events; and was perhaps concerned in drawing up a petition respecting the Reform Bill from some members of the University. Mr W. M. Praed, then member of Parliament for St Germains, writing to him on March 20, says: "I am much obliged to you for the copies of your petition, which I received this morning. It seems to me very admirably drawn up—temperate, grave, and forcible—just such a remonstrance as ought to come, and to come with effect, from such men at such a time." Mr Jones, in a letter of which the date is not quite legible, but which seems to be March 27, says: "Do you see yourself abused in yesterday's Times? I suppose we shall find at last that those who keep above or aloof from mere party politics had better keep aloof altogether, unless *in extremis;* for it is clear we shall get understood by neither party, and abused by both. My objections to the reconstructive part of the measure are I suspect more vital, and would therefore be more abused, than yours; while I rejoice at getting rid of the nomination Boroughs, and should be well abused by our friends of the opposite tack for that I suppose."

Writing about this time to his friend Mr Rose, Professor Whewell says: "It makes me glad in the worst of times to find that I am not likely to lose your regard and confidence, but I have lost my trust in my own guidance as to action, and should be well content to sit and speculate, if times and people would allow us. I send you a few sentences which I have printed, not for general circulation, but for the benefit of persons who, like myself, like to look at a matter on various sides. I do not conceive that you will agree with them, but you may perhaps bear to see them." No copy of these printed sentences has been found among the papers of Dr Whewell.

At the general election which took place in May, Professor Whewell supported Mr Cavendish as a candidate for the University; Mr Goulburn and Mr W. Y. Peel were elected. Lord Palmerston was also a defeated candidate.

In March, Professor Whewell was requested by Mrs Young to write the life of her late husband, Dr Thomas Young, or, in case of his own inability to undertake it, to suggest a suitable person; he seems to have advised an application to Mr Peacock, who, after more than twenty years, accomplished the work.

Towards the end of the year Professor Whewell was placed on the Council of the Royal Society. Mr Jones regretted this, as he thought that it might lead to the loss of the goodwill of some estimable persons—referring of course to the dissensions which had for some time disturbed the Society.

I now proceed to notice the various publications of the year.

An article by Professor Whewell, entitled *Cambridge Transactions. Science of the English Universities*, appears in pages 71...90 of the *British Critic*, No. 17, published in January, 1831.

The object of the article is to vindicate the English Universities, especially that of Cambridge, from the charge of neglecting modern knowledge and improvements. To repel the charge, the article, after a few preliminary remarks, reviews the contents of the third volume of the Transactions of the Cambridge Philosophical Society; it is maintained that this volume is a "proof of intellectual activity, of zeal for science, of perseverance and intelligence in its prosecution, of familiarity with the most valuable portions of recent discovery."

An extract from the early part of the article may be given:

"Indeed, it seems to be allowed, by some of the complaining voices, that the two habits—that of original discovery and that of academic instruction—are not very compatible; and that it is rather too much to expect that a person undertaking one of these offices is to be called upon for the performance of the other, precisely in proportion as it is impossible for him to give his thoughts and time to such a pursuit. A project, it appears, founded on the assumption of this incompatibility, has been proposed for the benefit of the country; the amount of which is this, that £400 or £500 a year being conceived to be sufficient for a lecturing professor, the surplus of the income, which is supposed to be of about equal value, shall be given to a philosopher whose time is to be employed in original research. We believe our English professors will look with admiring envy on the happy ignorance which could suggest such a plan as applicable to their condition. In what distant and favoured land these golden visions may exist as realities we know not. In the English University with which we are best acquainted, we lament to say, the *average* income of the professorships of physical science is under £200; and we believe, that if we were to add any emolument which may arise from lectures, it would very slightly affect this average. We include in this estimate the Plumian Professorship, which has been raised to £500 per annum in consequence of having the important and heavy labours of Observer added to the duty of Professor of Experimental Philosophy. It will easily be understood therefore, that several of these professorships do not exceed £100 per annum, as is in fact the case. We can imagine that the holders of these would be delighted with the operations of a reformer who should level all these offices *up to* the standard above mentioned."

The reader will observe in the preceding extract two phrases which have since become famous : *original research* is the motto inscribed on the banners of our contemporary academical reformers; and *levelling up* is familiar to us in the vocabulary of party politics.

The article includes a notice of a memoir by Professor Whèwell himself on the "Mathematical Exposition of some Doctrines of Political Economy." We extract a part of this :

"In dealing with the moral elements of our nature, the laws of space and number bestead us but little. If Professor Whewell, as a mathematician, chooses to lay claim to all the political economy which has nothing to do with such elements, we can only wish him joy of his acquisition : and we shall be greatly obliged to him to invent symbols, as we have no doubt he may do, which will condense the whole science into a single sheet of paper. We may point out to Professor Whewell's notice, that M. Dupin, in France, has been pursuing a somewhat similar train of speculation, and apparently in a much bolder spirit."

The first volume of Lyell's *Principles of Geology* appeared in 1830; it was reviewed by Professor Whewell in pages 180...206 of the *British Critic*, No. 17, published in January, 1831.

The review is a very lively and interesting notice of a book which has since become very popular. Professor Whewell awards high praise to the author, while holding opinions contrary to his on one important point. Mr Lyell maintained "in short, the well-known Huttonian doctrine, that the strata at the surface of the earth have been formed by the agency of causes which will continue to act in the usual course of the world." In opposition to this doctrine of *uniformity* Professor Whewell enforces that of successive *catastrophes*, and the opinions which he here delivers he advocated subsequently in his *History and Philosophy of the Inductive Sciences*. The contrast between the two opinions will be seen best from a few sentences of the review:

"Hutton, for the purpose of getting his continents above water, or of manufacturing a chain of Alps or Andes, did not disdain to call in something more than the common volcanic eruptions which we read of in newspapers from time to time. He was content to have a period of paroxysmal action—an extraordinary convulsion in the bowels of the earth—an epoch of general destruction and violence, to usher in one of restoration and life. Mr Lyell throws away all such crutches; he walks alone in the path of his speculations; he requires no paroxysms, no extraordinary periods; he is content to take burning mountains as he finds them; and with the assistance of the stock of volcanos and earthquakes now on hand, he undertakes to transform the earth from any one of its geological conditions to any other. He requires time, no doubt: he must not be hurried in his proceedings. But if we will allow him a free stage in the wide circuit of eternity, he will ask no other favour...."

And again,

"Common readers will probably not be disposed to consider as requiring any refutation, a theory which asserts that the elevation of the Andes from the bed of the Pacific is a phenomenon 'of the same kind' as those which happen in our times. But, to do Mr Lyell justice, it is to be taken into account, that having, as he thinks, eternity for his working-time, and volcanos and earthquakes for his tools, he naturally trusts much to the result of continued labour. He deems it an extravagance to raise a mountain 10000 feet at once, when he can gain the same object by raising it a yard at a time, only taking care to begin his work a few millions of years earlier. He treats with disdain the notion, 'that nature had been, at any former epoch, parsimonious of time and prodigal of violence.' To let off all his volcanos at once, would appear to him as contrary to the economy of nature, as to blow up his powder magazine would be to that of a good general. He brings out, on the contrary, a well-disciplined park of subterraneous artillery. ' Let a series of two hundred earthquakes strike the shoal, each raising the ground ten feet'—such is the word of command—'the result', he adds, 'will be a mountain two thousand feet high'."

Professor Whewell took a great interest in the progress of the work on Rent by his friend Mr Jones, and corrected the proofs for him. The work was published early in the year 1831 under the title, *An Essay on the Distribution of Wealth, and on the Sources of Taxation. Part I. Rent. By the Rev. Richard Jones, A.M., of Gonville and Caius College, Cambridge.*

The *London Literary Gazette* of February 19, 1831, contains a short notice which we may attribute to Professor Whewell on the authority of one of his own letters. The notice occupies a little more than three columns : it explains the different kinds of rents according to the views of Mr Jones. The work was reviewed by Professor Whewell in pages 41...61 of the *British Critic*, No. 19, which was published in July, 1831.

An article had been intended and accepted for the *Quarterly Review,* and Professor Whewell had looked forward to the great pleasure he should have in introducing to the public through the same number of the famous literary journal, both the work of Jones and Herschel's *Discourse on Natural Philosophy :* he anticipated with just pride this association of himself with his two distinguished friends. But the department of Political Economy in the *Quarterly Review* was claimed by another person, and Lockhart, the editor, expressed his regret at the breach of the engagement which had been made, at least implicitly, with Professor Whewell. The article prepared for the *Quarterly Review* seems to have remained unpublished; while the present article was gladly accepted by the editor of the *British Critic*.

Mr Jones divides Rents into various classes, namely, *Serf* Rents, *Metayer* Rents, *Ryot* Rents, *Cottier* Rents, and *Farmer's* Rents: according to Mr Jones and his reviewer, political economists had hitherto overlooked all forms of Rent except the last. Moreover they had adopted a *deductive* mode of treating the subject, instead of an *inductive*. The review gives a clear account of Mr Jones's views and method, and strongly recommends them; it may be said to be condensed into the earlier pages of the *Prefatory Notice* contributed by Dr Whewell to the *Literary Remains* of Mr Jones, published in 1859.

The reviewer maintains very strongly that the method hitherto adopted by political economists was bad; they had attempted to construct the subject on a few arbitrary definitions and axioms, instead of making careful inductions from the mass of actual facts; and it was consequently not surprising that different schools had arrived at different conclusions. He says then: "We do not think, however, that it is difficult to point out what influence it is which has thus sent dissension into the host of the Greeks. The names about which they contend, and which they wish to make to appear to be the marks of arbitrary notions limited by definitions, are in fact too goodly spoil to be so easily resigned or transferred. If the fair-cheeked Chryseis, and Briseis with the radiant eyes, had been mere names, and had not been the objects of desire and interest, it is not to be supposed that the king of men and the divine Achilles would have torn the camp asunder with their mutual ire. If rent and taxes, wages and capital, were things with regard to which we might assert our propositions without moving men's minds more than we do when we talk of *lunes*, or of the *angle of a semicircle*, we should no doubt be left to make of them what definitions we chose. But the fact is far otherwise. Both the professors of the science intend, and the public expect of them, that the propositions which they from their ample store put forth on such subjects, should employ the words in the sense in which they are commonly understood, should refer to the interests with which those grave realities are connected. Hence, however the political economist may begin, as he goes on he must necessarily endeavour to bring his language into a consistency with common apprehension and obvious facts. And hence arises a perpetual action and reaction between the practical feelings of mankind and the logical consistency of the school; and this intercourse is, we regret to say, very much to the disadvantage of both parties; since it ruins the only scientific merit of the theorists, while it introduces into the business of the world a number of dogmas and forms of language, which, as they are applied, are arbitrary and false, and could never, by any fair philosophical process, have acquired circulation or influence."

Mr Herschel's *Preliminary Discourse on the Study of Natural Philosophy* appeared in 1830; it was reviewed by Professor Whewell in pages 374...407 of the *Quarterly Review*, Number 90, published in July, 1831. Mr Lockhart says in one note to Mr Whewell, "The article appears to me very excellent," and in another, "I have just seen Murray, who is a good judge of the probable popularity of an article, and he says yours on Herschel would be greatly improved by some more quotations of an interesting character." The second note is dated March 1, and thus, perhaps, in the long interval which elapsed before the publication of the article, Murray's suggestion was adopted; for there are several quotations. As is the case with many of Professor Whewell's articles, we find anticipations of the matter afterwards produced in his great works; here especially we may notice his views on the nature of induction, and on the distinctive merits of Bacon.

We will extract a few characteristic sentences, beginning with a joke which seems sufficiently forgotten to bear revival.

"Some time back, the eminent person who now, as a scientific Lord Chancellor, is peculiarly Bacon's successor, expressed a hope that in the course of the schoolmaster's triumphs the day would come when every Englishman would be able to read Bacon: Mr Cobbett, who has long maintained that the true interest of Englishmen is to keep pigs and read his Register, observed somewhat contemptuously, that it would be more to the purpose to hope we might all come to *eat* bacon."

"It is ever thus in the progress of inductive philosophy, as in the wanderings of the Alpine traveller. The hill which at starting seemed the only barrier between him and his journey's end, serves but as a point from which he descries higher mountains to climb, and wider fields to traverse; but still every step is an ascent, every new prospect is a gain; and the august forms which surround him, and the pure atmosphere into which he rises, make his toil a sport and his perseverance a delight."

After speaking of the combination of induction and deduction which forms the entire scheme of a perfect science, the reviewer remarks that few sciences have completed this cycle; they are indeed in very varied points of their progress. He continues thus: "They have long been understood to form a sisterhood, and the different individuals of the group offer to us all gradations of growth, from the tottering girl to the full-formed matron. Some of them, indeed, have been long established in the world, and can look for no further advancement, except an increase of their progeny; others appear to experience all the restlessness of youthful expectation and hope of change; others, again, have not yet left the nursery and the spelling-book."

A curious illustration of the strange difference of opinion which exists as to the distinctive merit of Bacon is seen in the fact that Mr Herschel scarcely notices an element which Professor Whewell considers of vital importance: "But there is one warning of the inductive legislator respecting the vicious indulgence of this spontaneous impulse, which it would be wrong not to mention. We refer to his condemnation of the method of *anticipation*, as opposed to that of gradual induction; a judgment indeed which of itself almost conveys the whole spirit and character of his philosophy, and which therefore, we have been surprised to find not more distinctly touched upon in Mr Herschel's discourse."

In the review some notice is taken of "that portion of physical science which Bacon calls 'deductio ad praxim'; its application to the uses and needs of human life." Here, as in his great works, Professor Whewell looks coldly upon such application as quite distinct from science itself, and even somewhat unfavourable to it; he was however more tolerant in one of his latest works, namely, the *Novum Organon Renovatum.* He says in the present review:

"The true and worthy claim of knowledge is that which every lover of it feels in his own heart;—that it is valuable for its own sake; that truth is worth having because it is truth; the more worth having, the more pregnant and comprehensive the character which it possesses. Utilitarian moralists may maintain that we cannot have any wise motive of action except our own advantage; utilitarian philosophers may maintain that we cannot have any sufficient inducement for research except the tangible benefit of our expected discovery. The consciousness of every good man contradicts the former dogma; the irresistible impulse of every true philosopher—every man with the spirit of a discoverer—is inconsistent with the latter. Even if we were to confine ourselves to the pleasure produced, if we were to put the love of truth on a level with the love of turtle, still the former delicacy may probably be more widely and intensely relished, certainly more generally and equally diffused; and we do not see why the gratification which men may receive from knowing the laws which regulate the motions of light, is not as worthy our regard as that which they would derive from travelling from London to Brighton in an hour and a half, or from breakfasting on fresh strawberries every day in the year. But in fact the love of knowledge ought not to be degraded so far as to be weighed ounce for ounce against the pleasures of sense. It differs from them as a duty differs from an indulgence. Knowledge is followed because it is itself a good: it is an end, not a means. There are affections directed towards it as distinctly marked, and as elevated in their kind, as any other portion of our mental constitution. We are its suitors for itself, and not for its dower; and if we were allowed to borrow Mr Shandy's favourite quotation, and to translate as freely as he was in the habit of doing, we would say, 'Amicus Socrates,' utility is a pleasant companion, 'sed magis amica veritas,' but truth is a beloved friend."

The review of Herschel's *Discourse* is mentioned more than

once in Professor Whewell's letters to his friend Mr Jones. Thus he says it "took me a good fortnight on a subject quite familiar to me, and where I had nothing to read." And again: "You may perhaps recollect when you see the Q. R. that a passage about the Family Library near the beginning, and one about Sedgwick and others near the end, are interpolations of the worthy editors. With the latter however I by no means quarrel, except at being made to praise Leslie's style." Mr Jones directed his attention to various striking passages in the book which might be available for quotation.

Mr Herschel's own opinion of the review will be seen by the following extract from a letter, dated Sept. 29, 1831: "Had I known where you were I should not have let so long elapse without thanking you for the splendid review you put forth on my poor book, as well as for the high treat of your two reviews of the science of the English Universities and Lyell's Geology—the latter especially I have read over till I have almost got it by heart, and think more of it than. I will trust myself to say. As to my own share of your critical dispensations, I can only say that it seems to me to approach as nearly to the nature and value of posthumous fame, as any contemporary applause can possibly do, and that I should have envied the author of any work if a stranger, which could have given occasion for such a review."

Mr Herschel's *Discourse* was published in Lardner's *Cabinet Cyclopædia*, and he afterwards contributed to the same series an elementary Treatise on Astronomy. Professor Whewell was not quite satisfied with the channel which his eminent friend thus accepted for his writings. Dr Lardner was a man of scientific attainments, and of considerable ability for popular exposition; his importunity in urging the fulfilment of the promises which he obtained of cooperation in his Cyclopædia, and his name *Dionysius*, which it was conjectured he had himself modified from the more familiar Denis, naturally led to the appellation *tyrant*, which was given to him in a good-tempered manner by Southey and other literary men of the period. He made various attempts to induce Professor Whewell to join his staff, and in particular during the present year wished to engage him to write on Political Economy; but the applications were in vain. Professor Whewell, perhaps, mentioned the matter to Mr Jones, as we may conjecture from a sentence in a letter from him, " I should like to write a treatise for the tyrant if he would wait two or three years, but he shall not have the *prémices* of my speculations."

I may briefly notice an octavo pamphlet, entitled, *Reply to "Observations on the Plans for a new Library, &c. by a member of the First Syndicate." By a member of both Syndicates. Cambridge,* 1831.

This consists of 36 pages, besides the Title and the prefatory notice on four pages; it was published anonymously, but may be safely attributed to Professor Whewell. The "Observations" were, I believe, written by Mr Peacock; they advocated a design for the new Library which was proposed by Mr Cockerell. The "member of both Syndicates" was in favour of a design which was proposed by Messrs Rickman and Hutchinson. After the delay of a few years the University finally chose Mr Cockerell as the Architect for the new Library. The University seems to have been unfortunate in this and some other of its architectural efforts; for further discussion of the matter I may refer to a pamphlet, entitled: *Observations on some recent University Buildings...by Francis Bashforth, M.A., Fellow of St John's College, Cambridge,* 1853.

The year 1832 was full of work for Mr Whewell. He was busy in the construction or reconstruction of some of his mechanical text-books, in drawing up a Report on Mineralogy for the British Association, and in the composition of his *Bridgewater Treatise.* To these we must add his college duties, which had now probably reached their greatest extent, and his incessant efforts to rouse his able but languid friend, Mr Jones, to continue his labours in Political Economy.

We need not wonder that he resigned the professorship of Mineralogy; he was succeeded, to his great satisfaction, by Mr W. H. Miller. On retiring from the office Mr Whewell presented to the University a collection of books and specimens, together with the sum of £100. The specimens consisted of about 1000 articles, in a great measure arranged, ticketed, and catalogued:

the money was to be devoted towards the erection and fitting of a room for the reception of the books and minerals.

On Feb. 3, Mr Whewell says, in a letter to Mr Jones, "I am plunging into term-work, hurried and distracted as usual;—the only comfort is the daily perception of what I have gained by giving up the professorship. If I can work myself free so as to have a little command of my own time, I think I shall be wiser in future than to mortgage it so far. Quiet reflexion is as necessary as fresh air, and I can scarcely get a breath of it."

In April, Mrs Somerville with her husband, and Sir W. R. Hamilton of Dublin, paid a visit to Mr Whewell, who was much pleased with his distinguished guests. Mrs Somerville acknowledged in warm terms the satisfaction she had felt, "A week such as we passed at Cambridge does not often occur in the course of one's life, and I assure you the impression it has made will not be effaced." A few letters from this remarkable lady are preserved among Mr Whewell's papers, and it is curious in connexion with them to recall a circumstance connected with her childhood. Mrs Fairfax regretted the money spent on her daughter's early education, as she had not learnt even to write fairly and to keep accounts. The deficiencies were fully supplied in after life; for the authoress of the *Mechanism of the Heavens* could have found little difficulty in household arithmetic, and her hand-writing is conspicuous for its regular beauty. No letters in the large collection which Dr Whewell preserved, are finer specimens than those which conclude with the honoured name of Mary Somerville.

In June Mr Whewell attended the meeting of the British Association at Oxford. Fifteen years later Dr Hawkins, the Provost of Oriel, reminded him of the earlier visit, saying, "You were to have been one of my guests when the British Association first met in Oxford, but Buckland caught you away. Shall I be more fortunate on this occasion, and shall we also have the pleasure of receiving Mrs Whewell?" In August Mr Whewell made an architectural tour in Normandy in company with Mr Rickman; the materials which he collected were published in the second edition of his work on *German Churches*. In October he preached at Bury in behalf of some schools. In a Parliamentary election in December he seems to have supported Mr Lubbock, afterwards Sir J. W. Lubbock, who was not successful.

The correspondence of Mr Whewell with various scientific friends seems to begin in the present year—namely, with Professor De Morgan, Professor Phillips, Professor Rigaud, and Mr Eaton Hodgkinson. Professor De Morgan consulted him respecting an important memoir on Physical Astronomy, to which it was proposed to award a medal by the Astronomical Society. The business of the British Association led to his intimacy with Professor Phillips, who was then connected with the Yorkshire Philosophical Society. With Professor Rigaud he corresponded first on some points in the history of Mechanics, and afterwards on circumstances connected with the life of Newton. Mr Hodgkinson invited Mr Whewell's attention to the experiments he was then making on the impact of bodies; these afterwards formed the subject of a communication to the British Association.

I shall now notice some publications which belong to this year.

Mr Whewell reviewed the second volume of Lyell's *Principles of Geology*, in pages 103...132 of the *Quarterly Review*, Number 93, published in March, 1832. The review was written at the request of Mr Lyell, who supplied Mr Whewell with the sheets in the course of the printing. Mr Lyell was in some fear that the review would be finished before all the sheets had been received and read. This volume, like the first, is highly praised by the reviewer, and on one very important point he is on the same side as the author. I will extract a few sentences:

"His [Mr Lyell's] first volume contained a very masterly exposition of the present mode of action and the intensity of the moving forces of the earth, with a defence of the sufficiency of these to explain all the geological phenomena which belong to that part of the subject. The present volume is occupied with various discussions on the laws and limits of the variability of organisation, and is an estimate of the nature and amount of the alterations which causes, belonging to the animal and vegetable world, are now producing. The author's conclusion is, that the changes of this kind at present going on, are highly important towards the explanation of many of the facts of geology: but that the appearance of new species at successive epochs, which we learn from irresistible geological evidence to have repeatedly occurred, is a fact *not* belonging to the operation of that tendency to change in organised beings, which we see still brought into play."

The following sentences are in the same vein as some in the review of the first volume:

"The Huttonians, Mr Lyell's predecessors, thought, that when they had laid the materials of their calcareous or siliceous strata at the bottom of the sea, the action of subterraneous fire was requisite in order to convert them into sparry limestone or quartz rock. They deemed it necessary to bake their cake, when they had kneaded it; and all will recollect their exultation when Sir James Hall drew from his oven a marble loaf made of chalk flour. What does Mr Lyell intend to substitute for the Plutonic cookery of these elder assertors of the constancy of nature? Or is he prepared to maintain that this application of fire is superfluous, and that time alone, who does so much for him, will give the due solidity and structure to the stratified masses?"

The following allusion is to Sir R. Murchison and his former exploits in the hunting-field: "The same geologist has further had the joyous recollections of his earlier life revived by *viewing and digging out* of that locality a fine fossil fox, ..."

Two opinions on this review are contained in a letter from Mr Jones of March 20: "Herschel likes your review of Lyell much, but almost laments the jokes, good he says, but he does not like to spare your 'strong statement and sound and lucid philosophy' even for good jokes." "I like your last on Lyell better than any thing you have done in that way—the style is much the best."

A paper by Mr Whewell on *English Adjectives* occupies pages 359...372 of the *Philological Museum*, Vol. I, 1832.

Adjectives may be divided into two classes, which are called *adjectives of quantity* and *adjectives of relation;* the main purpose of the essay is with adjectives of the second class; "those, namely, which have a manifest and distinctly felt reference to some primitive: either a concrete substantive, as *wooden, fatherly*, or a verb as *tiresome, seemly.*"

The subject of the essay had a special interest for Mr Whewell on account of the demands which his scientific friends made upon him for the terminology they required in order to expound their researches. Some of the points noticed in the essay appear again in the *Aphorisms concerning the language of Science*, which form part of the *Philosophy of the Inductive Sciences*. Thus we have, here as there, *kallesthetics* suggested as the word to describe the doctrine of the perception of beauty. This word allows of a corresponding adjective; Mr Whewell says: " *Tasty*, as Mr Coleridge has observed, is a word which milliners only can venture upon."

The following is the last paragraph of the essay:

" Even in genuine English words it sometimes happens that the adjective has been derived from its root viewed under some particular association, so that the reference to the fundamental notion as commonly understood is by no means obvious. Of this we have a good instance in the excellent proverb, " home is home, be it ever so homely." A writer of Latinised English would perhaps thus separate the feelings, which the early framers of our language have mingled: " The scene of our domestic comforts must always have a peculiar charm, in spite of the inelegance which is often found among our familiar habits."

The present contribution to the *Philological Museum*, and another in 1833, have simply the letter W. subscribed. The W. W., which at a later period was so familiar to Cambridge men, might at this date more naturally have suggested William Wordsworth; and to him, probably, we may attribute a *Translation of the First Book of the Æneid*, subscribed with these letters, in the first volume of the *Philological Museum*.

A pamphlet of 8 octavo pages bears the title *Clipt Words;* it has neither date nor name, but I believe it is by Mr Whewell, and presume that it may belong to about the year 1832. I will extract the introductory paragraph, and a few examples:

" The strength and singleness of the principal accent in English words point out one part of each word or phrase as prominent and characteristic: and our love of speed and dislike to long words give us a tendency to select and employ this prominent part, rejecting the other syllables as a superfluous appendage. Such a tendency is shewn at first when expressions are used in a familiar or humorous manner : but the words thus *clipt* are often gradually established as respectable portions of the language, and end by being as completely recognized as any part of the vocabulary."

"*Phiz* is in common use for *physiognomy*, which was formerly called *physnomy*, and in Spenser is written *visnomy*, perhaps as though connected with *visage*."

"The old word, a *chapman* (*kaufmann*), for a tradesman, has been curtailed into a *chap*."

"The *drawing-room* was originally, what Mrs Margaret Bellenden calls it, the *withdrawing-room*."

"*Doll* perhaps comes from *idol, idole*."

In the year 1833 Mr Whewell's *Bridgewater Treatise* appeared, and also the first of a long series of papers on the *Tides* in the *Philosophical Transactions;* I shall speak of these after noticing some other matters of less importance.

In June 1833 the British Association met at Cambridge. Mr Whewell induced M. Quetelet of Brussels and Sir W. R. Hamilton of Dublin to attend the meeting; the latter had hesitated on first receiving the invitation on account of his approaching marriage; for he says : "this will make so sensible a perturbation in my orbit, that I do not venture to predict when I shall return to the perihelion of the British meeting."

Mr Whewell was one of the Secretaries of the meeting, and delivered an address similar in character to those which are now usually given by the President for the year. It is contained on pages xi...xxvi of the volume of the Association for 1833.

This Address is very vigorous and interesting, containing the brief exposition of principles which were afterwards developed in the author's works. Thus we have Astronomy recognized as "not only the queen of sciences, but, in a stricter sense of the term, the only perfect science." We are told that "a combination of theory with facts, of general views with experimental industry, is requisite even in subordinate contributors to science......Or if the word *theory* be unconquerably obnoxious, as to some it appears to be, it will probably still be conceded, that it is the rules of facts, as well as facts themselves, with which it is our business to acquaint ourselves." The superiority of science to art is strongly enforced, though with due gratitude to the latter. "Art has ever been the mother of Science; the comely and busy mother of a far loftier and serener beauty."

I do not know how the extreme antiquity of Astronomy is supported, which is implied in the statement on page xxiv, that for its success "the labour of the most highly gifted portion of the species for 5000 years has been requisite."

Early in June Mr Whewell made an architectural tour in Northamptonshire in company with Mr Rickman; later in the summer he accompanied Professors Sedgwick and Airy on a geological excursion.

In October Mr Whewell obtained relief from part of his College duties, particularly from all that related to financial matters, by engaging, as Assistant-tutor, Mr Perry, now Bishop of Melbourne. A report in consequence arose that he had himself resigned the tutorship, and he had to contradict this, and to explain the circumstance to various correspondents.

In July Mr J. C. Hare took up his abode in the Rectory of Hurstmonceux, and his departure from Cambridge was a great loss to his friends in the University. Mr Whewell succeeded to his rooms in College, which have always been much admired on account of their pleasant situation in the new Court, looking down the lime tree avenue. In speaking of one of the rooms to his predecessor, in a letter dated Sept. 22, Mr Whewell says: "it is really very pretty, and I think prettier with my furniture than it was with yours, as I hope you will have the sagacity to perceive and the candour to allow in November." Mr Hare was unable to pay the visit to Cambridge in November, and in a letter, dated Dec. 13, he says: "Thank you for your ample account of our rooms: but really you must suppose me to be the most supercandid person in the world, to fancy that I should allow that they look as well now as they used to do, when they used to be the admiration of everybody who entered them, were so often pronounced to be unique and *ganz einzig*, and when Landor declared that the only thing which would induce him to live in England would be to live in them. However, I have no doubt that in your hands they will be very well-looking, although those who remember the first temple might some of them feel inclined to regret it."

Late in the year Sir J. Herschel left England to take up his abode at the Cape of Good Hope, in order to survey the stars of the Southern Hemisphere; and during the next four years he wrote long and interesting accounts of his occupations to his friend Mr Whewell. He returned to England in 1838. On Dec. 10 Mr Whewell mentions in a letter to Mr Jones, that he had just began his book, that is, the *History of the Inductive Sciences.*

A paper by Mr Whewell, *On the Use of Definitions*, occurs on pages 263...272 of the *Philological Museum*, Vol. II, 1833.

The doctrine here enforced is that which he always maintained, namely, that definition is the last stage in the progress of knowledge, and not the starting-point: see the *Philosophy of the Inductive Sciences*, 2nd edition, Vol. II, pages 11...16. The illustrations given in the paper, for the most part, reappear in the author's later writings. I give one as an example:

" A lady who was describing an optical experiment which had been shewn her by a great philosopher, said: " He talked about increasing and diminishing the angle of incidence; and at last I found he only meant moving my head up and down." The philosopher's phraseology would have been far less commendable than the lady's, if he had not known that his terms referred to an essential, and her's to an accidental, condition of the experiment." See the *Philosophy of the Inductive Sciences*, 2nd edition, Vol. II, page 25.

The following interesting anecdote is not reproduced in the author's later works; though there is an allusion to it in the *Philosophy of the Inductive Sciences*, 2nd edition, Vol. II, p. 504 :

" At one of the meetings of the Geological Society of London, a memoir was read on " *The Green Sand,*" by an eminent member of the Society. At these meetings, the readings are followed by oral discussions, usually conducted with a rare mixture of acuteness and good breeding. On the occasion just mentioned, a distinguished geologist, well known both for the extent of his knowledge and the fastidiousness of his taste, stated that he had three objections to the *Title* of the paper: First, to the article *The*, since there are several green sands: second to the adjective *Green*, since the stratum spoken of is more commonly red: third to the substantive *Sand*, because in many cases it is more calcareous than siliceous. The subtlety of this criticism was applauded: but the name still keeps its ground, and is to this day a good and serviceable name, inasmuch as it is universally understood to designate certain members in a known and widely extended series of strata."

To this paper on Definitions, Mr Malthus seems to allude in a letter dated April 1, 1833: "I confess I was a little alarmed at it at first; and thought it was an attack upon my definitions in Political Economy, which I certainly do not consider as useless. I agree with you in thinking that new definitions of terms are not always necessary to get at truth; and that the most exact definitions are not so much the causes as the consequences from our advances in knowledge. At the same time, I should say, that in regard to this latter position, they act and react upon each other, and that without some understanding as to the meaning of the words used the advances in knowledge would be very slow, though it might still be quite true that you would not arrive at the very best definitions, till a very great progress had been made."

CHAPTER V.

THE Bridgewater Treatise was probably the work which first made Mr Whewell famous in the wide circle of general readers, as distinguished from the cultivators of science. It is well known that the eight thousand pounds left by the Earl of Bridgewater for the production of a work on the Power, Wisdom, and Goodness of God, was divided among eight selected writers; to Mr Whewell was assigned the subject of *Astronomy and General Physics considered with reference to Natural Theology.* The work was published in 1833; it was the first of the series, and perhaps has been the most popular of them on the whole. It is an octavo volume of nearly 400 pages; four other editions in the same size were issued between the present date and 1837; and subsequently two editions in a smaller size were issued, the last in 1864. There is I believe no variation of any importance in the successive editions, so that the work is one of the few examples of nearly permanent form in the author's career. I have used the third edition, which is dated 1834.

The work is dedicated to the Bishop of London, Dr C. J. Blomfield, by whom Mr Whewell had been selected as one of the writers. It was reprinted in New York, and translated into German.

The treatise is divided into three Books, which are entitled *Terrestrial Adaptations, Cosmical Arrangements,* and *Religious Views.* The first of these three Books seems to me on the whole the most striking; the way in which the laws and facts of nature work in harmony to secure the wellbeing of man and animals and plants is illustrated with great force and variety. The chapter devoted to the Atmosphere furnishes a very remarkable example; the diversity and the importance of the functions which the atmosphere discharges are exhibited in a most impressive manner, and the inference is irresistibly drawn that such exquisite adjustment testifies to the existence of an intelligent and beneficent Creator. The second Book presents less novelty than the first, except in the last chapter, where the advantages resulting from *Friction* are developed with great ingenuity: the chapter is somewhat out of place in this Book, as a note at the beginning of it admits. The third Book is remarkable for two chapters in which Inductive and Deductive Habits of Mind are contrasted; Mr Whewell holds the former to be of greater value in science, and to have a stronger tendency to religion than the latter: this opinion he never relinquished. I venture to doubt the propriety of the distinction he draws, and the accuracy of the judgment he pronounces; but the few remarks I shall offer on the subject will be more appropriate in connexion with his works of an exclusively scientific kind.

A few points of detail may be briefly noticed.

On page 83 Mr Whewell refers to the law according to which water contracts as the temperature diminishes till we come *near* the freezing point, and then by a further diminution of the temperature it contracts no more, but expands till the point at which it becomes ice. The important consequences of this law are traced and applied to lakes and seas. It is however now known that *salt* water, unlike *fresh* water, continues to contract as the temperature is lowered to the freezing point: but this does not affect the inferences drawn in the treatise to any appreciable extent.

With respect to sound we read on page 122 "The gravest sound has about thirty vibrations in a second, the most acute about one thousand." A correspondent in 1839 called Professor Whewell's attention to this passage, and pronounced the number *one thousand* much too small; although no change was made, there can be no doubt that the number ought to be between three and four thousand.

A paragraph from pages 202 and 203 has been transformed into blank verse by Mr J. E. Reade, and printed in the Notes to

Canto v. of his poem on *Italy*, 1838. Mr Reade says "That a simple relation of the operations of Nature form Poetry of the highest order,...the following passage will testify, being merely a passage, and not a selected one, from Whewell's Bridgewater Treatise, thrown into blank verse." The poem was abridged in later editions, and the passage does not occur in them; I will therefore reproduce it, and the reader may compare it with the original, and decide whether he prefers the prose or the verse.

> Say not man only perishes: he shares
> The lot decreed to all save God himself.
> The oak endures for centuries, and falls;
> The crumbling Mountains change, and earthquakes cast them
> From their foundations: even the sea retires,
> And the emerging green field smiles above
> The roar of weltering waves: the starry worlds
> Fall, and their place in heaven is known no more.
> The Sun and Moon have written on their foreheads
> The lines of age; that they must end: they have
> Only a longer respite given than man.
> Th' ephemeræ live their hour, man threescore years;
> Empires, too, have their centuries, their rise,
> Their spring and autumn; and volcanic fires
> Hurl the fixed Island from his Ocean throne!
> The very revolutions of the sky
> Which make our time, will languish, and stand still.

On page 233 the rotation of the Earth on its axis is very strangely given as an illustration of the first law of motion; this will probably be considered as unsatisfactory. For the law asserts that a particle in motion, if acted on by no force, will move uniformly in a straight line; and the rotation of a rigid body, so far from falling under this simple law, is a matter for profound mathematical investigation. It is true that if a homogeneous sphere be put in rotation round a diameter, and left free, it will rotate round that diameter uniformly and permanently; but such a result will not hold in the case of all bodies whatever, if put in rotation round any straight lines: though some writers not conversant with the subject have supposed that this would be the case in virtue of a principle of inertia.

In pages 269...280 Mr Whewell has occasion to allude to a question which he afterwards discussed in his *Plurality of Worlds*, namely, whether the various bodies of the sky are inhabited; it is curious to see that though he does not, as in his later work, reject the notion of such inhabitants, yet he speaks with cautious reserve. He says "No one can resist the temptation to conjecture, that these globes, some of them much larger than our own, are not dead and barren...." " To conjecture is all that we can do..." "...a few of the shining points which we see scattered on the face of the sky in such profusion, appear to be of the same nature as the earth, and may perhaps, as analogy would suggest, be like the earth, the habitations of organized beings..." On the whole his own opinion seems even now to have taken that position in which it subsequently remained fixed.

With reference to the Bridgewater Treatise, Sir John Herschel makes the following remark: "But that a great change in his views as to the origin of our fundamental axioms must have taken place between the production of these [the History and the Philosophy of the Inductive Sciences] and the last-mentioned work, may be inferred from a remarkable passage in that Treatise, in which he distinctly refers the origin of even the axioms of Mathematics to experience, *i. e.* to a slow process of inductive observation, growing with our growth, and not to any innate *à priori* intuition." A reference is given to *Bridgewater Treatise*, p. 336, ch. IX. *et seq.;* where IX is apparently a mistake for VI. But it seems to me, from the study of Dr Whewell's works and correspondence, that his main philosophical doctrines were completely formed at the epoch of the appearance of the Bridgewater Treatise, when he was nearly forty years old; and thus a few sentences, which might appear to bear a contrary interpretation, must not be opposed to the uniform current of his language. Sir John Herschel, I venture to think, gives an incorrect sense to the passage; the word *axiom*, which he employs twice, does not occur in the original, though it is true we do have *first principles.* Mr Whewell might perhaps have been more cautious in some of his expressions if his attention had been drawn to them, but even as they stand they may, without any forcing, be understood consistently with the opinions explicitly maintained in his later works. By *first principles* here he very probably meant the foundations of applied sciences; for example, in political economy the first principles may be certain facts as to population, wealth, rent, and the like.

In a biographical dictionary of merit the following strange statement occurs with respect to the Bridgewater Treatise, which has been obviously deduced by some perversion from the language of Sir J. Herschel: "In this work we have the idea which he afterwards elaborated, that the origin of even the axioms of mathematics is not due to any innate *a priori* intuition, but to experience, or a slow process of inductive observation." To make this agree with Sir J. Herschel's opinion, we must change *elaborated* into *opposed* or *contradicted*.

As I have already stated the last edition of the Bridgewater Treatise is the seventh, which was published in 1864; it is in small octavo, but corresponds very nearly page for page with the older editions: I have compared it with the third, and find that even the misprint of Fourrier, for Fourier, is preserved on page 76.

This edition contains a new preface in fourteen pages. After shewing how the argument from design had been employed by Socrates and by Cicero, Dr Whewell proceeds to notice the form which some of the arguments *against* design have assumed in modern times. It has been asserted that the structure of animals has become what it is by the operation of external circumstances and internal appetencies. Dr Whewell considers that there are two enormous assumptions which make such speculations a mere work of fancy.

"First, it is assumed, that the mere possibility of imagining a series of steps of transition from one condition of organs to another, is to be accepted as a reason for believing that such transition has taken place:

"And next, that such a possibility being thus imagined, we may assume an unlimited number of generations for the transition to take place in, and that this indefinite time may extinguish all doubt that the transitions really have taken place."

Dr Whewell quotes a passage relative to the eye from an advocate of the principle of natural selection; he does not give the name of the author.

There is an article in the Quarterly Review of October, 1833, on four of the Bridgewater Treatises, of which Mr Whewell's is one. The Reviewer, while admitting the eminence of the authors selected to write this series of works, expresses a dissatisfaction, which has been very generally felt, at the arrangements made for disposing of the Bridgewater bequest.

An article in the Edinburgh Review of January, 1834, is devoted to Mr Whewell's Treatise; we may safely attribute it to Sir D. Brewster, who wrote thus to the author on June 10th, 1833: "These distractions, however, have not prevented me from reading your volume, which has been my companion in two journies to the Highlands, and I need scarcely say that I never derived more pleasure or instruction from any other book. Did I not think the opinion would be general, I would not have ventured to say to yourself that I know of no other person who could have written such a work. Some objections, however, have occurred to me in reading the Chapters on Terrestrial adaptations, especially the one on the length of the year; and if they do not disappear on a farther examination of the subject, I may perhaps refer to them in a notice of your volume, which Mr Napier has requested me to draw up for the October number of the Edinburgh Review. It is probable that I have not seized the spirit of the Chapter under the cerebral agitation of a Mail Coach." The review, while praising the ability and eloquence of the treatise, does not consider it of great importance as a contribution to Natural Theology; Sir G. C. Lewis mentions it as a "dull attack on Whewell's Bridgewater Treatise:" *Letters*, page 32. Professor J. D. Forbes wrote to Mr Whewell: "It was in many respects an unfortunate article in the Edinburgh, and as to the tone in which it was written, I am sure the author was not aware of its causticity, as indeed I *know from experience he never is.*"

The Edinburgh Review agrees with the Quarterly in the dissatisfaction I have mentioned above.

Dr Whewell was always unwisely prompt in noticing the criticisms of reviewers; and he replied to the article in the Edinburgh Review, in pages 263...268 of the *British Magazine* for March, 1834: this periodical was edited by his friend Mr

Rose. One paragraph may be noticed as enunciating a principle which was developed and enforced by the author in later publications, especially in his Memoir on the *Transformation of Hypotheses*. The critic had charged him with founding his argument on precarious theories, as for instance, the undulatory theory of light, and the doctrine of a resisting medium. He replies, " But, it may be said, you found your arguments on transitory theories, and when the theory falls, the cause you have pretended to support with such buttresses is shaken. To answer this objection at full length, would lead me too far. I will only observe, that when a theory, which has been received on good evidence, appears to fall, the really essential and valuable part of it survives the fall; that which has been discovered continues to be true." There is nothing of special importance in the reply, but it is a welcome relief in the course of a dreary volume, which seems to have been devoted mainly to the purpose of excluding dissenters from the Universities.

Mr Whewell's Bridgewater Treatise afforded much gratification to his numerous friends and correspondents.

Mr Davies Gilbert, who, as President of the Royal Society, had been officially charged with the execution of the will of the Earl of Bridgewater, expressed himself as delighted in the highest degree. His letter has the following postscript : "I am very much pleased by your Inductive and Deductive Habits, but I hear that some mathematicians are quite violent against them ; and displeased with the note from St Helena, page 338. I once paid a visit to Dr Parr at his house near Warwick—it was forty years ago—and on that occasion he more than once went into the subject pretty much as you have treated it."

Mr Malthus says, "...I can assure you now with perfect sincerity that I have been quite delighted and much instructed by many parts of the work. Perhaps the very early portion of the volume is not quite so good as the rest; but the great mass is excellent; and on the whole it appears to me that you have brought forward very valuable materials for your purpose, and have arranged and applied them in a very masterly and striking manner. The proofs of design are indeed every where so apparent, that it is hardly possible to add much to the *force of the argument* as stated and illustrated by Paley; but still it is gratifying to contemplate the new illustrations which the almost infinite variety of nature furnishes, and these you have brought forward in abundance."

Mr H. J. Rose was enthusiastic in his praise, especially commending the beauty, eloquence, and ease of the style. He says, "I have often admired the richness and majesty of your sentences. *Now* they have the only thing they ever seemed to me to want, *perfect ease.*"

Mr Rose also remarks, "I see that you have contented yourself, and rightly I think, with asserting the strength of proof from contrivance to a contriver without arguing against the denial of that truth. Will you tell me where that is best done?" Judging from the tone of recent speculation, the point which Mr Rose considered to be rightly assumed would require to be fully discussed in any future work of the same nature as the Bridgewater Treatises.

Some sentences may be quoted from a letter written by J. Blanco White, so well known for his remarkable ability and for the strange vicissitudes in his life and in his religious opinions: "...I took up the book in the morning, and could not leave it out of my hands till I had read it through that very day. I was too well acquainted in my youth with the writers whom you refute, and having long and attentively weighed their evasions of the arguments for design and final causes in the creation, I found myself prepared for such a rapid perusal as I gave your book. The interest besides, which you have given to your argument, the masterly hand with which you manage the most extensive views of a vast subject, the truly philosophical arrangement which prevents vagueness and confusion in the statement of the subject— acted upon my mind like a charm, and made me read on with a more healthy and delightful glow of soul than I ever experienced from any Poem."

The Bridgewater Treatise was reviewed by the eminent German astronomer Littrow in the Vienna *Jahrbücher der Literatur*, 1838.

CHAPTER VI.

On the subject of the Tides Mr Whewell expended much time and labour; he published his results in fourteen memoirs in the *Philosophical Transactions*, the first in the volume for 1833, and the last in the volume for 1850. He seems to have originally taken up the subject with the idea of reporting on it to the British Association. In 1837 a royal medal was awarded to him by the Royal Society for his investigations; at that time eight of the memoirs had been published, and some account of them is given in connexion with the award of the medal : see the *Abstracts of the Papers...of the Royal Society*, Vol. IV. pages 6, 7, 24...27.

An article was published in the *United Service Journal* for January, 1838, entitled *Account of Mr Whewell's Researches on the Tides;* this also includes the first eight of the memoirs, giving a rather fuller notice of them than that in the Royal Society's *Abstracts*. The article is very good, and was probably written by Mr Whewell himself.

The well-known elaborate treatise on *Tides and Waves*, by the present Astronomer Royal, appeared in the *Encyclopædia Metropolitana* in 1843. In sections VII and VIII of this treatise will be found frequent reference to Mr Whewell, twelve of whose memoirs had then been published; with his name is associated that of another labourer in the same field, namely Mr Lubbock, whose work ranges between the years 1831 and 1837. In general great praise is awarded to both these investigators, but the regret is expressed that they did not follow a better guide than what is called the *equilibrium theory :* see *Tides and Waves*, Arts.

496 and 571.　Mr Lubbock in his *Elementary Treatise on the Tides*, 1839, refers frequently to Mr Whewell's contributions to the subject.

Speaking generally, we may say that the main object of Mr Whewell's researches is to obtain an accurate knowledge of the *facts* of the Tides; to employ an illustration which is familiar to us by frequent use, he occupied the position of Kepler, leaving that of Newton for some successor.　Hence there is little reference to mathematical formulæ in the memoirs; for such theory as Mr Whewell used he was content to follow Bernoulli, and he seems to have regarded the investigations of Laplace as mere speculations of no practical utility.　Perhaps both here and elsewhere Mr Whewell scarcely appreciated sufficiently the great ability of Laplace, which is the more remarkable because it was united with a penetrating sagacity in the discussion of natural philosophy, not always combined with such eminent mathematical power: the history of science seems to shew that while Laplace yields, as all must do, to the transcendent genius of Newton, he yields to that alone.　His researches on the Tides have received the emphatic commendation of his solitary successor in these abstruse investigations: see *Tides and Waves*, Art. 117.

I will now briefly notice the memoirs in order.　The first memoir is entitled *Essay towards a First Approximation to a Map of Cotidal Lines;* this was read May 2, 1833: it occupies pages 147...236 of the volume for 1833.

By *cotidal lines* is meant lines drawn on the surface of the ocean, and passing through all the points where it is high water at the same moment.　The memoir is accompanied by two charts, one exhibiting the greater part of the world, and the other the British Islands; the cotidal lines are drawn on these charts, which thus become an epitome of the whole memoir.　The charts are copied, with some modifications derived from Mr Whewell's later memoirs, in the treatise on *Tides and Waves;* the copies are on a smaller scale than the originals, but are perhaps quite as convenient for study: see Articles 524, 574...585 of the treatise. Sir G. B. Airy records in his Article 579 his dissent from one of Mr Whewell's opinions : " Mr Whewell has inferred...that the tides of the Atlantic are mainly of a derivative character, produced by the tides of the Southern Ocean and transmitted up the Atlantic in the same manner in which the tides of the Atlantic are transmitted up the English channel. We doubt this entirely." But even at this epoch Mr Whewell seems to have given up the opinion, and he formally retracted it subsequently: see the last page of the article in the *United Service Journal*, page 304 of the sixth memoir, and page 5 of the thirteenth.

The second memoir is entitled *On the Empirical Laws of the Tides in the Port of London ; with some Reflexions on the Theory.* This was received Nov. 13, 1833, and read January 9, 1834: it occupies pages 15...45 of the volume for 1834.

The tides depend principally on the moon. The epoch of high water, at a given place, follows that of the moon's transit across the meridian ; the interval between the two however is not constant, but depends on the positions of the sun and moon. Likewise the height of the tide is not constant, but depends on the positions of the sun and moon. Mr Lubbock had discussed the observations of tides which had been made during nineteen years in the port of London, and had given his results in the form of tables. Mr Whewell examines these tables in order to discover the laws which regulate the variations in the heights of the tide and in the interval which occurs between the moon's transit and the subsequent tide; and he succeeds in revealing these laws with considerable clearness, at least so far as they depend on the moon. He compares his results with Bernoulli's theory, and shews that there is a partial coincidence between them.

This is perhaps the most striking of the memoirs: Sir G. B. Airy says in his Article 492, "we confidently refer the reader to this investigation as one of the best specimens of the arrangement of numbers given by observation under a mathematical form." It would be very interesting to know what was actually the order of the process adopted : in the memoir the laws seem to be first deduced simply from the observations, and then compared with the theory; but the course *may* have been the reverse, that is, the theory may have been first consulted for suggestions, and the calculations made under their guidance.

The third memoir is entitled *On the Results of Tide Observations made in June* 1834 *at the Coast Guard Stations in Great Britain and Ireland;* this was received March 27, and read April 2, 1835: it occupies pages 83...90 of the volume for 1835.

Observations of tides had been made on the coast of England, Ireland, and Scotland, from June 7 to June 22 inclusive. These required to be submitted to calculation before conclusions could be drawn from them, and at this date only those relating to the south coast had been discussed. However, some interesting results had shewn themselves, among which may be noticed the fact that the sea between the Isle of Wight and the Downs is affected by two tides, one coming from the north and one from the west. The observations furnished some corrections for the cotidal lines in the British seas as drawn in the first memoir.

The fourth memoir is entitled *On the Empirical Laws of the Tides in the Port of Liverpool;* this was received Nov. 10, and read Nov. 19, 1835; it occupies pages 1...15 of the volume for 1836.

Mr Lubbock had discussed the observations of tides which had been made during nineteen years at the port of Liverpool; and the fourth memoir stands in the same relation to this discussion as the second memoir did to Mr Lubbock's discussion of the London tides: the conclusions support those obtained in the second memoir.

The fifth memoir is entitled *On the Solar Inequality and on the Diurnal Inequality of the Tides at Liverpool;* this was received Feb. 23, and read March 3, 1836: it occupies pages 131...147 of the volume for 1836.

In the second and fourth memoirs the laws which connect variations in the tides with the position of the *moon* had been investigated; in the present memoir Mr Whewell proposes to determine in like manner the influence of the *sun* on the tides at Liverpool: this is called the *Solar Inequality.*

"The *Diurnal Inequality* of the tides is that which makes the tide of the morning and evening of the same day at the same place, differ both in height and time of high water, according to a law depending on the time of the year." The existence of such an inequality had often been noticed by seamen and others, but its laws had never been correctly laid down. Mr Whewell was the first person who insisted on the importance of this matter, both in theory and practice, and from the date of the present memoir he gave great attention to it: if he had contributed nothing besides to the subject of the tides his exposition of the law and the facts of the diurnal inequality would entitle him to a distinguished place among cultivators of Physical Astronomy.

The sixth memoir is entitled *On the Results of an extensive system of Tide Observations made on the coasts of Europe and America in June* 1835; this was received June 2, and read June 16, 1836: it occupies pages 289...341 of the volume for 1836. Tide observations were repeated at the British coast-guard stations similar to those made in June 1834; and also numerous observations were made simultaneously on the coasts of North America, Spain, Portugal, France, Belgium, Holland, Denmark and Norway. This extensive work was accomplished by the zealous efforts of the governments of the various countries. The results furnished corrections of the cotidal lines of the first memoir. Accordingly a map is given containing the cotidal lines of the British seas and the German Ocean, and also a map containing the cotidal lines for the coasts of Europe; moreover the range of the high tide in yards round the British isles and the coasts of the German Ocean is recorded in a map.

Much valuable information with respect to the diurnal inequality was obtained from the observations.

The tides of the German Ocean presented great difficulties, and the view taken by Mr Whewell is thus noticed in Article 525 of the treatise on *Tides and Waves*. "Mr Whewell...has had recourse to the supposition of a revolving tide in the German Ocean, in which the tide wave would run as on the circumference of a wheel, the line of high water at any instant being in the position of a spoke of a wheel. Although our mathematical acquaintance with the motion of extended waters is small, we have little hesitation in pronouncing this to be impossible."

In a letter to Professor J. D. Forbes, dated Oct. 23, 1856, Mr Whewell spoke of this sixth memoir thus, "I look upon it as my great achievement in Tidology." Professor Forbes appears to have had in his possession a copy of the thirteenth memoir, but he must have forgotten to consult it; for in speaking on the Tides in his *Review of the Progress of Mathematical and Physical Science*, page 20, he lays down that statement of the propagation of tides in the Atlantic Ocean which, as we have seen, Mr Whewell retracted.

The seventh memoir is entitled *On the Diurnal Inequality of the Height of the Tide, especially at Plymouth and at Singapore; and on the Mean Level of the Sea;* this was received March 7, and read March 9, 1837: it occupies pages 75...85 of the volume for 1837.

This is a very interesting paper. The laws of the diurnal inequality are well established for Plymouth. It appears that the diurnal inequality vanished four days after the moon was in the equator. Another rule prevailed at Singapore, where this inequality is very conspicuous.

It is shewn in this paper that the mean height of the sea, that is half the sum of the heights at high and low water, is nearly constant at Plymouth; and similarly at Singapore.

The eighth memoir is entitled *On the Progress of the Diurnal Inequality Wave along the Coasts of Europe;* this was received June 14, and read June 15, 1837: it occupies pages 227...244 of the volume for 1837.

The observations made in June 1835 are used to exhibit in the form of Tables the diurnal inequality for nineteen places; and for seven of these places the results are also put in the form of curves: see *Tides and Waves*, Art. 563, where this is described as a "very admirable investigation." The memoir concludes with pressing on maritime nations the duty of systematic exertions in making out the laws of the tides.

The ninth memoir is entitled *On the Determination of the Laws of the Tides from short Series of Observations;* this was received June 8 and read June 14, 1838: it occupies pages 231...251 of the volume for 1838.

The object of the memoir is to shew that the observations of the tides for a single year at a given place may be sufficient to determine the laws of the variations produced by the position of the moon; such laws for London and Liverpool had been originally made out by the aid of observations continued for nineteen years. The materials used in this paper are four years observations at Plymouth and four at Bristol. Mr Whewell says that this memoir appears to be suited to wind up the series which during some years he had laid before the Royal Society from time to time; however we shall find that five others appeared after this. In a letter to Professor J. D. Forbes, already cited, Mr Whewell speaks of the eighth and ninth memoirs as not important.

The tenth memoir is entitled *On the Laws of Low Water at the Port of Plymouth, and on the Permanency of Mean Water;* this was received April 11 and read June 6, 1839: it occupies pages 151...161 of the volume for 1839.

Six years tide observations at Plymouth are here discussed; the main result is that the height of mean water, that is the height midway between high and low water, is constant: also the fluctuations in the height of low water are greater than those in the height of high water.

The eleventh memoir is entitled *On certain Tide Observations made in the Indian Seas;* this was received April 11 and read June 6, 1839: it occupies pages 163...166 of the volume for 1839.

The chief peculiarity here noticed is an enormous diurnal inequality of the heights at two places in the Gulf of Cambay.

A paper entitled *Additional Note to the Eleventh Series of Researches on the Tides* occupies pages 161...174 of the volume for 1840; this was received Feb. 7 and read April 2, 1840.

This note discusses some Russian observations which resembled those of the eleventh memoir. The most remarkable case is that of Petropaulofsk, where the tides "show more clearly than any that have yet been examined, the manner in which the diurnal inequality may be so large as to lead to the appearance of only one tide in the twenty-four (lunar) hours."

The twelfth memoir is entitled *On the Laws of the Rise and Fall of the Sea's Surface during each Tide;* this was received June 13 and read June 18, 1840: it occupies pages 255...272 of the volume for 1840.

The result obtained is stated thus: "If we suppose a point to move uniformly through the circumference of a circle in a tidal half-day, the height of this point above a horizontal line will represent the height of the surface of the water, supposing the velocity and the radius to be duly adjusted."

The investigation is founded on five months observations made at Plymouth, three months observations made at Liverpool, and twelve months observations made at Bristol. It is said that "this inquiry may be the more useful, inasmuch as the laws of rise and fall of the surface are nearly the same at all places; the differences being, for the most part, of such a kind as can be ascertained and allowed for without much difficulty." I do not know whether it is meant merely that the laws *ought* to be the same by theory, or that the laws really *are* the same in fact. According to the observations of Captain F. W. Beechey, the tides in the Bristol Channel and the Irish Sea do not follow the law stated by Mr Whewell: see *Philosophical Transactions* for 1848, page 110.

The thirteenth memoir is entitled, *On the Tides of the Pacific, and on the Diurnal Inequality;* this was received Nov. 11, and read Dec. 16, 1847, forming the *Bakerian Lecture* of the year: it occupies pages 1...29 of the volume for 1848.

This is an interesting paper; and taken in conjunction with the article in the *United Service Journal* to which we have already referred, will give a good account of Mr Whewell's investigations on the Tides.

This memoir is subsequent to the treatise on *Tides and Waves* by the Astronomer Royal, and refers to that treatise occasionally. Some parts of the first memoir are now retracted, so that we read: "I do not think it likely that the course of the tide can be rightly represented as a wave travelling from south to north between Africa and America."

Mr Whewell in his sixth memoir had arrived theoretically at the conclusion that at a certain spot of the North Sea the tide would be almost insensible; this had been verified by the observations of Captain Hewett, which are recorded in the volume of the British Association for 1841: see also the treatise on *Tides and Waves*, Art. 528. The present memoir alludes to this matter. Phenomena of a similar kind were afterwards discovered in other parts of the British seas; *Philosophical Transactions* for 1848, pages 5, 106, 107, 115.

The present memoir embodies the information respecting the tides in the Pacific, which had been gained since the appearance of the first memoir; and special attention is paid to the *diurnal inequality* which Mr Whewell first introduced to the serious notice of those who are practically or theoretically concerned with the tides.

There still remained a part of the ocean for which good observations had not been made, for Dr Whewell says: "I shall not entangle myself in the seas broken by innumerable large and small islands, which extend from Torres Straits to the coasts of India, Arabia, and Africa...."

It may be observed that the volume of the *Philosophical Transactions* for 1848 contains an important paper by Captain F. W. Beechey on the Tides in the Irish and English Channels; and there are papers on the Tides by the Astronomer Royal in the volumes for 1842, 1843, and 1845.

The fourteenth memoir is entitled, *On the Results of continued Tide Observations at several places on the British Coasts;* this was received Oct. 24, 1849, and read Jan. 31, 1850: it occupies pages 227...233 of the volume for 1850.

The results relate to the height of high water, and the variations which this height undergoes between spring tides and neap tides, and between neap tides and spring tides. "It is found, by examining the observations at 120 places and throwing the heights into curves, that the curve is very nearly of the same form at all these places."

Mr Whewell's investigations on the tides involved a large amount of numerical calculation, and he records his obligation to those who assisted him in this respect. The name of Mr D. Ross occurs most frequently, and next to his that of Mr Dessiou; Mr Bunt, Mr Boddy, and Mr Nayler are also mentioned. Letters from Mr Ross, Mr Dessiou, and Mr Bunt are preserved among the papers of Dr Whewell; and their number shews the extent of the labour involved both in the calculations themselves, and in the general direction and superintendence of them. Mr Dessiou and Mr Ross were employed under the Admiralty in the Hydrographer's office in London. Mr Bunt resided at Bristol, and was associated with Mr Whewell, not only in calculations on the tides, but in some important levelling operations undertaken for the British Association. Mr Bunt also contributed independently to the subject of the tides; he constructed a new Tide-Gauge, which is described in the *Philosophical Transactions* for 1838, and also in the treatise on *Tides and Waves*, Art. 474: and a paper by him entitled *Discussion of the Tide Observations at Bristol* was printed in the *Philosophical Transactions* for 1867. He also published a *Planetarium*, which consisted of a large diagram with accompanying Tables, designed to represent the planetary motions during a period of two centuries, ending in 1940: this was highly commended by Mr Whewell, Admiral Smyth, Sir J. Herschel, and Mr Airy. On Mr Bunt's merits Mr Whewell expressed himself very strongly in a letter to Mr Lubbock, Feb. 2, 1839.

Mr Whewell's memoirs are on the whole very clear, though, as might have been expected where processes of observation and numerical results enter largely, there are occasional obscurities. I have mentioned one example in the account of the twelfth memoir, and will notice another. In the second memoir, speaking of the rates of the lunar to the solar tide, we are told, on page 20, that Laplace from the Brest observations obtained 2·6157; while, on page 35, it is stated that the Brest observations give 2·6167. In a note to Art. 537 of the *Tides and Waves*, the Astronomer Royal says he does not know on what authority the number 2 6157 is ascribed to Laplace. Also, on page 35 of the memoir, it is stated that, according to the London observations, the ratio is 2·9887: the Astronomer Royal, in the place just cited, has a different result, and says he does not know on what authority the number 2·9884 is given: I presume the last 4 should be 7.

On pages 234 and 235 of the ninth memoir Mr Whewell quotes from Mr Bunt a description of the process of treating the observations so as to obtain the mean results from them; this description is difficult to understand, and fortunately it is reproduced in the *Tides and Waves*, Art. 494, with useful explanatory additions.

But the few obscurities which may be found in the memoirs do not detract to any appreciable extent from their substantial merits; the memoirs would indeed have been no inadequate return for the occupation of a long series of years if profession-ally devoted to Astronomy, and form a conspicuous example of voluntary devotion to science, rarely surpassed in England. The notion of drawing cotidal lines across the great oceans, which formed the starting-point of the investigations, proved ultimately to be of small value; and much of the earlier labour must have been expended with little direct benefit; but attention was thus called to the general subject, and the knowledge of the succession of the cotidal lines along the coasts of certain seas has been pro-ductive of valuable results both theoretical and practical. The most important parts of Mr Whewell's researches seem to me to consist in his discussion of the variations produced in the half-daily tide by the change in the declination of the moon and in her distance from the earth, and his exhibition of the reality and the nature of the daily inequality. The methods employed have obtained the high commendation of the Astronomer Royal, who expresses himself thus in Art. 496 of his treatise on *Tides and Waves*: "... viewing the two independent methods introduced by Mr Whewell, of reducing the tabular numbers to law by a process of mathematical calculation, and of exhibiting the law to the eye without any mathematical operation by the use of curves, we must characterize them as the best specimens of reduction of new observations that we have ever seen..."

The work, like so much that is accomplished in science, is however of a temporary and introductory character; the efforts of the early enquirers in any subject must be often confined to preparing the foundation and constructing the scaffolding, leaving the erection of the building to those who follow. This is recog-nised in the official award of a medal for the researches; we read in the Royal Society's *Abstracts*, Vol. IV. page 26:

"The interest which attaches to such investigations, which is so great during the progress of the structure which is to be raised upon them, ceases in many cases when the fabric is completed : a remark which is applicable to many of the most important researches and discoveries in philosophy, where we are accustomed to regard the last form only in which the theory is compared with the facts which are observed, and to forget or to neglect the series of laborious investigations which have led to its establishment, but which are no longer necessary for its explanation or proof."

I may remark that Mr Whewell seems to me less felicitous in this subject than he is elsewhere as to his technical language. The term *semi-menstrual* which is much used he himself disliked, and finally changed to *semi-mensual;* I should have preferred *half-monthly.* And not even his high authority can reconcile me to the barbarous compound *tidology.*

Mr Whewell issued various occasional papers connected with the Tides, principally bearing on the practical observations required. I will mention them.

Memoranda and directions for Tide Observations. Eight octavo pages.

Suggestions for persons who have opportunities to make or collect Observations of the Tides. Two quarto pages.

A paper in three quarto pages containing directions by which persons may *reduce* the Observations of Tides which they have made.

Directions for making a series of Tide Observations. Three quarto pages containing a diagram of a tide-gauge.

Queries for Nautical Men. A single quarto page dated Hydrographic Office, Jan. 10, 1837.

Directions for Tide Observations. The Establishment of the Place. Three octavo pages from the Nautical Magazine, No. XXIII.

Directions for Tide Observations. The Semimenstrual Inequality. Five octavo pages from the Nautical Magazine, No. XXIV.

Directions for Tide Observations. The Diurnal Difference of the Tides. Six octavo pages from the Nautical Magazine, No. XXXI.

Mr Whewell delivered a lecture on the Tides at Edinburgh in 1834 to the British Association, and also one at the Bristol

Institution about 1837; a newspaper report of the latter has been preserved.

The practical directions respecting the Tides were embodied by Dr Whewell in an article which he contributed to the *Manual of Scientific Inquiry,*—a volume edited for the Admiralty by Sir J. Herschel. The first edition of the work appeared in 1849; Dr Whewell's article is entitled *Directions for Tide Observations,* and occupies about 20 octavo pages; the third edition appeared in 1859, and in this Dr Whewell's article is slightly enlarged.

Besides the memoirs and papers already mentioned, there are a few brief communications by Mr Whewell to the volumes of the British Association; these I will notice in connexion with all his other scientific memoirs: his last article on the subject is contained in the volume for 1851, and bears the disheartening title *On our Ignorance of the Tides.*

There is a very good sketch of the history of the investigations respecting the tides in the *History of the Inductive Sciences;* see the 3rd edition, Vol. II. pages 91, 190, 470: by a misprint on page 190, which is in all the editions, D'Alembert is named instead of Maclaurin.

When the medal was awarded by the Royal Society to Mr Whewell for his researches on the Tides, the chair was occupied by the Duke of Sussex; he was intimate with one of the senior fellows of Trinity College at that period, and from this circumstance, as well as from the business of the Royal Society, was probably well acquainted with Mr Whewell. The present section may be appropriately illustrated by a quotation from the remarks made by the President on the occasion of the award of the medal by the Royal Society; see the *Abstracts of the Papers,* Vol. IV. page 7.

"When I read his essays on the architecture of the middle ages, on subjects of general literature, or on moral and metaphysical philosophy, exhibiting powers of mind so various in their application and so refined and cultivated in their character, I feel inclined to forget the profound historian of science in the accomplished man of letters, or the learned amateur of art; but it is in his last and highest vocation, whilst tracing the causes which have advanced or checked the progress of the inductive sciences from the first dawn of philosophy in Greece to their mature development in the nineteenth century, or in pointing out the marks of design of an all-wise and all-powerful Providence in the greatest of those works and operations of nature which our senses or our knowledge can comprehend or explain, that I recognise the productions of one of those superior minds which are accustomed to exercise a powerful and lasting influence upon the intellectual character and speculations of the age in which they flourish."

Mr Wilkinson, in a letter dated Dec. 10, 1837, alludes to the award of the medal thus: "I congratulate you on the honour of a Medalist, which the *Times* told me had been conferred upon you by the Royal Society. I am sure it is fairly earned. What a world of troubles you are involved in, or rather I should say a sea of troubles. Had I been of the committee I should have proposed, *more Romano,* a crown of sea-weed, with a motto *Mari devicto.*"

CHAPTER VII.

1834...1837.

In the year 1834 Mr Whewell was much engaged in correspondence with Professor Faraday, who required a good nomenclature for the exposition of his researches in Electricity, and naturally applied to the person generally recognised as the best authority on scientific language. Mr Whewell was able to satisfy his correspondent, who wrote thus on May 15:

"I have taken your advice and the names, and use *anode, cathode, anions, cations,* and *ions;* the last I shall have but little occasion for. I had some hot objections made to them here, and found myself very much in the condition of the man with his Son and Ass, who tried to please every body; but when I held up the shield of your authority it was wonderful to observe how the tone of objection melted away.

"I am quite delighted with the facility of expression which the new terms give me, and shall ever be your debtor for the kind assistance you have given me."

Mr Whewell attended the meeting of the British Association at Edinburgh in September; Dr Chalmers, with whom he was to have stayed, fell ill, and he became the guest of Professor Forbes. Previously to the meeting he made a tour in Scotland, and this furnished him with materials for a communication on architecture to his friend Mr Rickman.

During this year Mr Whewell invented his *Anemometer,* that is, an instrument for measuring the force and the direction of the wind; he described it in a letter to Professor Phillips, then resident at York, who took great interest in the subject, and employed the instrument. Mr Whewell's construction was also

adopted for some time at other places, but it has now been superseded by more convenient inventions.

Mr Whewell was much pleased with an essay, which belongs
to this date, on the *Analogy between Mathematical and Moral
Certainty*, by Mr Birks, a student of Trinity College, now Professor of Moral Philosophy. Mr Whewell distributed copies of
the essay among his friends, and found that his opinion of its
merit was shared by Dr Chalmers, Dean Conybeare, J. C. Hare,
Sir W. R. Hamilton, and Wordsworth. The essay is reprinted in
the volume which contains the Introductory Course of Lectures
given by Professor Birks at Cambridge.

In this year Coleridge died, and John Sterling, then living as
curate with J. C. Hare, proposed to raise a sum of money to found
a Prize Essay at Cambridge, in commemoration of the philosopher;
the subject of the Essay was to be the *Philosophy of Christianity*.
Mr Hare urged the subject on the attention of Mr Whewell,
but did not receive very much encouragement from him; and
finally the Heads of the Colleges declined to accept the proposition.

The work on the History and Philosophy of the Inductive
Sciences was now being prosecuted vigorously; a letter to Mr
Jones of July 27, 1834, explains the author's plan as then ar-
ranged:

"You are to understand that I am to consist of three Books.
Book 1, *History* of Inductive Science, namely, Astronomy, Mechanics, Physics, Chemistry and Botany, historiographized in a new
and philosophical manner. Book 2, *Philosophy* of Inductive
Science, which is what I want to shew you. It will be dry and
hard I fear, as it must contain most of the metaphysical discussions which have been attended to of late, but it must also contain all the analysis of the nature of Induction and the Rules
of its exercise, including Bacon's suggestions. Book 3, *Prospects*
of Inductive Science. The question of the possibility and method
of applying Inductive-processes, as illustrated in the philosophy of
Book 2, to other than material sciences; as philology, art, politics,
and morals."

Two other matters remain to be noticed under the present

date, namely the controversy with Mr Thirlwall, and a review of a book written by Mrs Somerville.

In the year 1834 some controversy arose in Cambridge respecting the admission of dissenters to the University; in this Mr Whewell and Mr Thirlwall took part. Mr Whewell published two pamphlets on the subject.

The first pamphlet is entitled *Remarks on some parts of Mr Thirlwall's Letter on the Admission of Dissenters to Academical Degrees;* it consists of 23 octavo pages. Mr Thirlwall was at that time a lecturer in Trinity College; he published a Letter to Professor Turton on the admission of Dissenters to Academical Degrees: the letter is remarkable for its vigour of thought and language. In the course of the letter Mr Thirlwall spoke unfavourably of the compulsory attendance on College Chapels. Mr Whewell's *Remarks* bear almost entirely on this point, and touch but slightly the main question which Mr Thirlwall had discussed; they are written in a grave and temperate tone.

The second pamphlet is entitled *Additional Remarks on some parts of Mr Thirlwall's two letters on the Admission of Dissenters to Academical Degrees;* it consists of 19 octavo pages. Like the former pamphlet this is devoted to combating the opinion of Mr Thirlwall, that the compulsory attendance at College Chapel was unfavourable rather than favourable to devotion. The question is one of great difficulty, and the existing practice at Cambridge is defended by Mr Whewell with calmness and discretion. I should have thought, according to such traditions as still linger in the University, that the College services at that date must often have been very far from creditable: Mr Whewell makes some admission to this effect on his page 15, but less distinct than I should have expected.

The discussion excited great interest at the time in Cambridge, which was naturally augmented by the fact that Mr Thirlwall was for his share in it deprived of his lectureship by Dr Wordsworth, then Master of Trinity College; against this proceeding Mr Whewell remonstrated urgently and repeatedly. Some of Mr Whewell's correspondents were strong in their disapprobation of Mr Thirlwall, others were much more moderate in their tone: Mr Rose was one of the former class, Mr Hare one of the latter.

The progress of events has practically settled in favour of Bishop Thirlwall the question he raised as to the admission of dissenters to the University; I confess I should have been glad if Mr Whewell had taken the same side. I had the gratification of hearing the opinion of Bishop Thirlwall quoted with approbation at the last meeting which it was found necessary to hold at Cambridge in order to extend to dissenters the full advantages of the College endowments; Professor Sedgwick was present to advocate the opinion he had long maintained, and all resistance was speedily overcome.

Mrs Somerville's work *On the Connexion of the Physical Sciences* was reviewed by Mr Whewell on pages 54...68 of Number 101 of the *Quarterly Review,* published in March, 1834. Mr Lockhart applied on Jan. 24th for the article; saying he did not know to whom to turn unless to Mr Whewell, for Herschel was away and "Brewster engaged to do the thing in the Edinburgh." On Feb. 1st Mr Lockhart returned thanks for the "spirited little paper."

The review is, as Lockhart had prescribed, "a lightish paper." Mrs Somerville's book is highly commended, and several extracts are given. Perhaps too much stress is laid on the fact, which is brought prominently forward, that such a work had been written by a *woman.* A few sentences may be quoted relating to this subject, which at the present time is much discussed.

"Notwithstanding all the dreams of theorists, there is a sex in minds."

"He [that is, man] learns to talk of matters of speculation without clear notions; to combine one phrase with another at a venture; to deal in generalities; to guess at relations and bearings; to try to steer himself by antitheses and assumed maxims. Women never do this: what they understand, they understand clearly: what they see at all, they see in sunshine. It may be, that in many or most cases, this brightness belongs to a narrow Goshen; that the heart is stronger than the head; that the powers of thought are less developed than the instincts of action. It certainly is to be hoped that it is so."

The article contains two specimens of verse "from the mint of Cambridge" respecting Mrs Somerville. The first is by Mr. Whewell himself, and is reproduced in his *Sunday Thoughts and other Verses,* 1847.

In the year 1835 Mr Whewell attended the meeting of the British Association at Dublin in August, and was one of the Vice-Presidents. After the meeting he gave some attention to Irish architecture, and became acquainted with Mr Petrie, from whom he derived information about the famous round towers of Ireland. Shortly afterwards he made a tour in North Wales. In December he paid a visit to Professor Sedgwick at Norwich.

From letters addressed to Mr Whewell in 1835 it appears that he took much interest in the experiments made by Mr Scott Russell on Waves and the Resistance of Fluids.

I pass to the publications of the year.

A Sermon was preached before the Corporation of Trinity House : this will be noticed with the other sermons.

A Report on the Mathematical Theories of Electricity, Magnetism, and Heat was communicated to the British Association at the Dublin meeting : this will be noticed with the other scientific memoirs. Mr Whewell's Report may have led to his acquaintance with the late Sir W. Snow Harris, from whom he received various long letters on Electricity.

A preface contributed to Mackintosh's Dissertation on the Progress of Ethical Philosophy belongs to about this time; it will be noticed with the other contributions to Moral Science. Mr Whewell's attention seems to have been drawn to the subject mainly by the circumstance of having to lecture his College students on Butler's Sermons.

In this year Mr Whewell published *Thoughts on the Study of Mathematics as a part of a Liberal Education;* it is an octavo pamphlet of 46 pages, including the title-leaf. About five pages are devoted to upholding the educational superiority of mathematics over logic, and the remainder to the discussion of some faults in the manner of teaching mathematics, by which the benefit of these studies may be seriously diminished. These faults are thus stated on page 8 :

"I would reply, then,—that if mathematics be taught in such a manner that its foundations appear to be laid in arbitrary definitions without any corresponding act of the mind;—or if its first principles be represented as borrowed from experience, in such a manner that the whole science is empirical only;—or if it be held forth as the highest perfection of the science to reduce our knowledge to extremely general propositions and processes, in which all particular cases are included;—so studied, it may, I conceive, unfit the mind for dealing with other kinds of truth."

The opinions thus stated are enforced with ability and moderation; they were permanently held by the writer, and reappear in substance in his *Philosophy of the Inductive Sciences*. The pamphlet is reprinted in the work *On the Principles of English University Education*, 1837, in which it occupies pages 143...181.

I may observe that, on page 30 of his pamphlet, Mr Whewell justly censured a certain faulty definition of a fluid. Ivory, whose conscience perhaps reproached him for his errors in hydrostatics, suspected some reference to his peculiar notions, and wrote to Mr Whewell on the subject. Ivory says: "Is the citation from any of my Papers? if so, have the goodness to specify the passage."

Mr Whewell's pamphlet was placed at the head of an article in Number 126 of the Edinburgh Review, published in 1836; the article was written by Sir W. Hamilton, and has become notorious as a wild and indiscriminate attack on mathematics by a person very slightly acquainted with them.

Mr Whewell replied to the Review in a letter dated Jan. 23, 1836; this letter was reproduced in pages 183...186 of the *Principles of English University Education*, and also in the *Edinburgh Review*, Number 127. The letter points out the mistake of the reviewer in representing a page or two of the *Thoughts* as containing all that could be said in favour of mathematics by an able advocate, whereas in fact the object of the pamphlet was quite different. A natural and modest wish is expressed for the titles of some works on Practical Logic and Philosophy, which the reviewer would recommend for their educational efficiency as rivals to the well-known mathematical treatises. The reviewer however, in a brief reply in Number 127 of the Edinburgh Review, declined to accede to what he called a "misplaced request."

A letter from a friend in Edinburgh addressed to Mr Whewell just before the publication of the notorious article is very moderate and reasonable in tone. "You will not be surprised, I dare say, to hear, that in this place, once celebrated for metaphysical and all intellectual studies, though it has unfortunately for some time ceased to be so, there are still a few who cling to those pursuits, and who do not view with much favour any attempts to give to mathematics a preference over them, as a means of developing and cultivating the mental powers. We allow to mathematics as a *Science* every perfection which you can claim for it; but we are very far from thinking that, as a *study*, it affords any thing like the best schooling for powers which must so much more largely and necessarily be employed in the field of contingent enquiry. In short, my dear Sir, we differ widely from you; and you will see *how widely* in an article of the Number of the Edinburgh Review to be published in a few days....You will, perhaps, revise your opinions, and either admit that we are right, or shew that we are greatly wrong."

In the year 1836 Mr Whewell's attention was drawn to a railway dispute, by Mr Barlow, Mr Drinkwater, and Mr Pringle; and he was requested to give evidence about it before a Parliamentary Committee. The question seems to have been respecting the relative advantages, for the South-Eastern Railway, of a shorter line with steep inclines, and of a longer line with gentle inclines. The answers of Mr Whewell to his correspondents have not been recovered; but it is obvious from their letters that he asked for experiments and observations on the subject, few of which had been made up to that time.

In August Mr Whewell attended the meeting of the British Association at Bristol.

In November Mr Whewell was a candidate for the Lowndean Professorship of Astronomy and Geometry, which became vacant by the death of Professor Lax; Mr Peacock obtained the appointment.

In the year 1836 Mr Whewell vindicated the character of

Newton against some attacks relating to his conduct towards Flamsteed. A large quarto volume entitled *An Account of the Rev. John Flamsteed, the first Astronomer Royal...* was printed by order of the Lords Commissioners of the Admiralty in 1835, under the editorship of Francis Baily. This contained Flamsteed's own version of his quarrel with Newton; and an article in the *Quarterly Review* strongly maintained Flamsteed's part. Mr Whewell published a pamphlet on the subject. The first edition is entitled *Newton and Flamsteed. Remarks on an Article in Number CIX. of the Quarterly Review.* In the second edition the Title has the following additional words: *To which are added two Letters occasioned by a note in Number CX. of the Review.* The pamphlet is in octavo; the first edition consists of 19 pages, and the second of 32.

Mr Whewell conducts his case with skill and moderation; he shews that the Quarterly Reviewer was not familiar with the scientific history of the period, and that he accepted Flamsteed's statements and opinions without exercising due caution in checking the evidence furnished by a witness in his own behalf.

An article on the subject of Newton and Flamsteed appeared also in the *Edinburgh Review;* Mr Whewell briefly alludes to it in connexion with one element of the controversy, saying, " The Edinburgh Reviewer, wiser than his brother, has pointed out this." Professor Rigaud, in a letter to Mr Whewell, ascribed the article in the *Edinburgh Review* to Mr Galloway.

Mr Whewell's pamphlet was favourably noticed in the *Philosophical Magazine* for February 1836.

Mr Whewell says in a letter to Mr Jones, " I send you a few copies of a pamphlet which I scribbled off at a sitting after reading Barrow's article in the *Quarterly.*" I have no doubt that this refers to the present matter, though the year is not recorded in the original letter, and Dr Whewell himself, when his letters to Mr Jones came again into his hands at a later period, put the date 1826 on it with a note of interrogation.

The most important circumstance in the year 1837 is the publication of the *History of the Inductive Sciences;* of this work I shall speak in the next chapter, and also of two letters connected with it, one to the Editor of the *Edinburgh Review*, and one to the Editor of the *Medical Gazette;* I proceed now to the other incidents of the year.

Mr Whewell became President of the Geological Society in February; this office was tenable for two years, and one of the most responsible functions attached to it always was the recommendation of a suitable person as successor. Mr Whewell had the honour of following Mr Lyell, and of being strongly urged, both by him and by Professor Sedgwick, to accept the office.

The Marquis of Northampton edited for a benevolent purpose a volume called the *Tribute*, consisting of poetical contributions from various writers; Mr Whewell was requested to assist by Mr T. Spring Rice, afterwards Lord Monteagle, and accordingly he sent a piece entitled *The Spinning Maiden's Cross;* this he afterwards reprinted in his *Verse Translations*. Some notice of the *Tribute* will be found in the *Edinburgh Review* of October, 1837.

In the summer Mr Whewell made a tour in France, Germany, and Holland. In September he attended the meeting of the British Association at Liverpool, and was one of the Vice-Presidents.

In November Mr Whewell preached before the University four Sermons on the *Foundation of Morals;* these were published, and will be noticed with the other contributions to Moral Science.

I will now advert to the *Letter to Charles Babbage, Esq., Lucasian Professor of Mathematics in the University of Cambridge.* This letter is printed on 7 small pages; it is dated *Athenæum, May* 30, 1837. A few words from the beginning shew the occasion of the letter.

"I have just read your 'Ninth Bridgewater Treatise'... I have still been unable to get rid of the persuasion, that displeasure at a sentence or two in my Bridgewater Treatise, had a considerable influence upon you, both as to the design and the execution of your book."

Mr Babbage thought that Mr Whewell had given support to the prejudice "that the pursuits of science are unfavourable to religion." The Chapter VI. of Book III., which is entitled *On Deductive Habits*, was probably the part of the Bridgewater Treatise which appeared to Mr Babbage open to objection. In defending

W.   7

himself Mr Whewell writes in a very friendly manner, and expresses his satisfaction at having Mr Babbage for a volunteer fellow-labourer. The letter concludes thus: "There have been in the recent literature of our country, many proofs how generally acceptable the subject is; but none in which the .sympathy of others with regard to it has given me more pleasure, and none in which it is treated in a more original manner."

In the year 1837 Mr Whewell published a work *On the Principles of English University Education.* This is on small octavo, and contains besides the *Title* and *Contents* 186 pages, of which 142 are new; the remainder consists of a reprint of the *Thoughts on the Study of Mathematics,* and the *Letter to the Editor of the Edinburgh Review:* to these I have already alluded. The work is divided into three Chapters, namely; of the Subjects of University Teaching, of Direct and Indirect Teaching, and of Discipline.

The first Chapter is a vindication of the paramount claims of classics and mathematics as educational instruments; it is, in my opinion, most interesting and valuable. The author is here perhaps somewhat of an advocate, while in his later writings he discharges the office of a judge; and he seems subsequently to have attached a higher value to the modern studies than in the present volume: but on the whole he adhered firmly to the educational creed here enunciated.

In the second Chapter, among other things, he treats on the subject of private tutors; he always had a great objection to the unofficial teaching furnished by this class of persons, so numerous and so popular at Cambridge. I am surprised that in his remarks he never once adverts to the great fact that on the whole the private tutors gain their position by their ability solely. The Colleges and the University offer very few permanent positions for teachers; and it is not extraordinary that those who devote themselves to the training of private pupils as a profession should be superior to the transitory and inexperienced holders of the official lectureships. I do not assert that the private tutors are superior in ability and attainments to the professors and others who occupy the posts of dignity; but it is not too much to say that for the union of knowledge, zeal, and didactic skill they are conspicuously eminent.

In the third Chapter, among other things, Mr Whewell alludes to the subject of compulsory attendance at College Chapels. He holds the same views as he did in 1834 on the occasion of the controversy between himself and Mr Thirlwall; and even those who may hesitate as to the best course in this difficult question will allow that the current practice is defended in an earnest and yet conciliatory manner. There is, however, no allusion to some defects which many excellent men have admitted to exist in these services; such as the monotony which accompanies the incessant repetition of the same devotional language, and the undue prominence of the Apocrypha in the course of the daily lessons: the latter defect has been remedied in more recent times.

I will quote the decided opinion expressed by Mr Whewell on a point which has been lately much discussed: he says on his page 46: "Euclid has never been superseded, and never will be so without great detriment to education."

Again, he says on his page 134: "As I have already said, I should be sorry to see Euclid lose his ancient place, or even his ancient form, in our system."

A second edition of the work appeared in 1838 with the following notice: "In this Edition, I have added a few Reflexions tending to illustrate further the nature of the intellectual training which the study of Mathematics supplies; and the mode in which it must be conducted, after the first stage, in order to answer its purpose. W. W." The part thus added consists of about eight pages; it is a reprint of the first half of the preface to the mathematical work entitled *The Doctrine of Limits*.

An eminent friend of Mr Whewell expressed a high opinion of the work on English University Education. "It has interested me so deeply that I have finished it within the second day, after receiving it. The whole argument is strong, or I would say, convincing—greatly needed moreover, at this particular season, when we are surrounded by rash novelties in speculation. You have, without concession, made a substantial peace with the mathematicians, and thoroughly thrown on his back my old acquaintance Hamilton the Edinburgh Reviewer...."

CHAPTER VIII.

THE work on which the reputation of Dr Whewell is perhaps generally supposed to rest is his *History of the Inductive Sciences, from the earliest to the present time.* I shall probably find myself in a very small minority when I say that I prefer the later work on the *Philosophy of the Inductive Sciences* to the earlier work on the *History.* Either opinion is consistent with that held by many persons that the Essay on the *Plurality of the Worlds* is the cleverest of all the author's numerous writings; but from the nature of the subject that essay does not present the remarkable combination of wide learning with great ability, which distinguishes the two works on Inductive Science, and in reading it we rather admire the skill of an advocate than bow to the decisions of a judge. It is conceivable that Dr Whewell himself may have considered his contributions to Moral Philosophy as the most valuable of his publications.

The first edition of the *History* appeared in 1837 in three octavo volumes, containing altogether about 1660 pages. The second edition, which is stated to be *revised and continued,* appeared in 1847, in three octavo volumes, containing altogether about 1830 pages. The third edition, which is stated to be *with additions,* appeared in 1857 in three small octavo volumes: these *additions* were also issued in the octavo size for the convenience of possessors of the second edition. Although in the second edition, and to a further extent in the third, we have some account of the more recent investigations, yet the words in the title-page, *to the present time,* became less applicable than they originally were. This arose, partly from the rapidity with which science advanced, and partly from the fact that the author had turned aside to other studies, and no longer endeavoured to keep himself on a level with the current researches.

There is something striking about the date of publication taken in connexion with the following passage written so many years previously: in a letter from J. C. Hare to Mr Whewell, dated Dec. 21st, 1819, after referring to the treatise on Mechanics which had just appeared, he says "Do you intend to keep your history of the science for the next volume, or to publish it separately,·or to keep it back to form a part of the history of mathematics from the creation of the world down to the year of the Lord 1836?"

The *History* is dedicated to Sir John Herschel, then at the Cape of Good Hope, engaged in astronomical labours. The dedication issues from 5, Hyde Park Street; Mr Hare objected to the honour thus transferred to a private residence which he thought belonged rightly to Trinity College: Mr Whewell explained the reason to be the pleasant recollections which the favoured place would suggest to Herschel; for it was the residence of their common friend, Jones.

The first edition of the *History* was reviewed in the *Edinburgh Review* for October, 1837, somewhat unfavourably. The article was written by Brewster, and his complaint seems to have been chiefly that *Optics* and *Scotchmen* were inadequately appreciated. Mr Whewell replied in a printed letter of four octavo pages: I shall return to this hereafter.

In the *Edinburgh Review* for July, 1838, there is an Article on the first two volumes of *Comte's Cours de Philosophie Positive;* it is undoubtedly by Brewster. The reviewer makes an opportunity for speaking unfavourably of Mr Whewell's *History:* it is curious that he seems to approve of the Nebular Theory, which he calls *Laplace's Cosmogony.* Mr Whewell refers to this Article in the preface to the first edition of his *Philosophy of the Inductive Sciences.*

An Article appeared in the *United Service Magazine* in 1847, and was reprinted under the title of *Strictures on Dr Whewell's History of the Inductive Sciences:* but the word *Strictures* is inappropriate, for the Article is merely an account of the work with very few criticisms.

The third edition of the *History* was noticed by Professor Forbes in *Fraser's Magazine*, for March, 1858, in an article on the *History of Science and some of its lessons.*

The *History* and the *Philosophy* were reviewed by Sir J. Herschel in the *Quarterly Review* for June, 1841. The Article— one of the most elaborate of the author's productions—is reprinted in his *Essays from the Edinburgh and Quarterly Reviews*, 1857, and occupies more than 100 pages of the volume. Sir J. Herschel awards high praise to the two works which he reviews, though he differs widely from the philosophical doctrines maintained in them. In a note to this Article Sir J. Herschel described a work by Dr Holland as "replete with profound philosophy," and Dr Holland in a letter to Mr Whewell said "no commendation of my own volume has so much pleased me as that which is contained in a note to this Article." I may observe that Sir H. Holland notices Dr Whewell on page 240 of his *Recollections of Past Life.* He thinks that the two works—the *History* and the *Philosophy*—might better have been embodied into one. In fact it is obvious that in the *History* there is much which is not narrative; for example, Book XVIII., is principally theory and speculation.

The *History and the Philosophy* are placed at the head of an Article in the *Quarterly Review*, Oct. 1861 ; but merely as a text. The Article does not review the books, but treats on the supposed immutability of the Laws of Nature. Dr Whewell himself thought this Article wanted point and compression.

The *History* was reprinted at New York. It was translated into German by the eminent astronomer Littrow; and to this circumstance Mr Whewell refers with proper satisfaction in the preface to the first edition of the *Philosophy.* An Italian translation was projected by Bishop De Luca, now a Cardinal, who had given an analysis of some of the Bridgewater Treatises in an Italian Journal; but it does not seem to have been published. Some intention was formed at one time of executing a Russian translation, by Professor Braschman, of Moscow.

The special characteristic of the *History of the Inductive Sciences* is the distribution of the course into various decisive Epochs, each Epoch having its Prelude and its Sequel: this the author claims as a novelty, and justly regards as of great value. The prominent parts of each science are well selected, and the whole is written with a vigour of language and a felicity of illustration rare in the treatment of such abstruse subjects. Thus the popularity which the work obtained was well deserved.

My own tastes having led me to explore some of the paths of scientific history, I may perhaps be allowed to offer a few remarks which have occurred to me in the careful perusal of Mr Whewell's work. The general accuracy and fidelity of the first edition are beyond question: perhaps only those who have endeavoured to throw into a continuous narrative the vast mass of details involved in any one line of historical study, are conscious how easy it is to fall into error; and such persons will readily admit the skill with which Mr Whewell executed the laborious task he undertook. But while awarding this testimony to the first edition, it appears to me that the second and the third editions do not exhibit the full amount of improvement which might have been anticipated. The author himself had wandered from science to moral philosophy; and even if he had not thus strayed, he might with Gibbon have "preferred the pleasures of composition and study to the minute diligence of revising a former publication." But still his position as the Master of the first College in England must have rendered it easy for him to find among the younger persons with whom he came into contact, some who would have regarded it as an honour and a pleasure to undertake the examination of selected portions of the work. Mr Whewell availed himself largely of the criticisms of his friends on his proof-sheets while they were passing through the press; but he does not appear to have systematically employed this method of improvement for his successive reprints, although as I have said his official station afforded him many opportunities for obtaining such revision: moreover, a competent reader can render much more effective assistance when he leisurely studies the whole work, than when he is limited to a rapid perusal of proofs for which the printer is waiting. It is true that in addition to his other literary and scientific occupations, the official duties of Dr Whewell concurred in absorbing his time; but these latter were not continuous. The University year consists practically of three terms of seven weeks each, thus leaving more than half the civil year available after satisfying all the claims of academical position.

I will mention a few examples as illustrations of the improvements which might have been effected in the later editions: in some of these, although it would be harsh to say there is absolute error, yet there is a tendency to produce an incorrect or imperfect impression: perhaps these may be partially excused on the ground of the necessity of compression in the prosecution of an extensive plan. I cite from the pages of the last edition of the *History.*

The title of Book II. is History of the Physical Sciences in Ancient *Greece;* but the writers to whom the Book chiefly relates are Euclid and Archimedes, who did not live in Greece.

In Vol. I. page 212 the superiority of Greek science to Arabian is attributed to the freedom of political institutions, and the national education, which "fitted the Greeks to be disciples of Plato and Hipparchus." But to this explanation there is the obvious reply that Plato lived at the close of Greek freedom, and Hipparchus long after its extinction; so that at most we can only regard freedom as a seed which yields fruit by the process of its own decay.

In Vol. I. pages 175, 258, 334 much prominence is given to a supposed Arabian discovery in the Lunar Theory; but before the second edition of the *History* appeared this had been contested by Biot, and some allusion to his adverse criticism ought to have found a place. Bertrand has recently followed on the same side as Biot. The supposed Arabian discovery is mentioned also without any hesitation in the *Philosophy of the Inductive Sciences;* first edition Vol. II. page 387, second edition Vol. II. page 225, *Philosophy of Discovery,* pages 42 and 123.

In Vol. II. page 172 we read "In 1827, Professor Airy compared Delambre's tables with 2000 Greenwich observations, made with the new transit instrument at Cambridge..." In all the editions we have the number 2000, whereas it ought to be about 1200. The curious blunder about *Greenwich* observations made at *Cambridge* is peculiar to the third edition.

In Vol. II. page 186 something is said to have taken place before "the return of the academicians:" but it really was *after* the return of those who went to Lapland, although *before* the return of those who went to Peru.

In Vol. II. page 190 an Essay on the Tides is attributed to D'Alembert instead of Maclaurin.

In Vol. II. page 198 instead of John Cassini we must read James Cassini.

In Vol. II. page 220 it is said that Lacaille was "four years at the Cape of Good Hope." Lacaille was indeed absent from France for nearly four years, but he did not spend quite two years at the Cape of Good Hope.

In Vol. II. page 362 the description of conical refraction omits half of the phenomena.

In Vol. III. pages 26 and 45 allusion is made to a memoir by Biot; the correct date 1801 is given in the former place, but 1811 in the latter: also in the note to the latter place for LI. read III.

Professor Sedgwick, in an amusing letter to Dr Whewell, objected to the frequent recurrence of the word *feature* in one of his scientific works: for my part I confess that the word which distresses me most is *already*, which is perpetually occurring in such phrases as "we have already seen." The exact place ought always to be indicated; even if we excuse an author for omitting such references in the excitement of a first edition, there can be no justification for the neglecting to insert them in the calm republication of a second or third edition. In all cases it is troublesome, in many it is very difficult, to ascertain the precise passage to which the *already* is intended to carry us back. For an example I may take the statement on Vol. II. page 145: "As we have already seen, even before they [Newton's discoveries] were published, they were proclaimed by Halley to be something of transcendent value." The reference *may* be to page 115, but this does not seem very obvious.

There are also occasionally vague allusions to what may or will be done in the sequel; these should all have been rendered precise in the second edition.

In the *Philosophy of the Inductive Sciences* the word *elsewhere* is often introduced, and then both the *History* and the *Philosophy* may have to be searched to discover the unknown place.

In Dr Whewell's later writings the tendency to allude to things which he had already noticed naturally increased : we often have the phrase *as I have said* when the real fact is that something of a *similar* kind may have been said, but that the similarity is rather faint.

I am a little surprised at the strong terms in which Mr Whewell condemns the labours of critics, commentators, and editors ; see the *History* Vol. I. pages 184 and 203. For he frequently undertook such labours himself and performed them with great success : as examples we may take his prefaces to Mackintosh, Butler, Sanderson, Grotius and Barrow, and his volumes on the *History of Moral Philosophy* and on the *Philosophy of Discovery*.

But leaving these matters of detail I may proceed to a few remarks of a more general character.

Mr Whewell is perhaps open to the charge, which so frequently applies to authors of works on science, that he did not form, or at least did not always maintain, a steady conception of the class of readers whom he wished to address. From the circumstance that no mathematical formulæ occur in a work of which a large portion is devoted to mixed mathematics, it might be supposed that the *History* is designed as a popular manual for the general reader; but it is on the whole far too difficult for the amount of knowledge and resolution which such a reader may safely be held to possess. On the other hand the references to the original authorities are not sufficiently numerous and precise to render the work adequate to the wants of the systematic student of science. Thus on the whole it seems that the position which it occupies is an unfavourable border land, which might well have been exchanged for one entirely within or entirely without the domain of exact knowledge. Perhaps also the plan might have been contracted with advantage, though in this case different readers would scarcely agree as to the portions which it would be desirable to omit : for my own part

I should be glad to see the Physiology and Comparative Anatomy withdrawn, and the space thus gained divided among the other subjects.

The work is called a History of the *Inductive* Sciences; and the writer here and in his other writings attaches supreme importance to *Induction:* he contrasts this not unfrequently with *Deduction* to the manifest disparagement of the latter. Thus in Book III. Chapter I. Section 8 he remarks that man is *prone* to become a deductive reasoner; and he so frequently speaks of *ascending* by induction and *descending* by deduction, that he seems to allow metaphorical language to impose upon his sagacious intellect, and to adopt the notion that induction involves something which elevates and ennobles, while deduction is grovelling and servile. In the *Philosophy* 2nd edition Vol. II. page 92. we find Induction *bounding* upwards, while Deduction has to move steadily and methodically downwards. It would seem, to use the language of the strongest and the fiercest spirit,

> That in our proper motion we ascend
> Up to our native seat: Descent and fall
> To us is adverse.

Now it has been remarked by J. S. Mill that "A revolution is peaceably and progressively effecting itself in philosophy, the reverse of that to which Bacon has attached his name. That great man changed the method of the sciences from deductive to experimental, and it is now rapidly reverting from experimental to deductive." Be this as it may with philosophy in general, it may well be asserted that the science which occupies the most prominent place in Mr Whewell's *History*, and probably in the whole range of human culture, is almost independent of that induction to which he attaches a preeminent value. The triumphs of Astronomy are essentially triumphs of deduction: Newton and Laplace are far removed from any Baconian influence. Bacon himself was behind his age with respect to this great science; for he does not appear to have become a Copernican.

I may illustrate this matter by some particulars.

Take for instance Book III. Chapter III. which is entitled *The Inductive Epoch of Hipparchus.* The first Section of the Chapter

is on the *Establishment of the Theory of Epicycles and Eccentrics.*
We are told as to this Theory that Hipparchus "not only guessed
that it *might,* but shewed that it *must,* account for the phenomena."
Now, as we learn from the same page, the *guess* had already been
made by Plato, and the *guess* is, as Mr Whewell maintains, the
main element of scientific induction; so that in the present case the
induction had been effected before the inductive epoch: see *Philosophy of the Inductive Sciences,* Book XI. Chapter V. Section 2.
But without laying undue stress on the word *guess* it is plain from
the account given by Mr Whewell that the merit of Hipparchus
consisted in a deductive process by which he established the
agreement of facts with an hypothesis already current.

Again Book V. Chapter II. is called the *Induction of Copernicus.*
But the text distinctly contradicts the title, by shewing that
Copernicus performed no feat of induction; he obtained his theory
as he says from the ancient authors, and then shewed deductively
that it explained the phenomena.

The same considerations apply with special force to what
Dr Whewell calls the Inductive Epoch of Newton, and the great
Newtonian Induction of Universal Gravitation; History Vol. II.
page 136. The word Induction is here used; and in the *Philosophy*
we have an *Inductive Table of Astronomy* suggesting successive
ascents until we arrive at the Theory of Universal Gravitation.
But the facts seem to be quite otherwise. The fame of Newton as
an astronomer reposes for ever on the Principia—the greatest work
known in the history of science—and this is a magnificent chain of
deductions almost as strictly such as Euclid's Geometry.

The terms *induction* and *inductive* have been used in modern
times with some fluctuations of meaning, and not unfrequently
without any precise meaning at all; so that panegyric and censure,
when very undiscriminating, are expressed by some vague appeal
to the "soundest principles of the inductive philosophy:" in fact the
words are used as loosely as *logical* and *illogical.* See De Morgan's
Differential and Integral Calculus, page 12. Hence some interest
attaches to the sense in which they are understood by a writer
who uses them so frequently as Mr Whewell does. I by no means
object to his exaltation of induction above deduction; for that is

a matter of scientific taste as to which every person may please himself. Nor do I deny that such a man as Newton possessed high inductive powers; for I yield to no one in my admiration of Newton. But I maintain that on Mr Whewell's own definition of induction and deduction it is the latter which is almost exclusively predominant in the fortunes of astronomy. The idea of an attractive force varying inversely as the square of the distance was not peculiarly Newton's guess; it was floating in the scientific atmosphere of the period, but Newton alone had the power of developing the idea in a series of transcendent mathematical investigations. See the *Philosophy of the Inductive Sciences*, Vol. I. page 255, and Vol. II. page 387.

One very just reason why the deductive labour of such men as Newton and Laplace is highly esteemed is that we can feel sure to whom it is due. It would be hard to apportion rightly the merit of the inductive guess about attraction among Newton's predecessors and contemporaries; while the glory of the deductive demonstrations of the Principia is all his own.

We must remember too, when we are appreciating the relative claims of induction and deduction, we must not compare some second rate performer of the latter with some first rate adherent of the former; we must take men of unquestioned eminence from each school. Or we may compare the deductive work of one man, say Newton, with his inductive guesses.

Mr Whewell maintains in a much later work his own opinions about induction and deduction against those of J. S. Mill; but I do not find any reason for modifying the remarks which I have made: see the *Philosophy of Discovery*, pages 283 and 284.

The fact I believe is that Mr Whewell first contemplated merely a History of Astronomy, or of Mechanics, and subsequently enlarged his plan to include what he considered the principal Inductive Sciences. But the unity which he supposed his work to retain when thus expanded seems more imaginary than real; it is not difficult to perceive the dissimilarity of the substances thus ideally combined—the head of gold, and the feet of clay.

The advantages which Mr Whewell enjoyed for the preparation of his work were very great in two respects : in the first place his habits of intimacy with so many of the scientific leaders of the day, gave him easy access to the earliest and best information ; and in the second place he was able to trace the course of some streams of knowledge almost from their sources. For example, geology, at least in any systematic form, may be said to have commenced within his recollection ; and so also may the undulatory theory of light. I ventured once in conversation with him to remark that Humboldt especially, and himself in great measure, had been in more favourable circumstances than any person could hope to be in future, now that the sciences had so vastly increased ; and to this he assented. On the same occasion I endeavoured to persuade him that he had estimated the late Dr Young too highly ; I maintained that if the conception of the undulatory theory was considered the main point, then Huygens was preeminent ; and if the development of the conception was the meritorious labour, then Fresnel claimed the praise. He was much more tolerant of the opinion thus held in opposition to that of his *History* than might have been expected, and seemed willing to admit on his own principles that more honour should have been ascribed to Huygens.

In fact, in science he alone is the true discoverer who reveals, not in dim oracles which enthusiastic votaries may subsequently interpret into truth, but in clear characters which contemporaries, even though hostile, cannot misunderstand or misrepresent. Dr Young was so obscure as to be scarcely intelligible ; and both in science and philology he maintained, with great positiveness, opinions which were afterwards shewn to be wrong. A sentence in a letter addressed to Mr Whewell, in 1826, by one of the most distinguished of his scientific contemporaries will confirm the opinion expressed as to Dr Young's obscurity : "I am glad that you could not comprehend Young's reasoning, as I found it utterly beyond me."

I do not think that the following sentence relative to Dr Young which we find in the *History*, Vol. II., page 353, gives a correct notion of the matter in question : "his office of the Superintendent of the *Nautical Almanac* subjected him to much minute labour, and many petulant attacks of pamphleteers." Another mode of statement might be given thus: Dr Young received the salary of £500 a year for editing the *Nautical Almanac;* he had no special qualifications for the duty, and obstinately resisted the requests for improvements urged by men, like Baily and Herschel, of the highest eminence in astronomy.

The note books of Mr Whewell attest the indefatigable labour with which he collected the materials for his *History;* his correspondence with his scientific friends affords additional evidence on the point, and shews that they did justice to the ability and integrity manifested in his work. For an illustration I may mention one point as to which his first judgment was challenged, namely, the share of merit due to Sir Charles Bell and some other persons, relative to a discovery in Physiology: the appeals made to him to reconsider the matter were earnest and respectful, shewing the high opinion which the claimants entertained of his fairness and skill. In consequence of a renewed study he made some changes in his original award: I shall recur to this point.

The *History* may rank, to say the least, on an equality with others of a similar kind in our language, such as the Dissertations of Playfair, Leslie, and Forbes, which are attached to the *Encyclopædia Britannica;* and it has the additional advantage of being readily accessible instead of being lost in a bulky compilation. When an author takes so wide a range there must necessarily be parts which he handles with more vigour than the rest; it seems to me that the subject of Mineralogy, which Mr Whewell had professionally studied, is treated with special ability. The progress of Astronomy is also very well narrated; but in this department the rivalry is strong, and it is difficult for a casual volunteer to compete with those who are almost absorbed in this attractive subject. Thus the sketches traced by Mr Whewell, though clear and impressive, are perhaps scarcely equal in interest and decision to such masterpieces as the article by Dr Bowditch in the *North American Review* for April, 1825, and the *Report* by the present Astronomer Royal in the first volume published by the British Association.

It is no slight testimony to the merit of the *History* that it rendered important assistance to J. S. Mill in the construction of portions of his *Logic;* the acknowledgment is made in the preface to his work, and it was emphatically repeated by Mr Mill in conversation with his friends.

We ought not to pass quite unnoticed the fragments of verse which occur in the volumes. Sir J. Herschel says towards the end of his review, "Among our author's various and brilliant accomplishments not one of the least remarkable is his poetical talent, of which we have specimens in the mottoes prefixed to the several books of his 'History', and in the following perfect little *bijou* from Goëthe, ... ;" and he proceeds to quote the lines which occur in Vol. III. page 360.

Sir H. Holland said in 1840, " If you translated into English verse the passage from Pindar standing as motto to your former work,·you ought to be called upon for translation of the Theban throughout—I looked at it by chance a few days ago, and thought it excellent." Sir J. Herschel said that he did not see the drift of the motto from Pindar.

A few sentences in a letter written to Mr Whewell by Mr Lyell, afterwards Sir Charles Lyell, not long before the publication of the *History* will represent the opinion of many of the most eminent men of the day on the range of his knowledge. " There was a time when I used to regret that you had not concentrated your powers on some one department of physical science and become a giant in that, or at least that you had been satisfied with some two or three of the Arts and Sciences, but I have for some years come round to the belief that you have been exercising the calling for which Nature intended you, and for which she gave you strength and genius, and that you have given a greater impulse to the advancement of science among us by being a Universalist, and by mastering so much of Chemistry, Mineralogy, Astronomy, Geology and other branches, than you would have done if restricted to the perfecting of any one alone."

I will notice two of the points in which the *Edinburgh Review* differed with the statements in Mr Whewell's *History*. One was respecting a story connecting Pythagoras with the origin of Harmonics, which is given in Book II. Chapter III.; the reviewer fairly complained of some awkwardness of expression, but he was himself unfortunate in the correction he proposed on his page 121 with all the emphasis of capital letters: the word *inversely* which he inserts ought to be omitted.

The other point is more curious, and is the origin of a long controversy, which at the risk of being tedious I will unfold. Suppose the light of the sun to be admitted through a very small aperture, and received on a screen at a considerable distance from the aperture; then whatever be the shape of the aperture, triangular or quadrangular, or any other, the luminous figure on the screen is round : there is no doubt whatever as to the truth of this statement of fact. In the first edition of the *History* the explanation of this fact given by Aristotle was condemned as an application of inappropriate ideas. In reviewing the *History* Sir D. Brewster dissented from this judgment, and ascribed Aristotle's failure to the circumstance that he did not examine the truth of his conjecture by new experiments. In Vol. II. page 186 of the *Philosophy*, published in 1840, Mr Whewell returned to the point, and expressly referred to the criticism of the Review, maintaining, however, his own opinion. Mr Whewell spoke of the hole as *triangular*; this does not affect the principle, and in the subsequent éditions of the History the hole is called *square*.

In the review of the *Philosophy* which appeared in the *Athenæum* of Sept. 12, 1840, the matter was noticed ; and Mr Whewell made the following remarks in his reply to the review :

" The critic in the Athenæum thinks that Aristotle made no error, but will not allow that his success was due to the clearness or appropriateness of his ideas. As the question is not concerning the character of Aristotle, but concerning the conditions of discovery, I will not here contest the critic's opinion that Aristotle's explanation is fundamentally right. But, says the critic, it is taking too short a method of reproving the Stagyrite (Stagirite) to say that it is demonstrable that the Sun's rays in passing through a circular hole, must by virtue of their rectilinear movement, form a circular image. 'The Sun's rays,' he adds, 'do no such thing.' The Sun's rays, I reply, do exactly this thing and no other, putting *nearly circular* for *circular*, which has of course been understood throughout the discussion of this question. And this is plain, from the critic's own account, notwithstanding his dogmatic denial of it..."

In the review of the *Philosophy* in the *Edinburgh Review* for Jan. 1842, Sir D. Brewster treated the subject with some detail; he concluded that Mr Whewell had not read the original account of the phenomenon, and its explanation. The following sentences occur:

" Aristotle uses a *quadrangular* not a *triangular* aperture. *He gives no such explanation as that which* Mr Whewell ascribes to him. He never mentions the *circular nature of the Sun's light*; and he gives an explanation of the phenomenon which it is manifest Mr Whewell could not have given, and which would not have done discredit to Newton himself."

Then Brewster proceeds to "the Aristotelian problem and its solution"; and in the course of this he says, "... and hence the geometrical solution of the problem is, *that a quadrilateral* image of the aperture is formed at all distances by the solar rays."

In an Examination Paper for Dr Smith's Prizes at Cambridge in the early part of 1842, Mr Whewell, with obvious allusion to the article in the *Edinburgh Review*, proposed the following question:

"It has been asserted that *by geometry* a quadrilateral aperture, through which the Sun shines, forms a quadrilateral image at all distances; (rays of light being rectilineal).

"Disprove this. Shew that when the distance is great the image is a circle. What is the form of the space illuminated by the *whole* disk of the Sun?

"Determine the intensity of the light in each part of the image."

In a communication to the Literary and Philosophical Society of St Andrew's on March 7th, 1842, Dr Anderson "laid before the Society an explanation of the nature of solar images, more especially of such as are formed when the Sun's rays are admitted through irregular apertures." Dr Anderson quoted and condemned Aristotle's opinions which he said had "been maintained with equal ability and zeal by a writer in the *Edinburgh Review.*" Dr Anderson sent to Mr Whewell a printed abstract of the communication, together with a long letter, on March 27, 1842. The communication and the letter consist of a satisfactory explanation of the phenomena by Geometry.

A printed paper entitled *Problem of Solar Images* dated St Leonard's, St Andrew's, April 7, 1842, is obviously due to Sir D. Brewster; this was sent to Mr Whewell with a note dated April 19th, 1842. The printed paper refers to the question in the Smith's Prizes Examination, and states that the opinion condemned in the question is not the Reviewer's but Aristotle's. A draft of a reply to Sir D. Brewster by Mr Whewell is preserved among the papers.

Finally in the second edition of the *History* Dr Whewell corrected his account of Aristotle's attempt at explanation; and now referred the failure rather to *indistinctness* than to *inappropriateness* of ideas. He says, Vol. I. page 92, " In the first edition I had not accurately represented Aristotle's statement." And in the *Philosophy* the passage to which we have referred as occurring in Vol. II. page 186 was omitted in the second edition.

Mr Whewell addressed a printed letter to the Editor in reply to the unfavourable Article on his *History* which appeared in the *Edinburgh Review.* The letter is dated Oct. 28, 1837; it occupies four octavo pages. It begins thus :

" My dear Sir. I have taken the liberty of answering the critique of my *History of the Inductive Sciences* which you have just published, through the same channel which I made use of for a reply to a former article on a much smaller work of mine— a reply which you were obliging enough to acknowledge in a succeeding number of the Review."

The " much smaller work" here referred to is I presume the *Thoughts on the Study of Mathematics;* but I do not know what *channel* is meant through which Mr Whewell replied: perhaps it might be a Cambridge newspaper, or some magazine.

The reply to the Article on the *History* consists in rejecting almost every correction which the Reviewer proposed. There are only two exceptions; an admission that perhaps sufficient attention had not been paid to the obligations of Alhazen to Ptolemy with respect to Optics; and a promise to re-examine the question of the relative claims of Sir Charles Bell and Mr Mayo with respect to the discovery of the distinction between nerves of sensation and of volition.

The Reviewer had foolishly expressed a wish to see the name of Mrs Somerville in the *History;* the reply is as follows:

"With regard to the excellent and accomplished lady whose name the critic has thought proper to introduce into his pages...I will only say, that if I had employed my office of historian for the purpose of complimenting her with a place among discoverers in astronomy, whatever others might have thought of such a step, I am persuaded that her clear sense and genuine modesty would have disapproved of the introduction of such a passage into my work."

The Reviewer spoke of the University of Cambridge as "the cloisters of antiquated institutions,...through whose iron bars the light of knowledge and liberty has not been able to penetrate." The reply is as follows:

"This wretched rant is the echo of a slavish tradition, handed down from the brighter and prouder days of the Edinburgh Review. That work, at all periods, has spoken of the English Universities with equal bigotry and ignorance; and the shallow and vulgar conceit thus generated, which tarnished some of the fairest pages of the better scientific critics of its former times, is well fitted to make it contemptible, now that it has no longer any such to boast of."

In the letter Mr Whewell promised to examine again the question as to the merits of Sir C. Bell, and Mr Mayo. There is a reprint of the letter, at the end of which is the following notice: "N.B. This letter is a repetition of one written in October last, with an alteration of the paragraph respecting Sir C. Bell and Mr Mayo." Accordingly he now gives the results he had obtained by further study of the question; these will be stated immediately. The reprint is dated Jan. 3, 1837, which must be a misprint for Jan. 3, 1838.

Mr Whewell published the results of his further study of the history of discoveries in the nervous system in the form of a letter to the editor of the *Medical Gazette*, dated Trinity College, Cambridge, Dec. 11, 1837: the letter is contained on pages 525...528 of the number of the Magazine for Dec. 30, 1837. By a curious chance it follows an article by Sir D. Brewster.

In the *History* Mr Whewell had associated the names of Sir Charles Bell and Mr Mayo; the Edinburgh Review objected to this as highly unjust towards Sir C. Bell. Mr Whewell now concludes that no injustice had been done by joining the name of Mr Mayo to that of Sir C. Bell; but this is not to be understood to imply that they had equal shares in the discovery. The name of M. Magendie is also introduced.

After treating of the discovery Dr Whewell proceeds to touch on the further *confirmation* of the doctrine by succeeding observers; and in this he assigns the chief merit to Professor Müller of Bonn. He sums up thus: "Looking upon the case as above stated, with the best judgment I can use, I find myself led to the conclusion that no injustice was committed by joining Mr Mayo's name to that of Sir Charles Bell, as I did; but that, as I have already stated, this is not to be understood to imply that they had equal shares in the discovery. If I should have occasion to reproduce this part of my history, I should wish to describe the discovery as having been 'made by Sir Charles Bell, Mr Mayo and M. Magendie; the latter two physiologists having corrected and completed the researches of the former;' to which might be added the above notice of its confirmation." In the second edition of the History the matter was noticed; but not quite in the manner of the article in the Medical Gazette; for a fourth person, Mr John Shaw, is now introduced. The third edition coincides with the second. Thus there are really three different judgments by Mr Whewell on the matter. This shews the difficulty of writing the history even of contemporary science; but perhaps the subject of physiology might have been omitted with advantage.

The opinion of the merits of the *History of the Inductive Sciences* which was formed at the time of its first appearance has been confirmed on more recent occasions. Thus Professor J. D. Forbes, in his *Review of the Progress of Mathematical Science,*

1858, referring to the History and the Philosophy of the Inductive Sciences, says "An English philosopher of wonderful versatility, industry, and power has erected a permanent monument to his reputation in a voluminous work bearing the preceding title." Professor Owen in his address as President at the meeting of the British Association at Leeds in 1858 spoke of Dr Whewell as "the ablest historian of natural science." And at the meeting of the same body at Belfast in 1874 the President, Dr Tyndall, in his Address alluded to the *History*, and especially commended the remarks regarding the spirit of the Middle Ages.

In the title-page of the first edition of the *History of the Inductive Sciences* the vignette was introduced which the author afterwards so constantly repeated—a hand is represented transmitting a torch to another hand, and a motto of four Greek words is placed beneath. The words are from Plato who in allusion to an Athenian ceremony says, *Holding torches they will pass them on one to another.* In the coat of arms which Dr Whewell used a similar emblem is introduced—a hand holding a torch, and the motto is *Lampada tradam.* Sir Francis Chantrey drew two designs for the vignette, which are still preserved; neither of them quite coincides with the form which was adopted. Dr Whewell, writing to his friend Mr Jones on August 24, 1846, drew his attention to the works of Sir Francis Palgrave, and especially to the book entitled *The Merchant and the Friar,* furnished, as he says, "with a vignette constructed in antithesis to mine:" this vignette represents apparently a person receiving a torch from heaven. The first edition of *The Merchant and the Friar* appeared in 1837; it alludes to the *History of the Inductive Sciences* as a "work combining the imagination of the poet with the precision of the mathematician, and perhaps, containing more materials for thinking than any other of the present day."

The proof-sheets of the original edition of the *History of the Inductive Sciences* are preserved, and I have had the opportunity of examining them since the present Chapter was sent to the press. These proof-sheets are very interesting, as they contain corrections and remarks by the eminent friends of the author, to whom he submitted them; and they would be consulted with advantage if a new edition of the work should be hereafter required. It is easy to conjecture that the author abstained from attending to some of the criticisms which he received, on account of the delay and the expense which the requisite changes would have caused in the publication; but it is strange that in his second and third editions he did not adopt various suggestions which, even if unimportant seem obvious improvements. Still more strange is it to find that decided errors, though corrected on these proof-sheets, survived in the second and the third editions. Thus in the case I have mentioned on page 104 the number 2000 is changed to 1200 by the astronomer who was most conversant with the matter, and the mistake is also recorded on a fly-leaf of the volume—and yet the wrong number is preserved in the subsequent editions. Precisely the same remark applies to the fraction $\frac{1}{80}$ which occurs in a note on page 169 of the second volume of the *History;* it should be $\frac{1}{50}$. Also on page 156 of the same volume the word *comprises* should be *compresses.* It would be unfair to conclude from such examples of the vitality of error that Mr Whewell was specially unteachable or unduly resolute in his own opinion ; the explanation is I think simply this, that he forgot the remarks he had received, or could not easily find them among the confused mass of his papers, and so in the haste of his incessant occupations they were neglected.

CHAPTER IX.

1838...1840.

IN the year 1838 Mr Whewell was appointed to what was strictly called the *Professorship of Moral Theology or Casuistical Divinity* at Cambridge; but he himself preferred to consider the subject committed to his charge as *Moral Philosophy.* Mr Whewell attributed his election principally to the encouragement which he received from Mr Worsley, Master of Downing College, and then Vice-Chancellor of the University. The appointment had been for a long time, perhaps always, a sinecure; the new professor stated in one of his lectures—"so far as I am aware, no duties pertaining to the office had ever been performed: at any rate none had been performed for an interval of a century..." It is almost needless to say that Professor Whewell did not follow the example of official inaction which he received from his predecessors.

Professor Whewell attended the meeting of the British Association at Newcastle in August, 1838. During the year he was much engaged together with Sir J. Herschel and Professor Peacock in communication with the government respecting the expedition to be sent for the survey of the Antarctic regions.

In the year 1838 he published a volume entitled *The Doctrine of Limits, with its applications ; namely Conic Sections, the first three Sections of Newton, the Differential Calculus.* This is an octavo volume containing xxii and 172 pages, besides Errata and Additional Errata. The design of the work is indicated by the title; portions of subjects usually discussed in distinct volumes are here collected into one. Besides the matters named in the title there is also a section devoted to the Integral Calculus. The preface, as is usual with the author, is interesting and instructive; it contains acknowledgments of two errors, connected with limits, into which he had fallen in previous publications. The book seems to have enjoyed less popularity than the other elementary treatises by Professor Whewell, and did not reach a second edition.

At the anniversary meeting of the Geological Society on Feb. 16, 1838, it devolved on Professor Whewell as president to announce the award of the Wollaston Medal, and to deliver an address. The whole matter occupies about 30 octavo pages in the *Proceedings of the Geological Society* for Feb. 1838. The medal was awarded to Mr Owen, who received it in person and returned his thanks. The address gives an account of the more conspicuous members of the society who had died during the past year: these were Professor Turner, Professor Farish, Henry Thomas Colebrooke, and Von Hoff.

Professor Turner occupied the chair of Chemistry in what was then called the University of London, now University College, London. Mr Whewell seems to have been very intimate with him, and speaks of him in warm terms. He says, "Dr Turner entertained a conviction (I am stating the result of many interesting conversations which I have held with him) that the time was come when the chemist could not hope to follow out the fortunes of his science, and to read in her discoveries their full meaning, without being acquainted with the language, and master of the resources of mathematics." The same opinion was I believe held by a chemist of still greater fame, namely Faraday.

In the account of the recent progress of geology attention is specially directed to the researches of Professor Sedgwick and Mr Murchison. In consequence of these it is suggested that "one-third of our geological map of England will require to be touched with a fresh pencil."

The subject of technical phraseology is noticed, and a metaphor introduced which Mr Whewell employed frequently in his writings, with great effect: "Is it not true, in our science as in all others, that a technical phraseology is real wealth, because it puts in our hands a vast treasure of foregone generalizations? And if we evade the difficulties which may occur in the application of this phraseology to new cases, by declaring that our terms are of little importance, is not this to deprive our language of all meaning and all worth ?   Do we not refuse to recognize as valuable the tokens which we ourselves circulate, and plainly declare ourselves bankrupts in knowledge ?"

Mr Whewell maintains the same opinion in this address as in his later works with regard to a very interesting point. "...I have no belief that geology will ever be able to point to the commencement of the present order of things, as a problem which she can solve, if she is allowed to make the attempt...   When we do thus comprehend in our view the whole of the case, it is impossible for us, as I have elsewhere said, to arrive at an origin homogeneous with the present state of things; and on such a subject the geologist may be well content to close his own volume, and open one which has man's moral and religious nature for its subject."

Towards the end of the address the President alludes to the state of the Society, and his own connexion with it: "...To be placed for a time at the head of a body which I look upon with such sentiments, I must ever consider as one of the greatest distinctions which can reward any one who gives his attention to science.   I trust by your assistance and kind sympathy, gentlemen, I shall be able to preserve the spirit and temper which I so much admire;—to hand that torch to my successor burning as brightly as it has hitherto done.   And there is one consideration which will make me look with an especial satisfaction upon such a result.   I have not myself the great honour of being one of the members of the Society who are connected with it by an early interest in its fortunes, and by long participation in its labours.   I may consider myself as only belonging to its second generation. Now if there be a critical and a perilous time in the progress of a voluntary association like ours, it is when its administration passes out of the hands of its founders into those of their successors.   It is like that important and trying epoch when the youth quits the paternal roof."

In attempting to estimate the progress of a flourishing science there must be room for difference of opinion as to the value of recent contributions; and it is not surprising that Mr Greenough, an eminent geologist, wrote to the President of the Society respecting two points in his address. Mr Whewell had traced up to Von Hoff the merit of an important method of investigation; the correspondent thought that the two Delucs ought to have been mentioned, since they had adopted this method at a still earlier period. He also considered that Mr Whewell had been hasty in concluding that the researches of Professor Sedgwick and Mr Murchison in the north of Devon would necessarily lead to great changes in the geological map of England.

In the year 1839 Professor Whewell printed anonymously a translation of Göthe's *Herman and Dorothea* into English Hexameters. It was reproduced in 1847 in the *English Hexameter Translations;* and I shall speak of it hereafter.

In the months of June and July, 1839, Professor Whewell made a' tour in Germany; he was furnished with letters of introduction by Bunsen.

At the anniversary meeting of the Geological Society on Feb. 15, 1839, it devolved again on Professor Whewell, as President, to announce the award of the Wollaston Medal, and to deliver an address. The whole matter occupies about 40 octavo pages in the *Proceedings of the Geological Society* for Feb. 1839. The medal was awarded to Professor Ehrenberg of Berlin, and was received for him by the Chevalier Bunsen, who returned thanks. This may have been the occasion on which Professor Whewell first became acquainted with Bunsen. The address gives an account of the more conspicuous members of the Society who had died during the past year: these were Sir Abraham Hume, Benjamin Bevan, Nathaniel John Winch, William Salmond, Count Montlosier, Anselme-Gaëtan Desmarest, and Count Kaspar Sternberg. To this list was added Baron Von Schlotheim, who died in 1832, but of whom no notice had as yet been taken in the annual addresses.

With reference to Sir Abraham Hume we are told, "Indeed he may in a peculiar manner be considered as one of the Founders of the Society. English geology, as is well known, evolved itself out of the cultivation of mineralogy,—a study which was in no small degree promoted, at one time, by the fame of the mineralogical collections of Sir Abraham Hume and others."

After speaking of some meetings of cultivators of mineralogy, the address proceeds thus. "Out of the meetings to which I refer this Society more immediately sprung." The connection of mineralogy with geology is somewhat of the nature of that of the nurse with the healthy child born to rank and fortune. The foster-mother, without being even connected by any close natural relationship with her charge, supplies it nutriment in its earliest years, and supports it in its first infantine steps; but is destined, it may be, to be afterwards left in comparative obscurity by the growth and progress of her vigorous nursling. Yet though geology now seeks more various and savoury food from other quarters, she can never cease to look back with gratitude to the lap in which she first sat, and the hands that supplied her early wants."

The notice of Count Montlosier is very interesting. The Count was active in French politics in the early days of the first revolution, became an exile, and resided for some years in London, where he was the editor of a royalist journal. " ... Under the empire he returned to France, and was employed in the Foreign Office of the Ministry, but recovered little of his property except a portion of a mountain, which was too ungrateful a soil to find another purchaser. The situation however could not but be congenial to his geological feelings; for his habitation was in the extinct crater of the Puys de Vaches. The traveller, in approaching the door of the philosopher of Randane, had to wade through scoriæ and ashes; and from the deep basin in which his house stood, a torrent of lava, still rugged and covered with cinders, has poured down the valley, and at the distance of a league, has formed a dike and barred up the waters which form the lake of Aidat;—a spot celebrated by Sidonius Apollinaris, Bishop of Clermont in the fifth century, as the seat of his own beautiful residence, under the name of Avitacus." Some account of Montlosier as a politician will be found in the *Revue des deux Mondes* for Dec. 1874.

In the sketch of the recent progress of geology especial attention is given to the labours of Sedgwick and Murchison. Referring to the liberality with which our corner of the world had been supplied with groups of strata Professor Whewell observes, " As if Nature wished to imitate our geological maps, she has placed in the corner of Europe our island, containing an *Index Series* of European formations in full detail." Professor Whewell also speaks very highly of the researches which Mr Hopkins had communicated to the Royal Society, respecting the problem of precession and nutation on the supposition that the interior of the Earth is fluid.

With regard to the prospects of the science the President observes: " I confess, indeed, for my own part, I do not look to see the exertions of the present race of geologists surpassed ·by any who may succeed them. The great geological theorizers of the past belong to the *Fabulous Period* of the science; but I consider the eminent men by whom I am surrounded as the *Heroic Age* of geology. They have slain its monsters, and cleared its wildernesses, and founded here and there a great metropolis, the queen of future empires. They have exerted combinations of talents which we cannot hope to see often again exhibited, especially when the condition of the science which produced them is changed. I consider that it is now the destiny of geology to pass from the heroic to the *Historical Period.* She can no longer look for supernatural successes, but she is entering upon a career, I trust a long and prosperous one, in which she must carry her vigilance into every province of her territory, and extend her dominion over the earth, till it becomes, far more truly than any before, an universal empire."

In the concluding paragraphs of his address, Professor Whewell thanks the Society for the honour conferred on him by the office of President, and alludes to his own special pursuits: "For it has ever been one of my most cherished occupations, and will, I trust, long be so, to trace the principles and laws by which the progress of human knowledge is regulated from age to age in each of its provinces. To have had brought familiarly under my notice in a living form, the daily advance of a science so large and varied as yours, has been, as it could not but be, a permanent and most instructive lesson;—perpetually correcting lurking mistakes, and suggesting new thoughts."

Professor Whewell found much difficulty in obtaining a successor in the office of President; he applied in vain to

Sir J. Herschel, Sir P. Egerton, Leonard Horner, and Faraday: ultimately Dr Buckland, who had already discharged the duty, was persuaded to undertake it again.

Professor Whewell contributed to the *Philosophical Magazine* of January, 1840, the matter which occupies pages 65...70; this consists of a review of Professor Miller's *Treatise on Crystallography*, and of a review of the *Transactions of the Cambridge Philosophical Society*, Vol. VII, Part 1. The *Treatise* is highly praised with the slight reservation contained in the following sentences. "We cannot help thinking, however, which we do with regret, that this book, mathematically so admirable, will be a sealed book to a large body of crystallographical students. It is written with a rigorous brevity, worthy of the ancient mathematicians: a quality, in itself, doubtless, a beauty, but one of those stern beauties which repel, rather than attract, common beholders." The review finishes thus: "We cannot conclude this brief notice without expressing our satisfaction, that this subject of crystallography, after being put in so many forms for the last half century, has here assumed a shape which, so far as mathematical simplicity and symmetry go, leaves us nothing to desire, and therefore no reason for further change." The review of the part of the *Cambridge Philosophical Transactions* consists of short notices of various abstruse mathematical memoirs.

I may observe that Professor Whewell was the correspondent the justice of whose remark is allowed on page 104 of the *Philosophical Magazine* for February, 1840.

In this year Professor Whewell printed for private circulation some English Hexameters, entitled *The Isle of the Sirens*: it was reprinted in 1847, and will be noticed hereafter.

He seems to have taken some interest in a philanthropic society at Paris, called the *Institut d'Afrique*, of which he was appointed an honorary Vice-President about this time: the secretary, M. de Saint Anthoine, assures him that the Society would always see with pride among its foreign notabilities the name of the distinguished writer and illustrious Professor of Moral Philosophy of Cambridge. A letter from Dr Livingstone, dated Oct. 1859, giving an account of his travels in Africa, shews that Dr Whewell

long retained his sympathy with the exertions made for the civili-
zation of that quarter of the globe.

The *Philosophy of the Inductive Sciences* was published in
1840; it will be noticed in the next Chapter.

In the summer of 1840 Professor Whewell visited Holland and
Germany in company with Mr Thurtell of Caius College. In
September he was with Professor J. D. Forbes in Scotland, and was
I believe present at the early part of the meeting of the British
Association at Glasgow; but he was obliged to hasten back to
Cambridge to take part in the examination for fellowships in his
college. Towards the end of the year he and his friend Mr Hare
were very earnest in supporting the claims of Lord Lyttelton as a
candidate for the office of High Steward of the University; but
Lord Lyndhurst was elected.

CHAPTER X.

THE PHILOSOPHY OF THE INDUCTIVE SCIENCES.

In 1840 Professor Whewell published his important work entitled *The Philosophy of the Inductive Sciences, founded upon their History;* it consists of two octavo volumes containing altogether about 1230 pages. The second edition appeared in 1847 in two octavo volumes containing altogether about 1430 pages. The increase of bulk in the second edition is due mainly to three causes, namely, a fuller Table of Contents, the addition of some Philosophical Essays previously published, which occupy rather more than 100 pages, and the use of larger type than before for that part of the work called *Aphorisms.* The work was some years later separated into three, called respectively the *History of Scientific Ideas,* 2 vols. 1858, the *Novum Organon Renovatum,* 1858, and the *Philosophy of Discovery,* 1860; the last of the three contains large additions to the portion of the original work which it includes. These three publications are in small octavo, ranging with the third edition of the *History of the Inductive Sciences.* This separation of the original work into three cannot be considered a fortunate operation; it seems to shew a want of stability in the author which shakes the faith of his readers: Professor Whewell had already in his treatises on *Mechanics* exemplified the same process of rearrangement with unsatisfactory results. He says himself in a letter to Mr Jones, June 11, 1846: "I have more and more come to the conviction that to alter a book very much in the second edition spoils it. You may avoid some errors, but you lose the vitality and meaning of the work." In speaking of the *Philosophy* I shall refer to the pages of the second as the standard edition, reserving until a later period any remarks on the final issue of the work in three divisions.

The work is dedicated to Professor Sedgwick. The original Preface might have been advantageously reproduced in the second edition, but instead of it something quite different is substituted. The Preface to the second edition says that very slight alterations are made in the first edition, "except that the First Book is re-modelled with a view of bringing out more clearly the basis of the work;—this doctrine of the Fundamental Antithesis of Philosophy." The alterations occur in pages 1...50 of the first edition; they might be described by a friendly critic as merely natural modifications of the original views, while an enemy might exaggerate the changes so as to find evidence of great unsteadiness and fickle-ness. The want of an Index is a great defect in the book; in the *History* there are two lists, which, though far from being suf-ficiently extensive, do serve in some measure the purposes of an Index.

The *Philosophy* was reviewed in the *Athenæum*, Number 672, Sept. 12, 1840. To this review Professor Whewell replied in a pamphlet, which was probably printed only for private circulation. The pamphlet contains six very closely printed pages, besides the introductory paragraph, dated Sept. 22, 1840, which I will re-produce.

"Perhaps when you see the subject of this paper, you will at first think that I shew a needless sensitiveness in replying immediately to a review of my 'Philosophy,' in a weekly journal. If, however, you will read to the end, you will find that I have not written with any discontent against my Critic. His remarks af-forded me an opportunity of explaining a little further the objects of my Work; which I am afraid, with all explanation, my readers will find in many parts somewhat abstruse; and I have taken advantage of the opportunity. I have answered my Critic as I should have answered a fair objector in conversation, bluntly, but I hope not rudely."

One of the points discussed is the philosophy of perception. The critic held that Professor Whewell borrowed Kant's essential doctrines and stated them badly: Professor Whewell maintained that what was borrowed was explained nearly in Kant's own words, but that much was added for which he, and not Kant, was responsible.

As I have already said in Chapter VIII the work, in conjunction with the *History*, was reviewed by Sir John Herschel in the *Quarterly Review*. The *Philosophy* was also reviewed in the *Dublin University Magazine* for February and November, 1841; the first of these articles relates principally to Kant, the second is interesting, though rather too rhetorical: the signature, B, suggests that they may possibly be due to Professor Archer Butler.

The *Philosophy* was reviewed by Sir D. Brewster in the *Edinburgh Review* for January, 1842, in an unfavourable manner. Brewster, pre-eminent as an experimentalist in Optics, was no mathematician, and, what is rare among his countrymen, he seems to have disliked metaphysics; and consequently there was much in the work quite beyond the range of his sympathy. One thing which offended him much was the warmth with which Professor Whewell supported the claims of the Undulatory Theory of Light; the conclusion of the review will shew the opinion of the critic on this point: "It [the Undulatory Theory] utters predictions, and contrives to fulfil them; and not content with the dignity of a prophet, it wields the sceptre of a king in attempting to crush the spirit of experimental philosophy, on which the scientific glory of our country can alone repose. Thus armed with inquisitorial powers, it has enjoyed a temporary triumph. But its doom, as a physical theory, is sealed, and when it has lingered for another century as a mathematical hypothesis, the true cause of the phenomena of light will reward the diligence and genius of those who, in the spirit of genuine induction, have advanced in the straight and narrow way that leads to the Temple of Truth."

An article by Mr Herbert Spencer, entitled *The Universal Postulate*, in the *Westminster Review* for Oct. 1853, has the *Philosophy of the Inductive Sciences* among the works placed at its head. So far as the article bears on Dr Whewell, it relates to the controversy between him and Mr Mill on necessary truth, and on inconceivableness as a test of impossibility. Mr Spencer does not agree with either of the disputants, but is nearer to Dr Whewell than to Mr Mill.

As in the case of the *History*, I think that the later editions of the *Philosophy* ought to have shewn evidences of more careful revision than they actually received. A few examples of mistakes or difficulties which appear in all the editions may be noted; I cite the pages of the second edition.

Vol. I, page 68: "if equals be taken from equals the wholes are equal." Instead of *wholes* we must read *remainders*.

Vol. I, page 84 : "the mere formal laws by which appearances are corrected." Probably instead of *corrected* we should read *connected*.

Vol. I, page 99 : "the boundaries of an interminable space." This seems to me to involve a contradiction, namely the boundaries of a space without boundaries. It recalls to the memory a sentence almost at the end of Leslie's *Dissertation* in the *Encyclopædia Britannica*, where he speaks of the "shadowy visions which flicker along the horizon of Illimitable Space."

Vol. I, page 282: "the principles which form the basis of our mechanical reasonings,—that every change must have a cause, and that bodies can act upon each other only by contact." It is difficult to accept the statement that "bodies can act upon each other only by contact:" it seems contradicted by the fact of attraction, according to the author's own opinion on his page 260.

Vol. I, page 399. Here the doctrine of chemical compounds is stated in such a manner that A is made to neutralise n, without any condition as to the relative weights of A and n.

Vol. I, page 698 note. Instead of No. cxxIII, p. 126 we must read No. cxxxIII, p. 127.

A few remarks may be made on some of the topics discussed in the *Philosophy*.

The first Book is entitled *Of Ideas in General*. Dr Whewell maintains that there is a considerable number of *Fundamental Ideas* in all the sciences hitherto most successfully cultivated. He states them at the end of this Book, and proceeds in subsequent Books to discuss them in detail. Thus in the second Book he treats of the Ideas of *Space, Time* and *Number*, which are the foundations of Geometry and Arithmetic; in the third Book he treats of the Ideas of *Force* and *Matter* on which the Mechanical Sciences more peculiarly rest; and so on. Whatever may be the amount of conviction which the reader can obtain from the first Book, he will at least admit that the author's doctrines are developed and illustrated with great vigour and clearness. It has been said that many thinkers were unable to accept the ultra-Platonic hypothesis on which the *Philosophy* was based; but it would be more correct to speak of it as *ultra-Kantian* than *ultra-Platonic*. Although Professor Whewell paid much attention to Plato, yet it was chiefly at a later period of his life; and rather as a moralist than as a metaphysician.

In pages 55 and 59 Dr Whewell gives a criterion of necessary truth, namely that we cannot distinctly conceive the contrary. But if this be the case it becomes a puzzle to conjecture how we are to follow many of the indirect demonstrations in Geometry. For instance in Euclid I. 6 we have to take a triangle with two equal angles and conceive that the corresponding sides are unequal, and trace the consequences of this conception. Dr Whewell seems to admit that children and savages could form an indistinct conception in this case, but that persons of mature intellect could not: thus according to him Euclid's demonstrations may half convince a child or savage, but fail completely when presented to a trained mathematician.

In Vol. I, page 156, and Vol. H, page 359, a statement is made to the effect that if the properties of the conic sections had not been demonstrated by the Greeks the discoveries of Kepler could hardly have been made. It cannot be said that Dr Whewell himself attaches undue weight to the statement, but it has become popular and has sometimes been loaded with a weight of suggested inference which it will not bear. It is not true that any large amount of familiarity with the conic sections is required for the discoveries of Kepler; a very small fraction of the treasures accumulated by the Greek geometers would suffice for this purpose: probably a dozen pages would supply the necessities of a student who wished to master even the Principia of Newton. The notion that the exuberant developments and refinements now so characteristic

of pure mathematics constitute a capital which will hereafter produce an abundant return in physical applications is not warranted
by the tenour of scientific history, for the exigencies of natural
philosophy lead rather to the creation of new methods than to the
employment of those already current; this is well exemplified in
the theories of Attraction and of the Figure of the Earth. I am
little likely to undervalue the cultivation of pure mathematics,
but I do not wish the claims of this department of knowledge to
be placed on untenable grounds.

The third Book, which treats of the *Philosophy of the Mechanical Sciences*, is very interesting; especially the Chapter on the
General Diffusion of clear Mechanical Ideas. A few sentences
which occur on page 271 well deserve to be extracted. "The most
familiar words and phrases are connected by imperceptible ties
with the reasonings and discoveries of former men and distant
times. Their knowledge is an inseparable part of ours; the present generation inherits and uses the scientific wealth of all the
past. And this is the fortune, not only of the great and rich in the
intellectual world: of those who have the key to the ancient
storehouses, and who have accumulated treasures of their own;—
but the humblest inquirer, while he puts his reasonings into
words, benefits by the labours of the greatest discoverers. When
he counts his little wealth, he finds that he has in his hands
coins which bear the image and superscription of ancient and
modern intellectual dynasties; and that in virtue of this possession, acquisitions are in his power, solid knowledge within his
reach, which none could ever have attained to, if it were not that
the gold of truth, once dug out of the mine, circulates more and
more widely among mankind."

Nothing is more remarkable in the history of mixed mathematics
than the extraordinary inaccuracy and fluctuation of the language
of the earlier writers when measured by modern standards.
Although there has been a gradual improvement in this respect
much remains to be done, and perhaps some of the expressions now
current in our books will hereafter be proscribed. Dr Whewell's
language does not seem to be quite satisfactory, though it is far
superior to much that had survived to the date of his work. With-

out entering into details, I will repeat what I have said in Chapter II, that one great blemish is the frequent use of the word *pressure* instead of the word *force*. This is the more curious because in many cases the better course seems instinctively followed. Thus on page 185 there are some very good remarks on *Force*, and on page 203 the fundamental proposition of Statics is very properly entitled the *Parallelogram of Forces*. Many writers instead of the last phrase use the *Parallelogram of Pressures*; that is they seem to contemplate one species of mechanical action alone, to the exclusion of others: it would be as reasonable to speak of the Parallelogram of *Squeezes*, or the Parallelogram of *Thrusts*, or the Parallelogram of *Pulls*. Perhaps the word *pressure* is occasionally used instead of *force* for the mere rhetorical purpose of varying the expression; there cannot be a graver offence against scientific accuracy, and sometimes Dr Whewell may be guilty of it, but this will not account to any appreciable extent for the recurrence of the word *pressure*. Take an example from Vol. I, page 238, where a weight hanging by a string over the edge of a smooth level table draws another weight along the table; Dr Whewell speaks of the "pressures which restrain the descent of the first body and accelerate the motion of the second:" an examination of the context will shew that the word *forces* would be far preferable to the word *pressures*. In fact here the weights are not *pressed* but *pulled*. As another example we may take two sentences from the *History*, Vol. II. On page 37 we have "the Velocity which any force generates in a given time when it puts the body in motion, is proportional to the Pressure which the same force produces in a body at rest;" and on page 38 we have "in the same body, the velocity produced is as the pressure which produces it." It is obvious that in the second sentence *pressure* should be changed into *force*, in order to correspond with the first sentence; and in the first sentence there is something very awkward in having pressure produced *in* a body.

In Vol. I, page 267, Dr Whewell alludes to the opinion of Hegel that Kepler's glory had been unjustly transferred to Newton; this matter had been already noticed in the *History*, Book VII, Chapter II, Section 5. Hegel's opinions respecting Newton were considered by Dr Whewell with more detail in a paper printed in the *Cambridge Philosophical Transactions*, Vol. VIII, and reprinted in the *Philosophy of Discovery*. He should have remarked that the extravagances which he condemns in Hegel had been maintained by one who might have been expected to revere Cambridge and Newton. Coleridge in his *Table Talk* is reported to have made some remarks on the subject, in the course of which he assures us that "it would take two or three Galileos and Newtons to make one Kepler." I do not know that there is any evidence of Coleridge's competence to appreciate Galileo and Newton; and I presume he merely borrowed from Hegel without acknowledgment. Dr James Hutchinson Stirling has I believe published a work in which he discusses the question between Hegel and Dr Whewell; but I have not seen it.

The second Book of the *Philosophy* is called the *Philosophy of the Pure Sciences;* it treats in fact of the philosophy of the mathematical sciences. Now these are not inductive sciences at all, as Dr Whewell himself says on his page 83: their peculiarity is that they involve the process of deduction in a most remarkable character. The third Book of the *Philosophy* is called the *Philosophy of the Mechanical Sciences;* it is much more closely connected with that which precedes than with those which follow. The second and third Books form indeed one division of the whole work: it may be said that the lessons gathered from the study of Mathematics and Mechanics are applied in the remainder of the work to other sciences. This involves some important peculiarities with respect to the whole *Philosophy.* The mathematical sciences are avowedly deductive, and the mechanical sciences, so far as they are here considered, are more deductive than inductive; thus we really have principles mainly gathered from a survey of *deductive* sciences afterwards applied to subjects of a different kind, as chemistry and biology, which may be truly *inductive.* The second and third Books of the *Philosophy* are more precise and definite than the rest; the latter however are more novel and difficult. It is curious that no special Philosophy is ascribed to the subject which occupies the most prominent place in the *History,* namely Astronomy.

The peculiarities of Dr Whewell's treatment are seen then

most distinctly in the Books IV to X, which treat successively of
the Secondary Mechanical Sciences, the Mechanico-Chemical Sciences, Chemistry, Morphology, the Classificatory Sciences, Biology,
and Palætiology. Speaking generally we may say that in these
Books, instead of starting from an idea, as he starts from the idea
of cause in his third Book, he proceeds in search of an idea; and
what is finally presented to us as an idea is such a filmy abstraction that it can scarcely be considered as admitting of steady contemplation, and certainly not of any fertile application. In Book V
the idea of Polarity is propounded as that which is prominent and
distinctive. It is there shewn that in Electricity, Magnetism,
and Light, the word Pole and its derivatives Polarity and Polarisa-
tion have been usefully and extensively employed : but it seems
only a verbal generalization to say that the idea of Polarity is the
characteristic of these sciences and the guide to future progress in
them. We are directed to strip off from the idea all that is superfluous, and to arrive with Mr Faraday at the idea of polarity as
"an axis of power, having [at every point] contrary forces, exactly
equal, in opposite directions." Vol. I, page 355. The result is so
unsubstantial that it would scarcely be intelligible to any person
but Mr Faraday himself, and to him it would be superfluous.

There is something arbitrary in the collocation of the *Ideas.*
For instance *Cause* is allotted to the *Mechanical Sciences;* but it
would seem to be also quite as prevalent in the so-called *Secondary
Mechanical Sciences.*

In Chemistry we are told at first that the appropriate ideas
are those of *Element* and *Substance,* Vol. I, page 376; but subsequently the idea of *Chemical Affinity* is discussed at length ;
perhaps however this is meant to be only another name for the
idea of *Element:* see Vol. I, pages 376 and 388. An interesting
summary is given on page 546 of the attempts made to explain the
phenomena of Chemical Combination, and made in vain ; until at
last we find men "finally acquiescing in, or rather reluctantly admitting the idea of *Affinity,* conceived as a peculiar power..." It
might perhaps be said with more justice that men abandoned their
attempts at explanation and finally acquiesced in the *name* Affinity,
as simply a description of the phenomena without further analysis.

In Biology we are told that the peculiar idea is that of *Life*. But it does not seem that there is any agreement among Biologists as to what this idea really is. Dr Whewell himself works vigorously at a definition—a practice which he is usually far from commending. His final result is that *Organic Life is a constant Form of a circulating Matter, in which the Matter and the Form determine each other by peculiar laws (that is, by Vital Forces)*. Vol. I, page 588. If in Dr Whewell's own words (Vol. I, page 356) his idea of Life is "divested of all machinery," the residuum is merely the tautology that Life is something determined by Vital Forces. He seems indeed subsequently to allow that very small success had attended the pursuit of the idea or even of the definition: see Vol. II, pages 7 and 22.

The proof-sheets of the parts of the *Philosophy* which relate to Biology and Physiology were submitted to the criticism of Professor Clark of Cambridge, Professor Owen, and Sir H. Holland; the latter suggested that there was a little too much reference to foreign authorities, to the comparative neglect of English physiologists. Sir H. Holland objected to the word *sum* as an unsatisfactory translation of the French word *ensemble* in Bichat's definition, which Dr Whewell on his page 574 presents thus: "Life is the sum of the functions by which death is resisted." It seems strange that Dr Whewell did not yield to this objection.

In connexion with Chemistry the idea of *Substance* is discussed, and we are told that the "weight of the whole compound must be equal to the weights of the separate elements." See Vol. I, page 413, and Vol. II, pages 431, 659. Of course there is no practical uncertainty as to this principle; but Dr Whewell seems to allow his readers to imagine that it is of the same nature as the axiom that "two straight lines cannot enclose a space." There is, however, a wide theoretical difference between them, depending upon a fact which Dr Whewell has himself recognised in another place : see Vol. I, page 224. The truth is that *strictly* speaking the weight of the whole compound is not equal to the weight of the separate elements; for the weight depends upon the position of the component particles, and in general by altering the position of the particles, the resultant effect which we call weight is altered, though it may be to an inappreciable extent. Or we may say that we can always *make* the weight of the elements equal to the weight of the compound, if we are allowed to change at our pleasure the place where the experiment is performed, that is to weigh the compound at one place and the elements at another. Moreover even the *time* at which the weighing is performed is theoretically important; for weight changes with the changing positions of the sun and the moon in the sky. Trifling as may be the practical uncertainty which thus attaches to weight, it is quite sufficient to deprive the assumed principle of that absolute *a priori* certainty which the book seems to ascribe to it. The same considerations will form a powerful argument against the doctrine of Essay III in Vol. II, pages 624...634, which seeks to demonstrate that all matter is heavy.

Both in the *History* and in the *Philosophy* Dr Whewell gives much attention to Botany. His remarks are very interesting, but they seem to leave the impression that only the simpler and less valuable parts of the subject are treated. There is scarcely any account of vegetable physiology, which ought perhaps to receive the chief attention. The late Professor Henslow used to say that Botany offered us the best prospect of finding out what Life really is, from the ample facility which it supplies for experiments; probably Dr Whewell would have held that Zoology is the most promising study in this respect: see his *Philosophy*, Vol. I, page 540. It is worthy of notice that one of the most eminent of Dr Whewell's scientific antagonists, J. S. Mill, also paid considerable attention to Botany; and one point on which the two differed was the method of natural classification. Perhaps both writers looked upon the problem of such a classification as simpler and nearer to a solution than it really is. I extract a few words with respect to Mr Mill, which appeared in the *Examiner* of May 17, 1873. "Most botanists agree with Mr Mill in his objections to Dr Whewell's views of a natural classification by resemblance to 'types' instead of in accordance with well selected characters; and indeed the whole of these chapters are well deserving the careful study of naturalists, notwithstanding that the wonderfully rapid progress in recent years of new ideas, lying at the very root of all the natural sciences, may be thought by some to give the whole argument in spite of its logical excellence, a somewhat antiquated flavour." In the *Academy*, Oct. 1, 1873, a writer asserts that Mr Mill did not understand the proper end of classification.

The second volume of the second edition of the *Philosophy* is very interesting; it consists of Book XI, On the Construction of Science, Book XII, Review of Opinions on the nature of Knowledge and the method of seeking it, Book XIII, Of Methods employed in the Formation of Science, and the Aphorisms concerning Ideas, concerning Science, and concerning the Language of Science. An Appendix is formed by six Essays which had been previously published. The Review which constitutes Book XII may be especially commended. Dr Whewell always shines as a critic. His great power of acquiring knowledge and presenting it in a vivid manner is shewn to the best advantage in these historical sketches.

Books XI and XIII are, however, the most important in the volume; their design is legislative rather than historical and critical. Although full of instruction and interest, they do not seem well fused into an independent system. In the first place, it is not easy to see the mutual relation of the two Books—to distinguish between the *construction* of science, and the methods employed in the *formation* of science; this difficulty obviously occurred to Dr Whewell himself, for in the *Novum Organon Renovatum*, which consists mainly of a reprint of these two Books, the first is entitled simply *Of Knowledge;* but the mere change of title does not constitute distinction of substance. In the next place, there seems to be some discrepancy between the statements scattered over the two Books. I will examine the point with a little detail.

According to page 5, the processes by which we arrive at science are two, namely,

> Explication of Conceptions...A.
> Colligation of Facts ....B.

I place the capital letters here and throughout for the sake of reference.

It appears from page 46, that the combination of A and B constitutes Induction. Then it is not immediately obvious what part is performed by the Chapters of Book XI, which come after page 46. But it would seem from the last sentence of Chapter IV, that Chapter V, which is entitled, *Of certain Characteristics of Scientific Induction,* is to aid us in determining when the task has been rightly executed. Then Chapter VI, which is entitled, *Of the Logic of Induction,* may perhaps be said to be a comparison of Induction with Formal Logic. The remaining three Chapters of this Book may be described roughly as remarks on the nature of the results obtained by Induction. But there is a want of a well-defined plan in the Book; the connexion is not very obvious between the later Chapters and the earlier Chapters in which the two principal processes of the first Chapter are developed.

Now pass to Book XIII. We are told, on page 335, that there are *three* main points of the process by which science is constructed.

Decomposition and Observation of Complex Facts...C and D.

Explication of our Ideal Conceptions ....   A.

Colligation of Elementary Facts by means of these

    Conceptions ....   B.

Here, in addition to the A and B of the original statement, we have another process which, for the sake of convenient reference, I resolve into two parts, denoted by C and D respectively.

On page 336, "we have the following series of processes concerned in the formation of science."

    (1)   Decomposition of Facts .... C.

    (2)   Measurement of Phenomena ....D.

    (3)   Explication of Conceptions..............A.

    (4)   Induction of Laws of Phenomena ......E.

    (5)   Induction of Causes.......................F.

    (6)   Application of Inductive Discoveries...G.

I assume that *Measurement of Phenomena* is equivalent to what was before called *Observation of Complex Facts;* and from pages 336 and 337 I infer that *Methods of Observation* is another phrase for the same thing. Moreover, from page 336, it seems that E, F, and G together constitute what was formerly denoted by B.

Finally after the introduction the Chapters of Book XIII bear the following titles:

II. Of Methods of Observation.

III. Of Methods of acquiring clear Scientific Ideas; and first of Intellectual Education.

IV. Of Methods of acquiring clear Scientific Ideas, *continued*. Of the Discussion of Ideas.

V. Analysis of the Process of Induction.

VI. General Rules for the Construction of the Conception.

VII. Special Methods of Induction applicable to Quantity.

VIII. Methods of Induction depending on Resemblance.

IX. Of the Application of Inductive Truths.

X. Of the Induction of Causes.

It would seem that II corresponds to our D; that III and IV correspond to our A; that V, VI and VII correspond to our E; that IX corresponds to our G; and that X corresponds to our F. But still there are doubts which suggest themselves. It appears for instance from page 380, that V, VI and VII are equivalent to B; so that E is equivalent to B: whereas we formerly concluded that E, F and G together were equivalent to B.

Chapter V is called *Analysis of the Process of Induction;* but the Chapter really analyses Induction into three parts and discusses the first part. Then, very awkwardly, Chapter VIII treats of a process of Induction which was not included in the so-called Analysis of Induction.

By comparing pages 380, 389 and 395 it seems that the following phrases are to be considered equivalent: *Determination of the Magnitudes* and *Special Methods of Induction applicable to Quantity.* There is great perplexity caused by such needless variation of language.

We were formerly told that Induction consisted in the combination of A and B; but we are afterwards told that Induction consists of B alone: see pages 380 and 473.

It will be seen that Chapters IX and X are equivalent to the (5) and (6) which were quoted from page 336, but with a change of order. The change was doubtless prompted by rhetorical considerations, for a more impressive termination is thus secured for the Book; see page 440.

Although the Books XI and XIII are full of valuable remarks

and suggestions, yet they would have been much improved by condensation into one Book, and by attention to a consistent uniformity of doctrine and expression.

The first Aphorism concerning Ideas is the following: "Man is the Interpreter of Nature, Science the right interpretation." This is founded on Bacon's first Aphorism, omitting however all allusion to *Power;* for which Dr Whewell gives reasons in his Vol. II, pages 246 and 430. In the preface to the first edition Dr Whewell said respecting Bacon's first Aphorism; "Thus I have ventured to separate his first Aphorism into two;" but this statement is incomplete, for he should have added, *and to omit the second part.* In the first edition of the *Philosophy* the Aphorisms were placed at the beginning of the work; but I apprehend that this was merely for the convenience of making the two volumes of about the same size, and the paging seems to shew that they were printed after the rest. Moreover we read on page 51 of this edition, with respect to the Language of Science, "its rules and principles I shall hereafter try, in some measure, to fix." The *hereafter* here refers to matter which is really placed *before* in the volume.

The remarks concerning the Language of Science are expressed with great modesty; it could not have been inferred from them that the names adopted by some of the eminent masters in science, to whom Dr Whewell refers, were originally suggested by himself or at least submitted to his judgment. Such however was the case, as we learn from his correspondence: and nothing puts in a stronger light the eminent attainments of Dr Whewell than the respect with which he was consulted by so many of his contemporaries: among these we find Bell, Faraday, Lubbock, Lyell, Murchison and Owen.

The general structure of Dr Whewell's *Philosophy*, and also various subordinate parts, have been much criticised. I may in particular mention the article by Sir John Herschel in the *Quarterly Review,* to which I have alluded on page 102, the *Treatise on Logic* by J. S. Mill, and the *Biographical History of Philosophy* by G. H. Lewes.

Dr Whewell replied to Sir J. Herschel in a paper which was printed and circulated privately, and afterwards published in the second edition of the *Philosophy*, Vol. II, pages 669...679. The reply takes the form of a letter to Sir John Herschel, in which the Reviewer is alluded to as if he were some third person; this little fiction could I presume never have concealed from any person the real author of the Review. Dr Whewell replied to Mr Mill and Mr Lewes in his *Philosophy of Discovery.*

I shall not venture to join here in the controversy carried on by such distinguished men, except so far as to say that I cannot personally agree with Mill and Herschel to refer my conviction of the truth of the foundations of geometry to experience. I can venture to disclaim most emphatically any trace in my memory of obtaining certainty by trial and experiment as to the axiom that two straight lines cannot enclose a space. Whether there may not be some compromise between the views of Sir J. Herschel and Dr Whewell may furnish matter for conjecture. Why it may be asked should we all arrive at our convictions in the same way? Is it not possible that one person may have recourse to experiment, while another finds that he can do without it? Or may we appeal to the doctrine, now receiving much attention, that our faculties are modified by descent, so that we really *inherit* what appears at first sight to be *innate?* Thus we may imagine that one philosopher springs from a long line of ancestors, who have made sufficient experiments in their generation, so that their descendant enters into possession of results which he unconsciously supposes to be innate. Another philosopher owing to the neglect of experiment by his progenitors has to execute such labours for himself before he can arrive at the idea of Space in Dr Whewell's sense. He has in fact to acquire a fortune instead of succeeding to an entailed estate. Sir J. Herschel seems to me practically to surrender his own position in the words which are quoted with emphasis by J. S. Mill in his *Logic,* Vol. I, page 282, "including always, be it observed, in our notion of experience, that which is gained by contemplation of the inward picture which the mind forms to itself in any proposed case, ..." What is thus included by Herschel and Mill in the notion of experience, if allowed to be sufficiently prominent, will be nearly sufficient for Dr Whewell's demands. After all I suppose that Sir J. Herschel and Dr

Whewell would quite agree in the fact that we do ultimately possess certain fundamental ideas; and this is of great practical importance, even if the speculative problem as to the origin of such ideas remains unsolved.

There are slight indications in the correspondence of Dr Whewell that in very early years his philosophical views may not have been quite identical with those which he afterwards maintained. Thus his friend Hugh James Rose in 1822 seems to charge him with undue depreciation of Plato; and a letter of his own to Richard Jones of the same period is perhaps not quite consistent with his mature opinions. But on the whole it would appear that his philosophical principles soon became fixed, and are maintained with great consistency throughout his long series of publications: I have noticed an apparent exception in speaking of the *Bridgewater Treatise.*

It is according to Goldsmith a safe criticism to say that the picture would have been better if the artist had taken more pains; at the risk of appearing to repeat such an obvious remark I cannot forbear from expressing the wish that Dr Whewell had continued during the second half of his mature life the pursuits of the first. The quarter of a century between 1816 and 1840 produced the *History* and the *Philosophy;* if the next period of the same duration had been employed in a similar manner we might have obtained other works of a kindred character, or at least large additions and improvements to these. But whatever objection may be urged against specific portions of the *Philosophy*, it remains still a noble design executed with rare ability. The author proposed to traverse the wide fields of physical science, to extract the most important results from the past, and to indicate the principles which should guide the future. He did not despair of the fortunes of knowledge; he held that real connexions must subsist among the apparent infinite mass of details, and he encouraged investigators to seek for such general ideas in all their peculiar occupations. The lessons which he explicitly taught and implicitly suggested seem peculiarly useful at the present day when every pursuit is specialised. Mathematics alone form more than sufficient occupation for one student, and have to be separated into distinct parts; and the natural and experimental sciences in like manner are minutely subdivided. The claims of a candidate for scientific distinction are urged for instance, not on the ground that he is eminent in Natural History, nor even that he is familiar with Zoology, but that he is well versed in the *Lepidoptera*. The study of such a work as the *Philosophy* may well counteract the narrowness which the exclusive devotion to one selected branch of knowledge is apt to produce; to use language attributed to Professor De Morgan, it will encourage a man to know something about everything as a balance to knowing everything about something.

Perhaps however the desire for unity in science amidst the seeming endless diversity is again manifesting itself; and books appear among us which discuss the same subjects as in the preceding generation were treated in Herschel's *Discourse*, Whewell's *Philosophy*, and Mill's *Logic*. Without however depreciating any recent attempt, I may be permitted to recommend the older writings as still deserving of study. I will not attempt to make a comparative estimate of these three; but only observe that the merit of such general surveys of sciences consists rather in the aids they afford to thought than in the rules for guiding manipulations. I find the following remarks in a recent article relating to Mr Mill, but I cannot say that they seem to me satisfactory. "Sir John Herschel very much improved the rules of inductive enquiry, but they still remained empirical rules unsupported by theory. Whewell collected a large mass of facts connected with the progress of inductive science; but his work bears the same relation to a true philosophy of induction as Wombwell's menagerie did to a natural history museum. Neither Whewell nor Herschel solved the problem of induction and so distant did such a solution appear to be that not only Macaulay, but even Whately pronounced it impossible." The article maintains that Mr Mill did solve the problem which had baffled his predecessors.

More recently the following sentence appeared in another review. " We may state however our firm belief that Mr Jevons's mode of treating the subject has given the *coup de grâce* to Mr Mill's theory of induction."

I ought not to pass without notice a volume entitled *Exploratio Philosophica,* published in 1865, by Professor Grote, who succeeded Dr Whewell in the chair of Moral Philosophy at Cambridge. Sixty pages of this volume are devoted to an examination of Dr Whewell's Philosophy of Science, following as many which treat of Mr Mill's Logic. The volume bears ample testimony to the power of its author, and will recall, to those who had the high privilege of knowing him, many excellent traits of a character which they justly regarded with reverent affection. There are, however, some peculiarities which weaken the impression the work might produce, in spite of the great ability which frequently appears in incidental remarks. Perhaps few persons will follow the author in the high value which he assigns to Professor Ferrier's *Institutes of Metaphysic :* for while as a clever academical exercise that volume can scarcely be overestimated, it seems to have slender claims to be regarded as a serious contribution to the subject on which it treats. Professor Grote paid little attention to the systematic development of his views, and neglected the graces of composition; the latter is a fatal defect with respect to a science which has been adorned by some of the finest writers in the English language. Hume and Berkeley, Dugald Stewart and J. S. Mill, by their conspicuous charms of style have rendered students of metaphysics almost unreasonably fastidious. I will advert to two points of interest noticed by Professor Grote. On his page 203 he offers an explanation of the relations which the four parts of Dr Whewell's whole work, as finally divided, bear to each other—a matter which is by no means obvious. The History of the Inductive Sciences forms the first part; the Philosophy of this History is contained in the History of Scientific Ideas, which forms the second part; the third part is the Philosophy of Discovery, which is to be regarded as the History of Scientific Method; the Philosophy of this History is contained in the Novum Organon Renovatum, which forms the fourth part. This is ingenious ; but it will be seen that it reverses the chronology of the third and fourth parts: that is, according to this scheme the Novum Organon Renovatum should have appeared last, and not the Philosophy of Discovery. On his page 225 Professor Grote gives a note which involves the suggestion that while Dr Whewell and Mr Mill both put forward the name of Bacon, they were really rather followers of Descartes.

It is known from the correspondence of Dr Whewell that his labours on his *History* and *Philosophy* were carried on simultaneously; and thus at the time when the former was published he was far advanced in the preparation of the latter: indeed, if this had not been the case it would have been impossible even for his untiring energy to have produced the *Philosophy* within three years after the *History*, in the midst of his very numerous occupations. Eleven bound note-books have been preserved which contain abstracts and references connected with the *History* and *Philosophy*, and also drafts of portions of these works. Another note-book which has been kept with these bears a similar relation to the *Bridgewater Treatise*.

This Chapter may be fitly closed with the letter in which Humboldt acknowledged the receipt of a copy of the *Philosophy of the Inductive Sciences*.

"Monsieur,

"J'ai été extrêmement sensible Monsieur à l'honneur que vous m'avez fait de m'envoyer votre grand et important ouvrage de "Philosophie des sciences exactes." Je n'ai pu mériter cette distinction que par la vive admiration que depuis bien des années j'ai marquée pour votre "History of the Inductive Sciences." Je vous ai lu non dans la pâle traduction de M. Littrow, mais dans la langue dans laquelle on aime à lire ceux qui ont la puissance de la parole. Je savois que vous nous prépariez la jouissance d'un ouvrage philosophique propre à apprécier la sûreté des méthodes et la certitude sur laquelle reposent nos connoissances du monde intérieur; j'en avais parlé avec chaleur au Silurien M. Murchison, et comme nous respectons en lui, et dans l'excellent et respectable M. Sedgwick un ami commun, je suis tout satisfait de la bienveillance avec laquelle M. Murchison vous aura révélé la curiosité dont j'ai été tourmenté. J'aime à me rappeler de plus une soirée bien agréable que nous avons passée ensemble chez l'illustre Cuvier, dont l'esprit étoit plus logique que philosophique et qui blâmoit un peu trop tout ce qui tendoit au delà "de mettre de l'ordre dans les idées." Vous ne me rendez pas justice, Monsieur, si vous craignez que j'appartienne à cette secte de mes compatriotes qui dédaignent Bacon et le trouvent un peu *rococo*. Je suis innocent de ce dédain et aucun de mes travaux (si toutefois il en existent) montre de l'indifférence pour le "Novum Organum." J'ai du commencer par me justifier devant celui qui honore de nos jours par son nom le "collége de Bacon et de Newton." Quant à l'ensemble de votre vaste ouvrage, Monsieur, il m'est pour le moment impossible d'entrer dans des discussions partielles. Je vous écris ces lignes du haut d'une petite "colline historique," dont le nom peut vous rappeler bien des distractions et des devoirs de position qui ne sont pas tout à fait littéraires. Je ne connois pas d'ouvrage depuis 10 ans qui excita à un plus haut degré mon vif intérêt. Les chapitres Philosophy of the Pure Sciences, la Morphologie, les forces vitales, (un des mythes nombreux de nos physiologies), la mer ténébreuse des traditions palætiologiques que vous avez navigué avec une noble indépendance de l'esprit, les méthodes d'induction et le résumé historique que vous avez donné Vol. ii p. 283...469, m'ont offert des lectures pleines d'intérêt et de charme dans les jardins de Sans Souci! Après vous avoir suivi dans vos ingénieuses recherches sur les mouvemens de l'Océan, on aime à vous suivre à travers ces fluctuations de l'intelligence, variées selon l'espace et le tems et soumises cependant à des retours périodiques. Lorsqu'on voit les choses de si haut, comme vous Monsieur, on permet le doute à ceux, qui comme moi, ne savent que glaner dans les spécialités. Je n'ai encore vu votre grande composition que comme ces forêts vierges (sylvas sylvarum) dont on découvre la surface ondoyante lorsqu'on est placé sur le sommet des Cordilleras. Je n'ai encore pu jouir que du coup d'œil général, mais votre grand ouvrage m'accompagnera sous peu vers le nord, devant suivre le Roi à la prestation des hommages à Königsberg, où je trouverai deux savans illustres bien dignes de vous lire, M. Bessel et le grand géomètre M. Jacobi.

"Je saisis avec empressement cette occasion Monsieur pour vous renouveler l'hommage de ma haute et respectueuse considération.

LE B^N DE HUMBOLDT.

"Veuillez bien excuser mon écriture illégible. J'ai un bras malade des exhalaisons des forêts du Cassiguiaré.

À SANS SOUCI près POTSDAM,
  ce 14 Août, 1840."

CHAPTER XI.

1841...1853.

AT the beginning of the year 1841 Professor Whewell felt some inclination to accept the college living of Masham in Yorkshire, which was then vacant; and he consulted his friend Mr Hare on this proposed change in the nature of his occupations: ultimately, however, he declined the preferment. In this year he published a pamphlet entitled *Two Introductory Lectures to two Courses of Lectures on Moral Philosophy:* this will be noticed hereafter in connexion with the other works on Moral Science.

In the year 1841 the British Association assembled at Plymouth on July 29th; Professor Whewell was the President, and gave an address which occupies pages xxvii...xxxv of the Report of the meeting.

It is frequently the practice in such addresses to offer an estimate of the scientific work of the preceding year, and it might perhaps have been expected that one who had traced the history of several branches of knowledge through many centuries should have followed this course. But Professor Whewell speaks of such a task as "always difficult and sometimes long;" and instead of it proposes to occupy a few minutes with a slight sketch of the general aspect which the Association now appears to offer to a thoughtful spectator. Accordingly he adverts to the scheme of a Philosophical College which Bacon had drawn up in the *New Atlantis,* and called *Solomon's House.* The British Association may be considered to correspond in some degree with this scheme of Bacon's, the difference consisting in its voluntary character instead of its subjection to the control of the State. Professor Whewell warns the members of the Association against an evil which we see from his correspondence that he strongly deprecated, namely a tendency to introduce questions of legislation and politics into their discussions. He concludes his reference to this topic in the following words: "Knowledge *is* power; but for us, it is to be dealt with as the power of interpreting nature and using her forces; not as the power of exciting the feelings of mankind, and providing remedies for social evils, on matters where the wisest men have doubted and differed."

Two extracts may be given which contain interesting personal references to the speaker himself. With respect to his occupying the position of President he says: "On one account, at least, I may venture to undertake such a ministry as this: I have been a faithful attendant upon the meetings of the Association ever since its first institution, and there is scarcely any subordinate office of labour or dignity in the constitution of the body which I have not at one place or other discharged, with such zeal and care as was in my power." Adverting to the scientific zeal and intelligence of the miners of Cornwall, Professor Whewell says: "Perhaps I have had very unusual opportunities of becoming acquainted with their merits, for in two different years (1826 and 1828), in the prosecution of certain subterraneous experiments, undertaken in conjunction with the present Astronomer Royal and other persons, I lived four months the life of a labouring miner, and learnt how admirable for skill and conduct is the character of all classes of the mining population in that region."

On October 12, 1841, Professor Whewell married Cordelia Marshall, at Watermilloch Church, Cumberland, near Hallsteads, the residence of John Marshall, Esq., her father. On October 16 letters from Dr Peacock and others reached Professor Whewell at Coniston, informing him that Dr Wordsworth had resigned the mastership of Trinity College, and according to the advice given to him he came to London immediately. There he found a letter from Sir Robert Peel offering him the vacant post; he accepted it on the 18th, had an interview with Sir Robert Peel on the 19th, and returned by the night mail to Coniston. On the 23rd he sat down to write the first lecture of a new course on Moral Philosophy.

The selection of Professor Whewell for the Mastership was

approved with a hearty unanimity which shewed the general conviction of his great power and attainments. His claims were urged upon the consideration of members of the Government by various well-qualified judges, but Sir Robert Peel was justly proud of the circumstance that he recommended Professor Whewell to the Queen spontaneously before any of the favourable testimonies reached him.

In the year 1842 the British Association met at Manchester in June. Professor Whewell seems not to have been present, though it is usual for the retiring President to introduce his successor in office formally to the meeting. The new President, Lord Francis Egerton, in his address alluded in graceful terms to his predecessor: "You met last year, indeed, under different auspices. I cannot forget—I wish for the moment you could—how your Chair was then filled and its duties discharged. Could you forget the fact, it were hardly to my interest to awaken your recollection to it, that such a man as Professor Whewell filled last year at Plymouth, an office which I now hold at Manchester. I do so for the purpose of remarking that he, more able, perhaps, than any man living in this country to give you a concise and brilliant summary of all that he and his fellow labourers are doing, forbore in his discretion from that endeavour. If he, then, who is known in matters of science to have run

'Through each mode of the lyre, and be master of all,'

abstained from that undertaking, I may now be excused, not for my own silence, which would require no apology, but for not calling on one of your other functionaries to supply my place for the purpose."

In September, 1842, the inhabitants of Lancaster gave a public dinner to their eminent fellow-townsmen, Professor Whewell and Professor Owen. On Nov. 5th, 1842, Professor Whewell preached before the University, and naturally alluded to the Gunpowder Plot and to the glorious and blessed deliverance effected by William III. His great predecessor Bentley had preached before the University on Nov. 5th, 1715, and had attacked the errors and the crimes of Popery with all his vast ability as a scholar, a theologian, and a controversialist. Professor Whewell was appointed Vice-Chancellor of the University in November, 1842, and held the office according to custom until the November of the following year.

In 1843, during his Vice-Chancellorship, a proposal was offered to the Senate of the University for the temporary construction of a Board of Mathematical Studies. The Board was to consist of the Lucasian, Plumian, and Lowndean Professors, and of the moderators and examiners for the current year and the two preceding years. A report was to be presented by the Board to the Senate in 1844, and again in 1847. The function designed for the Board was to guide and control the mathematical studies and examinations of the University. The proposal however was rejected by the Senate of the University, as one had been some time previously of a similar kind, but having the Master of Trinity College as a proposed member. A printed paper of three quarto pages was circulated in the University recommending the proposed Board; the only copy I have seen has the date T. L. April 13, 1843, and the signature W. Whewell, both added in manuscript. Another printed paper of two quarto pages also recommended the proposed Board; it is dated April 3, 1843, and is unsigned: it may have been written by Dr Peacock. A Board of Mathematical Studies was established in the University by a regulation of the Senate on Oct. 31, 1848; see Dr Whewell's *Liberal Education*, Art. 289.

Among the matters which engaged his attention during his Vice-Chancellorship may be mentioned—a petition from the University to Parliament in favour of retaining the Welsh Bishopricks undiminished in number; the election of a Regius Professor of Divinity to succeed Dr Turton, when Dr Ollivant was the successful candidate; and the opposition of the University to the scheme of the Railway Company with respect to the site of a station in Cambridge. Professor Whewell's official engagements did not prevent him from taking an interest in other matters which were brought before him. The Bishop of London had suggested the formation of a committee to enquire as to what good school-books were at present in use for the schools of the poor; and the committee consulted Professor Whewell on the subject of English Grammar. He must have replied to the committee, as the receipt of a "very kind letter" is acknowledged. In the same year Bunsen, then occupied with his work on Egypt, wished to submit his views on the Philosophy of Language to Professor Whewell.

In the year 1844, Professor Whewell took the degree of Doctor of Divinity. He preached on June 30th before the University. He was present at the meeting of the British Association at York in September.

On January 30th, 1845, Dr Whewell preached before the University; the subject of his sermon was Christian patriotism, and he naturally alluded to the special services then connected with the death of King Charles the First.

In the year 1845, Dr Whewell published his *Elements of Morality*. He says in a letter, dated March 7, to Mr Jones, "I am to get my larger book out in May if possible: for this purpose I must print and correct at the rate of a sheet a day, which is fast going." The work will be noticed hereafter.

The British Association met at Cambridge in June, 1845. This was not according to Dr Whewell's wish; he thought that the trouble and excitement attendant on such a large concourse of visitors ought not to be imposed for a second time on the University while there were so many places which had not borne the burden once. But he acquiesced in the arrangement when made, and endeavoured to secure by personal invitation the presence of some distinguished foreign men of science.

In this year perhaps he went for the first time to Cliff Cottage, Lowestoft, an attractive residence near the sea, to which henceforward he frequently resorted. He purchased the place in 1851. He was much pleased with Fairlight Down in Sussex, and the shore near it, but, as he wrote to Mr Jones in September, 1845, he "did not discover its beauties till it was too late to make any enquiries."

In September, 1845, Dr Whewell attended the meeting of the Archæological Association at Winchester. In October, 1845, Dr Daniel Wilson, Bishop of Calcutta, wrote a long letter to

Dr Whewell respecting the offer to the University of Cambridge of a Prize of £500 "for the best Refutation of Hinduism, and Statement of the Evidences of Christianity in a form suited to the Hindus." The liberal donor of the Prize himself drew up an admirable paper, explaining the nature of the Essay which was required; this was issued to the University on May 27, 1846, in the form of an official document on seven quarto pages. It is addressed, "Rev. W. W. D.D.;" this I presume refers to Dr Whewell. The prize was awarded to the Rev. Rowland Williams, of King's College. Bishop Wilson, through whom the offer was made to the University, was at this time on a visit to England; he spent a few days at Trinity Lodge in November, 1845, and was much gratified at the kindness shewn to him.

Dr Whewell presented a statue of Bacon to Trinity College; the resident Fellows in November, 1845, requested Professor Sedgwick to convey their thanks to the Master for his gift; the interesting letters which were written on this occasion have been preserved.

I proceed to notice two publications of the year 1845.

In the year 1840 a work was published anonymously, entitled *Vestiges of the Natural History of Creation*, in which the old theory of Lamarck, respecting the development and transformation of the species of animals, was revived. The work was written in a captivating style, and thus excited considerable attention, while its doctrines were viewed with some alarm. In the course of the year 1844 Dr Whewell was urged by various friends to reply to it, and was requested to write an article on it in the *Edinburgh Review;* but he declined to do more than to issue some extracts from his former publications, in which, as he considered, the opinions of the author of the *Vestiges* had been anticipated and condemned. In his correspondence he attached great importance to the fact that the passages thus selected had presented themselves spontaneously in the course of his philosophical reflections, and had not been produced with any specially controversial object. Accordingly he issued in 1845 a slender volume in small octavo entitled, *Indications of the Creator*, consisting of extracts from the *History* and the *Philosophy of the Inductive Sciences*, together with one from the *Bridgewater Treatise*. The volume is dedicated to W. Smyth, the Professor of Modern History at Cambridge. The preface alludes in general terms to the subject which had been brought into notice by the *Vestiges*, but does not specifically mention that work. In 1846 a second edition of the *Indications* appeared with the addition of a new preface, which is very polemical in tone, and refers frequently to a publication entitled, *Explanations: a Sequel to the Vestiges of the Natural History of Creation*. Little remark on the subject is needed, for the facts are familiar to those who take an interest in science or literature. The popularity of the *Vestiges* was partly owing to the mystery which hung over the authorship, and may have been augmented by a vehement attack in the *Edinburgh Review;* the work went rapidly through several editions, but finally almost disappeared from public notice. Recently Mr Darwin has again drawn attention to the same theory of development, though in a more cautious and moderate spirit.

I pass to the work entitled, *Of a Liberal Education in general; and with particular reference to the leading studies in the University of Cambridge*. This is an octavo volume; the title, dedication, preface, and contents, occupy xviii pages, and the text 248 pages; it is dedicated to G. B. Airy, Esquire, Astronomer Royal. The book is of the same character as that on English University Education, which was published in 1837; but of much larger extent. It consists of four Chapters entitled respectively, Of the subjects of Educational Study, Of the method of teaching in Classics and Mathematics, Of the recent and present condition of Mathematical and Classical Education at Cambridge, and Plan of a standard Cambridge Course of Mathematics. It will be seen that the authority of Dr Whewell may be quoted in favour of the word *educational*, which has been condemned, I think, by Archbishop Trench.

In the first Chapter we have educational studies divided into the permanent and the progressive; the former being those which connect men with the past, and the latter those which connect them with the future. The paramount place in education is claimed for mathematics and classics. The whole discussion is very interesting, and is perhaps even more important at the present day than when it was first published, inasmuch as there is now a stronger tendency than formerly to dangerous heresies with respect to the principles of education. If I may venture to indicate what appears to me a defect, I should say that there is a want of practical direction as to the *extent* to which the training is to be carried. In his Art. 115, Dr Whewell strongly condemns the practice of allowing a student to confine himself to classics alone, or to mathematics alone, according to his choice; but the programme which is recommended as suitable with respect to these two departments of knowledge will appal even the most resolute youth. For the mathematical course is to include the *Principia* of Newton, the *Mécanique Analytique* of Lagrange, and the *Mécanique Céleste* of Laplace; while in the classical course a thorough acquaintance with the Greek and Latin languages, and a familiarity with the best classical writers, are indispensable. The demands thus made on the exertions of a student may remind us of the long list of the qualifications necessary for a poet, according to Dr Johnson, and may well draw forth the despairing exclamation—"Enough, thou hast convinced me I can never acquire a liberal education."

A few points of detail in the first Chapter may be noticed. In Art. 19 we read, with respect to the interest of the Greeks in Philosophy, "Such speculations formed a large portion of their Philosophy; and such Philosophy has occupied every succeeding generation up to the present time; and most, the generations of greatest intellectual activity." But the last clause of the preceding sentence seems scarcely consistent with a long and vigorous passage on pages 21...25 of the work on English University Education. I will extract from this one sentence, which is printed by the author himself in italics: "The progress of science corresponds to the time of practical teaching; the stationary, or retrograde, period of science, is the period when philosophy was the instrument of education."

In Art. 28 Dr Whewell assigns to the ancient classics a vast superiority over modern authors; and, in fact, distinctly implies that none of the latter approach the former in extent of influence.

It may be doubted, however, whether this is a safe or a fair estimate. In such a comparison we can frequently trace the habit of
putting, perhaps half unconsciously, all the Greek and Latin
writers in one scale, against some single modern writer in the
other scale. But take distinct examples; it would be scarcely
possible to find a single ancient poet whose influence for
range and power conspicuously surpasses that of Shakespeare or
Dante.

A statement in Art. 93 may be commended to the notice and
judgment of classical scholars; it seems rather hazardous: "It is
not likely that we have in modern times, any one who knows
Latin better than Erasmus did, or Greek better than Bentley."
The word *knows*, as applied to a language, however, is susceptible
of various meanings, and thus in some one or other of these the
opinion may be sound.

Dr Whewell is strongly in favour of presenting mathematics,
for the purpose of education, rather in the geometrical than in the
analytical form. He considers, and very justly, that the cultivation of the higher departments of analysis is rather the business
of the professional mathematician than the discipline of the general
student. But the same remark would apply equally to much of
the geometrical learning which he recommends; the *Principia* of
Newton, for example, at least in any approach to its whole extent,
may be suitable for a mature philosopher but scarcely for a youth
under educational training. In connexion with his comparison of
mathematical methods it may be observed that Dr Whewell does
not seem to appreciate sufficiently the value of new conceptions.
Take for example the subject of conic sections, which he strongly
recommends to be studied geometrically, and not by the method
of coordinates. Now it is obvious that when a youth has gone
through a course of Euclid he is only continuing essentially the
same kind of discipline in passing on to Geometrical Conics; he
gains indeed more facts and results, but no additional power
and no fresh principles; or at least any such acquisition bears no
proportion to the expenditure of time and attention which the
subject demands. But the study of the method of coordinates
introduces him to a region entirely unknown before; he has to

master the peculiarities of a new language of curious idioms and marvellous power; and the difficulty which young students usually experience in acquiring some familiarity with the strange tongue is a fair measure of the importance of the task, and the healthy vigour which it promotes.

An interesting opinion is recorded in Art. 65, though I may venture to say that my own would be exactly contrary to it: "I have no doubt that in any application of geometrical, mechanical, or hydrostatical principles to a problem of moderate difficulty, supposing the problem new to both of two students; one, a geometer of the English school of forty or fifty years back, the other a modern analyst, instructed in equal degrees; the former would much more accurately and certainly obtain a definite and correct solution."

In Art. 52 we read, "The general belief, for undoubtedly it is a general belief, that Mathematics is a valuable element in education, has arisen through the use of Geometrical Mathematics. If Mathematics had only been presented to men in an analytical form, such a belief could not have arisen. If, in any place of education, Mathematics is studied only in an analytical form, such a belief must soon fade away." The last sentence in the preceding passage seems contradicted by experience ; take for instance the mathematical education in the celebrated *Ecole Polytechnique* of France.

In Art. 31 we read with respect to the *Calculus of Variations*, which has been a favourite pursuit with some mathematicians: "Such steps of wide symbolical abstraction, however beautiful as subjects of contemplation to persons of congenial minds, are out of the range of any general system of Liberal Education.''

The second Chapter of the work on Liberal Education is entitled *Of the method of teaching in Classics and Mathematics.* This contains much that is interesting and valuable, as might have been expected from the long experience and advantageous position of the author; his opinions on *Examinations* especially deserve attention. The main element of the problem, so far at least as Oxford and Cambridge are concerned, seems however almost forgotten, though practically it is of paramount influence; namely, the fact that valuable fellowships are offered for attainments to be tested by competition in some precise and definite manner. At Cambridge on the average perhaps twenty students annually gain fellowships, and twenty more throughout their course may have reasonably indulged the hope of such success. Thus a disturbing force, tending to deflect from the direct pursuit of knowledge, acts on a large fraction of the men of each successive year; and on that fraction which may be supposed to include all the students of conspicuous ability. No discussion of University teaching and examinations can be complete which does not distinctly recognise this prominent fact, and endeavour to suggest remedies for the evils which it undoubtedly involves, even if we admit predominant benefits.

Dr Whewell alludes in his Articles 169 and 170 to the disputations which were formerly a part of the regular academical exercises for degrees; he regrets their extinction although he considers their restoration hopeless. I remember to have heard him state in conversation that very acute arguments were sometimes produced in these contests; in particular he said that Sir John Herschel, when passing through this stage of his undergraduate career, advanced a most subtle objection against one of the received propositions in science. Dr Whewell promised to give me the particulars of this on some future occasion; and I much regret that the opportunity never occurred.

It must be observed that notwithstanding the familiarity of the author with Cambridge practices his generalisations are sometimes too wide for perfect accuracy. Thus he says in his Art. 156 that students will select as their private tutors, "...if the constitution of the University allows it, those who have been, or are to be, Examiners;" but it is sufficient to say in reply that the late Mr Hopkins, who for more than twenty years was almost unrivalled as a private tutor of Mathematics, never held the office of Examiner in the University.

Dr Whewell recommends the combination of oral examinations with paper examinations; his advocacy of the former, which have almost disappeared from Cambridge, is moderate and judicious. I have discussed the subject in some published Essays and will only say that the practical difficulties which beset any scheme of searching oral examinations in mathematics seem insuperable. Mr Jones, in writing about the present work to Dr Whewell, says, amidst high general praise of it. "You have not converted me into any liking for oral examinations. Most of the men here [Haileybury] however side with you to some extent, that is, they value them as means of detecting imposture—but I once passed a morning in the schools at Oxford and came away with a profound conviction of the intense injustice of using oral trials for the purpose of assigning relative rank for which men have toiled for years, and I do not think that conviction will leave me on this side the grave." Dr Whewell must have grieved to see a philosopher whom he had so constantly exhorted to walk in the paths of inductive virtue, and whose eminent merit in this respect he had himself attested and praised, thus announce that he had formed a conviction, to be for ever unshaken, on a *single* instance.

The Articles 128...146 form a curious episode in the second Chapter of the work; they consist of a reply to some criticisms published in Lyell's *Travels in North America* on the opinions advanced in Dr Whewell's volume on *English University Education.* The criticisms and the reply seem both to have been out of place in the books where they respectively appeared; and the difference of opinion between the two authors is such as might naturally be expected when one fixes his attention mainly on the faults which he wishes to correct, and the other on the merits which he hopes to retain. From drafts preserved among Dr Whewell's papers it appears that he wrote privately to his critic in 1847, urging the same considerations as occur in the printed volume; and he received a long and most able reply. It is satisfactory to find from their correspondence in later years that the old friendship of the two eminent writers was not destroyed by their controversy. The logic of facts seems to have decided in favour of Sir Charles Lyell, who lived to see the accomplishment of many of the reforms which he advocated and the assured prospect of the rest.

The third Chapter of the work is entitled *On the recent and present condition of Mathematical and Classical Education at Cambridge.* This gives the history of Examinations at Cambridge, quoting in many cases the official language which prescribed the successive changes. Suggestions are offered with the view of improving the examinations, and in particular the plan for conducting the Mathematical Tripos examinations is recommended which was substantially adopted by the University, and lasted from 1848 to 1872, both inclusive. The establishment of a Board of Mathematical Studies is proposed in Art. 253; the suggestion had been previously made by Dr Peacock; the Board was established in 1848 and still exists, but it may be doubted whether its course has been on the whole judicious and the results of its interference salutary. Some remarks on Private Tuition in this Chapter seem more moderate in their tone than might have been expected from the indications of opinion in the earlier part of the volume.

The fourth Chapter consists of a *Plan of a Standard Cambridge Course of Mathematics*. This Course begins with Arithmetic, and then passes on to Algebra and Plane Trigonometry; so that very strangely Geometry is omitted. Specific works are recommended on the various subjects, but scarcely any of these survive in the present mathematical teaching of the University.

A friend of Dr Whewell writing to him about the book expressed the opinion that the word *Reason* was employed in various senses. He says: "In Art. 10 you must use it very largely for *Rational Thought* and something more probably; in Art. 17 you must mean the reasoning faculty simply, even to the exclusion of *Judgment*, which is not exercised in the study of Euclid." The writer mentions as containing excellent remarks, Copleston's *Reply to the Edinburgh Review*, p. 104 *et seq.;* and the article by Davison on Edgeworth's *Professional Education* in the *Quarterly Review* for Oct. 1811.

The remarks made by Dr Whewell in his Articles 100 and 101, ascribing small importance and value to Greek composition, were once quoted by Lord Stanhope in the House of Lords, and received with much attention.

I will give extracts from two letters which record opinions on some of the matters discussed in this work.

Professor Chevallier of the University of Durham, writing to Dr Whewell on Dec. 9, 1845, says:

"Many of the points which you touch upon have of course been brought prominently before our notice here; and many of your suggestions have been already brought into practice for some years, and with very good effect.

"The introduction of a vivâ voce examination in classical examinations for honours has been accompanied with great advantage: and I have long wished that a similar plan should be adopted in Cambridge. A vivâ voce examination is valuable for its influence on the examiners as well as on the candidates: and there are some branches of scholarship, such as an extended acquaintance with history, especially with the philosophy of history, as well as a knowledge of the subject-matter of all the books, which can scarcely be tested by any other means.

"We have applied vivâ voce examination also in mathematics, and found it perfectly easy of application, and very advantageous in those parts of an examination which relate to natural philosophy."

Mr Jones in the letter to which I have already referred says: "For myself I feel sincere pleasure in seeing your effort to purify and amend the mathematical training of the place—I have long been convinced that, as a matter of training, exclusive habits of symbolical reasoning are not merely useless but deleterious, and I see very often instances of their bad effects on men of very acute minds."

Dr Whewell's work on Liberal Education was reviewed in the *Westminster Review* of July, 1848. The work was reprinted in 1850 almost unchanged, except by the addition of new matter which will be noticed under that date.

In the year 1846 Dr Whewell published a book entitled *Lectures on Systematic Morality;* he also contributed to *Blackwood's Magazine* three letters on English Hexameters. The book and the letters will be noticed hereafter. Dr Whewell was present at the meeting of the British Association at Southampton in September, 1846. Sir John Herschel says in a letter dated Sep. 2, 1846, "I am very glad you are coming to Southampton and with a paper in your pocket. The subject is one which will not fail to be of great interest." There is nothing attributed to Dr Whewell in the Report of the meeting, except a short communication on a *Method of Measuring the Height of Clouds;* and to this Sir John Herschel's language seems scarcely applicable. Mr Joseph Kay of Trinity College, who had held the appointment of Travelling Bachelor in connexion with the University of Cambridge, publish-ed a work on the Education of the Poor in England and Europe; some controversy connected with this publication between him and Dr Whewell is I think in print, but I have not seen it.

Dr Whewell seems to have contemplated the establishment of a Magnetic Observatory in his house at Lowestoft, and letters alluding to the subject from Sir J. Herschel, Humboldt, and Professor Phillips have been preserved. Perhaps the design was abandoned as too expensive for a private person to carry out. Some important remarks by Sir J. Herschel, in relation to the contemplated observatory, ought not to be omitted: "In the first place I would observe generally that if I were asked how I thought an amateur of Magnetism and a good philosopher and mathema-tician could do most good for the subject presuming all lines of proceeding equally consonant to his private tastes I should say— to set to work one or more good calculators under his directions and furnished by him with formulæ and combinations to try and get some definite results from the mass of magnetic and meteorological observations already on hand—rather than add to the stock."

I alluded on page 41 to the communications which Dr Whewell sometimes received from ignorant men; as another example I may state that, in 1846, he received from a land-surveyor a diagram to support the assertion that the diameter of a circle is to the circumference precisely as 7 to 22. The writer says, "I shall feel much obliged if you will investigate it and give me your Candid opinion on its merits, or if your time is fully occupied you will I doubt not have the kindness to hand the paper to some Friend of yours who will give it his early attention."

I pass on to two mathematical publications of the year 1846. The first of these is entitled *Conic Sections: their principal properties proved geometrically.* This is an octavo pamphlet containing in all 47 pages besides a page of errata. It was doubtless occasioned by the University legislation of the period, which made great changes in the system of examination for mathematical honours, and explicitly prescribed Geometrical Conics as one of the subjects. Dr Whewell in his preface strongly maintains the importance of this subject, but his arguments will not secure universal assent: especially when the nature of the geometry here exhibited is considered, for it consists almost entirely in the laborious combination of ratios. The publication was found useful; the second edition appeared in 1849, and the third in 1855: the errata are corrected but substantially no change was made in the original work.

The other mathematical publication of the year 1846 is entitled *Newton's Principia: Book I. Sections I. II. III. In the original Latin; with explanatory notes and references.* This is an octavo pamphlet containing in all 74 pages. The preface complains that students under examination frequently deviated widely from Newton's own methods while professing to reproduce his demonstrations; and recommends attention to the original text as a standard which ought to be maintained. There are in the notes numerous references to the author's *Doctrine of Limits.* Some directions drawn up by Newton himself about the books to be read as preparatory to his *Principia* are here published for the first time. In the list of works advertised at the end of this pamphlet as *By the same Author* there is one entitled *Statics (Elementary); or a Treatise on the Equilibrium of Forces in One Plane. 8vo. Plates, 4s. 6d.* I believe that this is a mistake, and that Dr Whewell did not publish such a work.

An article in the *Edinburgh Review* for April 1847 treated on the subject of Arabian Philosophy; in the course of it a quotation was given advocating an extreme opinion on intellectual intuition, and a foot-note was added in these words: "This passage is at the service of Dr Whewell, in support of his arguments for that independence of all experience which he claims for certain truths." The matter is of no consequence, but it so happened that this was the first Number issued under the superintendence of Professor

Empson, who was a friend of Dr Whewell, and thus it seemed a curious time for a renewal of the old attacks on him by the *Edinburgh Review.* The new editor expressed his regret for the "saucy look" of the note, and explained that the article was printed off before he had accepted the charge of the *Review.*

In June, 1847, Dr Whewell attended the meeting of the British Association at Oxford. In November he was appointed one of the Visitors of the Royal Observatory at Greenwich in the room of Dr Pearson who died in September; the appointment was made I believe by the Marquis of Northampton, as President of the Royal Society. In December Dr Whewell visited Paris, provided with letters of introduction from Dr Hawtrey, the Provost of Eton.

Three publications of this year will be noticed hereafter in the appropriate Chapters; namely, *English Hexameter Translations,* a sermon entitled *The Christian's duty towards Transgressors,* and a volume of *Sermons preached in the Chapel of Trinity College, Cambridge.*

I proceed now to notice two other publications of the year 1847, both poetical.

In the early part of the year 1847, a volume appeared anonymously entitled *Verse Translations from the German: including Bürger's Lenore, Schiller's Song of the Bell, and other Poems.* The volume was published in London, by John Murray; it is in octavo. The translations and the notes occupy 87 pages; the title, preface, and list of contents, occupy 8 pages. The preface is peculiar on this account; two paragraphs take an impersonal form, and then the writer proceeds to say "I have added...", while no name is subscribed. The authorship of the volume is known from Dr Whewell's correspondence.

As a specimen of the translations we may take a passage from the *Song of the Bell;* in this case and in some others the original was annexed in order to supply "the opportunity of immediately seeing the degree of fidelity which the translations attain."

> Proud from the Girl the Boy must dart ;
>   To plunge in life his bosom burns,
> A pilgrim he roams earth's every part,
>   And a stranger back to his home returns.

And there before his startled eyes,
  In radiant virgin bloom arrayed,
Like some fair vision from the skies,
  Before him stands the blushing Maid.
Then nameless yearning fills his breast;
  In lonely paths he steals along;
Tears fill his eyes and break his rest;
  He shuns his brethren's noisy throng.
He follows, blushing, where she goes;
  Her greeting makes all nature smile;
He culls for her the fairest rose
  — To deck her bosom, all too vile!
O thoughts that thrill! O hopes that bless!
  O joys to first love only given!
The heart is drunk with happiness,
  The earth is brighten'd into heaven.
O that the time might endless prove,
  That happy time of early love!

The following piece is called *The Traveller's Evening Song:*

Lo! the fading gold of the west!
Darker and darker the purple unfolds its vest:
  The earth is silent growing—
  Louder the stream is flowing
    Lulling to rest.

See the last gleam dies in the west!
Still shades cover all things on earth's wide breast:
  Serene and tender the feeling
  Over the senses stealing
    Soothing to rest.

Under every covert is rest!
The branches quiver silently over the nest:
  The birds all slumber securely;—
  Wait awhile, wait awhile,—surely
    Thou too shalt rest.

A note says: "This is a translation of the German words which accompany a very pleasing musical air." The fact however is, that the first two verses are not a translation, but Dr Whewell's own; while the third verse is a translation, though not by him. The original of the third verse is due to Göthe; it will be found on page 101 of the first volume of the edition of his works in thirty volumes issued in 1850: the translation, however, is not very close. A lady has kindly recorded the circumstances connected with the composition of the English song, which Dr Whewell himself must have forgotten when it was printed. She was one of a party assembled at Trinity Lodge on May 4th, 1844, and sang among other things the words which form the last verse of the piece as it appears in the *Verse Translations*. Dr Whewell and Lord Houghton who were standing near her both liked the song, but said it was too short; the lady said she could not lengthen it for there were no more words, but she hinted that perhaps they might supply her with some. Lord Houghton said that the verse was complete in itself, and so a *second* could not be added; but Dr Whewell rejoined that a *first* might be put before it. The subject then dropped, but the lady took an opportunity of returning to it a few days afterwards, and sent her book to Dr Whewell; he soon returned it with the two original verses. A letter from Hermann Kindt a German friend to Dr Whewell, dated Oct. 26, 1863, praises very highly the last verse of the song as a translation.

One of the pieces in the *Verse Translations* is called *The Spinning Maiden's Cross;* we are told in a note, with respect to this, that it "has already been published at least three times, and each time has been mutilated in such a manner as to render it unintelligible." It appeared first in the volume called *The Tribute,* as I have stated on page 97; I do not know the other two publications in which the piece was included, but it seems from Dr Whewell's papers that one of them was a volume issued by Mr J. Burns. I am not sure whether the piece is entirely original or a translation; the Marquis of Northampton proposed this question to Dr Whewell, and his answer seems to leave the matter doubtful. A very able German friend of Dr Whewell, named Heinrich Fick, however, treats it as a translation. He says, in a letter dated May 29, 1847: "Having, as a most loyal interpreter of our great Schiller's genius, followed him through the most beautiful and varied lore of his muse, how delightfully you finished your task with the sweet ballad, 'The Spinning Maiden's Cross,' which you rendered with most charming touching simplicity." H. Fick resided for several years in England; his letters shew that he was a person of great literary ability, well acquainted with English, as well as with German, poetry. In an address, delivered at the distribution of Prizes at the College of Civil Engineers at Putney, he alluded to the *English Hexameter Translations*, and to the *Verse Translations*, and said, "they occupy the foremost place amongst the translations ever attempted from the German in point of fidelity and finish." As a specimen he quoted the piece entitled *The German Muse*.

In the later part of 1847 Dr Whewell printed for private circulation a volume entitled *Sunday Thoughts and other Verses*. The volume is in octavo, and contains 112 pages, besides the half-title and the list of contents on 4 pages. There is no preface, and no explicit record of the date of issue, or of the author's name: it is known to be Dr Whewell's from his correspondence. The *Sunday Thoughts*, which stand first in the volume, consist of eight pieces of a devotional nature; there are also other compositions of a serious cast, as a poem in blank verse, called *Gothic Architecture*, a sonnet to Mrs Somerville, and a sonnet to Miss W., that is, Miss Wordsworth. But the greater part of the volume is of an amusing character, including five Riddles, three Charades, and translations of thirty-five short German Fables. One piece is called *Nugæ Bartlovianæ*, and alludes in a very pleasing manner to the excavation in 1835 of some large barrows, near the village of Bartlow, in Essex; this had been printed before, at least in part, in a quarto form. One of the persons most interested in the excavations was Mr John Gage, the Antiquary; the curiosities obtained were deposited at Easton Lodge, the seat of Viscount Maynard, the proprietor of the Bartlow Hills. Dr Whewell introduced a Roman soldier remonstrating against the disturbance of the remains in the barrow:

> But you'll soon discover, as I've a notion,
> That we are not pleased with our promotion;
> And from Easton's shelves you'll a groaning catch,
> At the dark still hours of the midnight watch.

Dr Whewell sent a copy of his *Sunday Thoughts and other Verses* to one of those who had been present at the excavation; a few sentences may be taken from the reply:

"In reading them once again, the recollection of the joyous and interesting meetings which were held here connected with the

Barrow Exploration, was not unmixed with regret. Because it was impossible to forget that poor Gage, whom you so well describe as 'main cause of all this ill,' should have been the first, and I believe the only one of the party, who has been taken from us.

"But the Roman's prediction is still more extraordinary, Easton having been burnt at midnight, and every vestige of the contents of the Barrows consumed."

The last piece in the volume is called *The Isle of the Sirens;* this had been printed before in a quarto form : it is in hexameters, and consists of an adaptation of some passages of Carlyle's *Chartism,* published in 1840.

In August, 1848, Dr Whewell attended the meeting of the British Association at Swansea.

In the year 1848 Dr Whewell published three things which will be noticed in subsequent Chapters, namely, a review of Longfellow's *Evangeline* in *Fraser's Magazine* for March, a sermon entitled, *The Bulwarks and Palaces of Zion,* and an edition of Butler's *Three Sermons on Human Nature.* He was requested to review Mrs Somerville's work on *Physical Geography;* but he declined, saying, "I have not been conversant of late with the departments of literature and science to which it refers, and should either have to read a good deal bearing upon the subject, or to write a very shallow and scanty review; and my labours at present lie mainly in other fields." The task which Dr Whewell thus avoided was undertaken very appropriately by a great traveller—Sir Henry Holland : see his *Recollections,* page 248.

Dr Whewell delivered a lecture at the Royal Institution in London on Jan. 22, 1848, on the *Use of Hypothesis in Science.* A rough draft of the lecture in manuscript is preserved, and also a printed abstract of it, which is probably taken from some Journal of the Proceedings of the Institution. It is easy to see from these that the lecture coincided substantially with what Dr Whewell published elsewhere : see the *Philosophy of the Inductive Sciences,* 2nd edition, Vol. II, pages 54...74; and the memoir on the *Transformation of Hypotheses in the History of Science* reprinted from the *Cambridge Philosophical Transactions* in the *Philosophy of Discovery.* Mr Faraday said in a letter on Nov. 7, 1848, in reference to this lecture, " You must have done good to all who heard you, but I can testify of myself that the truth you laid down, that truth can more easily emerge from error than from confusion, has been to me practically useful, and a source of continual pleasure in observing that, in other things besides those I meddle with, it is so."

The Count of Habsburg appeared on pages 232 and 233 of *Fraser's Magazine* for February, 1848. It consists of 12 stanzas, each of 10 lines. I quote the introductory notice : " This celebrated Schillerian ballad has lately been set very expressively to music by Andreas Homberg. The following translation preserves the metre and the rhythm of the original, so as to be suited to the same music." The piece was translated at the request of the family of Sir G. B. Airy.

In the year 1849 Dr Whewell published some magazine articles on the subject of English Hexameters, which will be noticed hereafter in the Chapter on that subject. He examined Mr Mill's Logic in a pamphlet entitled, *Of Induction with special reference to Mr J. S. Mill's System of Logic;* this was afterwards embodied in the *Philosophy of Discovery.*

A lecture was delivered by Dr Whewell at the Royal Institution in London, on Jan. 19, 1849, on the *Idea of Polarity.* A printed abstract of the lecture is preserved, probably taken from some Journal of Proceedings of the Institution. We may infer that the lecture in great measure coincided with Book V of the *Philosophy of the Inductive Sciences.*

In the autumn Dr Whewell spent some time at Kreuznach in Germany, as Mrs Whewell had been recommended by her physicians to use the baths of that place. The same cause led in subsequent years to visits to the same place, but without any permanent benefit to Mrs Whewell.

On November 5, 1849, Dr Whewell delivered a lecture in London before the Royal Institute of British Architects; the subject was " The Gothic and After-Gothic of Germany." An abstract of the lecture is in print in the *Transactions* of the Institute, and occupies rather more than a quarto page : it is apparent from this abstract that the lecture must have coincided

in substance with a paper which Dr Whewell published sub-
sequently in Volume VII of the *Archæological Journal*, entitled
"Remarks on the complete Gothic and After-Gothic Styles in
Germany."

It remains to notice two magazine articles which belong to the
year 1849.

Sacred Latin Poetry. This is a review of the selection of
Sacred Latin Poetry edited by Professor Trench, now Archbishop
of Dublin; it was published in *Fraser's Magazine* for May, 1849,
pages 527...532. Dr Whewell gives translations into English
verse of some of the Latin pieces; these are very remarkable for
their fidelity and beauty. The papers of Dr Whewell contain in
addition a few lines which were omitted in the magazine, probably for want of space. The Latin Hymn to the Holy Spirit has
been recently translated into English verse by Dean Stanley, in
Macmillan's Magazine for June, 1873; it will be found interesting
to compare this version with that given by Dr Whewell.

On Mr Macaulay's Praise of Superficial Knowledge. This is
a review of a pamphlet published by Professor J. D. Forbes under
the title of *The Danger of Superficial Knowledge;* the review
appeared in *Fraser's Magazine* for August, 1849, pages 171...175.
In an address delivered at Edinburgh Mr Macaulay had ventured
to contradict the well-known line, "A little learning is a dangerous
thing." Professor Forbes held that the opinions of Mr Macaulay
were hurtful, at least to students, and gave a lecture to his class
on the danger of superficial knowledge: he submitted the lecture
in manuscript to Dr Whewell, who concurred, though not very
decidedly, in the author's opinion of the propriety of publication.
The review discusses the question raised by Macaulay, and in
the last paragraph alludes to the lecture by Professor Forbes.
The main point which Dr Whewell urges is that although more
may be known on a given subject at present than in preceding
times, yet it does not follow that any individual knows more than
the wise and great men of old. This was directed against
Macaulay's principle that the knowledge of former ages was insignificant compared with that of our own times. Perhaps the
review acquiesces in an extravagant estimate of the value of mere

knowledge without any specific reference to *thinking* and *reasoning;* this however was probably owing to the form which the controversy had already taken, and not to any real doubt as to the relative importance of these elements. It is interesting to find the extent of knowledge in past and present times discussed by two persons, each so famous for encyclopædic attainments as Macaulay and Whewell. The conclusion of the review is very ingenious: "And we are too magnanimous and too consistent to be discontented, if any reader, convinced by our reasons, is still of opinion that a little of such reasoning is a dangerous thing, and should determine to draw from Professor Forbes's pages a deeper draught of antidote to the siren strains in which Mr Macaulay sang his *Encomium Moriæ.*"

In the year 1850 Dr Whewell contributed an article on translations of *Herman and Dorothea* to *Fraser's Magazine* for January; this will be noticed in the Chapter on English Hexameters. The other publications of the year to which I now proceed are two articles connected with Architecture, a continuation of the work on Liberal Education, and a translation of a German novel.

The Lamps of Architecture. This is a review of the well-known work by Mr Ruskin; it appeared in *Fraser's Magazine* for Feb. 1850, pages 151...159. Dr Whewell commends the eloquence of the work, but does not admit the justness of all the opinions it contains; in particular, as befits a Cambridge man, he protests against the strong condemnation pronounced on King's College Chapel.

Remarks on the Complete Gothic and After-Gothic Styles in Germany. This paper was published in No. 27 of the *Archæological Journal* in 1850. Dr Whewell states that he formerly put forward a *theory of Gothic architecture* as illustrated by the churches of Germany; and the doctrine had been regarded with favour by eminent architectural authorities. He says: "The countenance thus given to the theoretical or ideal view of architecture justifies us, I think, in attempting to apply it in other cases also, at least in the way of trial." "Having before proposed a theory of the *formation* of the Gothic style, I wish now to suggest the theory of its *dissolution.*"

The work on *Liberal Education* which first appeared in 1845 reached a second edition in 1850. To the general title is now added *Part I. Principles and Recent History.* It is said in the preface that only slight and unimportant corrections are made in this Part. The second edition is in small octavo, and the Part I contains xx and 236 pages. Then follows Part II; to the general title is added *Discussions and Changes, 1840—1850*: it contains viii and 144 pages. The paging is distinct from that of Part I, but the numbering of the Articles is continuous. Part II was issued shortly before the new edition of Part I.

In Part II Dr Whewell passes in review the changes which had taken place in Cambridge education and examinations during the preceding ten years; many of these were such as he had himself warmly urged. He discusses at length some important matters, among them especially one which he did not commend, namely the emancipation of the classical students from the compulsion of taking mathematical honours. He has an admirable section on *The Great Classical Schools*, and expresses the strongest conviction "that those who wish to improve English education ought to direct their efforts to those quarters much rather than the Universities, as points on which their action, if successful, will produce a much wider and deeper effect." A very curious passage occurs in Art. 368: Dr Whewell says that as far as he can judge boys in general are more slow in understanding any portion of Mathematics than they were thirty years ago; and conjectures a cause for this decline.

Dr Whewell offered some criticisms on a mathematical work which had been recently published by Mr Harvey Goodwin, the present Bishop of Carlisle. The author replied in a pamphlet entitled *Defence of certain portions of an Elementary Course of Mathematics.* He maintains, I think with complete success, the advantage of what may be called the modern treatment of Trigonometry and Mechanics over the ancient which Dr Whewell preferred. The reply acknowledges the "kind tone and spirit of Dr Whewell's criticisms," and is itself of the same temperate character. There is an allusion to the work in the preface to the second edition of Part I, which as I have said was issued some little time after Part II.

It will be convenient to finish our account of the present work, and so I mention that in 1852 Dr Whewell issued Part III; to the general title is added *The Revised Statutes*. It is in small octavo and contains viii and 100 pages. The paging is distinct, but the Articles are numbered in continuation of those in Part II. After long discussion a scheme of revised statutes had been drawn up, and was about to be submitted to the Senate for adoption or rejection; the publication was designed by the author to assist the Senate in forming a correct judgment on the matter. But the questions probably never excited much attention elsewhere, and at Cambridge they have lost their interest in consequence of the legislation which followed the Royal Commission. It is remarkable to see how earnestly Dr Whewell labours through nearly 20 pages to extol and support an institution called the *Caput,* especially obnoxious to the ardent academical reformers of the epoch, and now extinguished. A few words occur incidentally in Art. 410 which involve such a happy metaphor that even the general reader may like to notice them. "When a man has, to use the expression familiar in the University, thus 'got up his subjects,' he can carry them about with him, and set them down at a moment's warning...". Examinations have rarely been so neatly condemned as by the last phrase, which compares the passing through them with the laying down of a burden, to the permanent relief of the bearer.

The translation of the German novel is entitled, *The Professor's Wife, from the German of Berthold Auerbach.* London, John W. Parker, 1850. This is a small octavo volume, containing altogether 200 pages. The translation is most skilfully executed; but English readers have access to so large a supply of original fiction that it is difficult to engage their interest in the comparatively inferior productions of foreign authors. The story is very slight; an artist marries a country girl, takes her to live in a town where he has an appointment as director of a picture-gallery, neglects her, and finally comes home one night intoxicated: whereupon she leaves him and goes back to her mother. A friend of the artist, who is an assistant librarian and is called the Collaborator, is present throughout the narrative, and delivers various moral and philosophical reflections. The manuscript and the proof-sheets of the translation have been preserved; they conclude with the words *Trinity Lodge, Sep.* 16, 1850, which were struck out in the impression.

There are a few allusions to this publication in the correspondence of Dr Whewell. A distinguished prelate asked for a copy. A lady whose own writings have gained her an honoured place among English novelists, says, " I am deep in the interest of the natural, exquisite little story, and heartily thank the translator who has given me the power of reading it in strong simple English." A German friend, with national profundity, discovered what the author probably never contemplated—"One of the features of the novel which strikes me as particularly good is that sort of poetical parallelism inherent in the composition between the causes of the fate both of Lorlie and Reinhard—she being crushed by her marriage with him and he simultaneously by his marriage—as if it were—with his new employment. For this reason I think the title so very well chosen." A Swiss friend wrote, " Mais j'avoue qu'en général je trouve cette vie rustique que nous n'avons que de trop près autour de nous d'un petit intérêt poétique. C'est différent en Angleterre où le véritable paysan d'autrefois a presque disparu."

In the year 1851 Dr Whewell published a sermon entitled *Strength in Trouble,* preached in Trinity College Chapel on February 23; it will be noticed hereafter. He was requested to preach in St Martin's in London, during the season of the Great Exhibition, on the Connexion between Science and Faith; but he does not seem to have complied with the request. He attended the meeting of the British Association at Ipswich in July, 1851.

The Great Exhibition, to which I have just alluded, was, as is well known, the first example in England of these extensive undertakings; the object of it was to bring together all that could illustrate the recent progress and the actual condition of arts and manufactures. His Royal Highness Prince Albert took a warm interest in the scheme, and contributed greatly to the success which attended it. After the close of the Exhibition a course of Lectures was arranged by the Society of Arts, at the suggestion of Prince Albert, for the purpose of drawing out the most valuable lessons which the whole proceedings could suggest. Dr Whewell began this course and his Lecture was published under the following title: *Inaugural Lecture, Nov.* 26, 1851, *the general bearing of the Great Exhibition on the progress of Art and Science.* The copy of this Lecture which I have read forms a pamphlet of 16 closely printed octavo pages. The Lecture was also issued in a volume, with others, by various persons, relating to the Exhibition.

Dr Whewell illustrates in a very striking manner some points which he had already noticed in his *Philosophy of the Inductive Sciences.* Thus he shews that in general Art has preceded Science; "Art was the mother of Science: the vigorous and comely mother of a daughter of far loftier and serener beauty." Nearly the same words as these occur in the lecturer's address to the British Association at Cambridge in 1833. He lays stress too on the lesson in Classification which was supplied by the method of arranging the immense contents of the Exhibition. He contrasts most impressively the case of a supposed traveller spending his life in examining the arts of various countries of the world with the inspection of the whole simultaneously, which the Exhibition rendered practicable. He alludes to a remark with which, as appears from one of his letters, he had been struck in early life, "You have most of you probably heard of the careful and economical critic, who proposed to reduce the extravagance of the wish of the impatient separated lovers, that the gods would annihilate space *and* time; and who remarked that it would answer the end desired if one of the two were annihilated."

Dr Whewell seems to have been at first reluctant to undertake the Lecture, for the following draft of a letter is preserved among his papers; it was probably addressed to the Secretary of the Council of the Society of Arts:

"TRINITY LODGE, CAMBRIDGE, *Nov.* 12, 1851.

" Sir,

"I have had the honour to receive your letter in which you propose to me on the part of the Council of the Society of Arts, that I should on Wednesday the 26th deliver an address on the tendencies of the age which have led to the Great Exhibition and on its probable effects.

"I am deeply sensible of the honour which is done me by such an application, and it is a matter of extreme regret to me not to be able to attempt, however imperfectly, the very animating task thus proposed to me. But my intercourse with the eminent men who were put in motion and brought together by the Exhibition, and who must be in a great measure the channels and exponents of its influence on the nation and on mankind, has, unfortunately for me, been so limited, that I do not feel myself capable of tracing and estimating that influence in a worthy manner.

"In returning this reply to the very flattering proposal of the Council, I am supposing that the suggestion of H. R. H. Prince Albert, of which you speak, referred only to the class of persons who should be requested to undertake the proposed lectures, and not to myself in particular. The admiration which I entertain for the whole of the conduct of H. R. H. relative to the Exhibition is such that I could not easily bear to treat any wish of his on that subject as impracticable."

In the year 1852 we have the following matters to notice ; a Letter to Edwin Guest, Esq., a Letter to the author of *Prolegomena Logica,* a speech at the opening of the New Grammar School at Lancaster, the evidence contributed to the Blue-book issued by the Royal Commission for the University of Cambridge, and an obituary notice of the Marquis of Northampton. To these I now proceed.

In Vol. v, pages 133...142 of the *Proceedings of the Philological Society,* there is a letter from Dr Whewell to Edwin Guest, Esq., Secretary to the Society. The letter is dated Trinity Lodge, Cambridge, Feb. 6, 1852; it was read to the Society on Feb. 20. The first paragraph explains the purport of the letter:

" My dear Sir. You are aware that an Etymological Society was formed at Cambridge, at a period a little previous to the establishment of the Philological Society in London. Many of the original members of the latter Society are aware of the existence of the former, from having taken a leading part in its proceedings; but some account of the plans and some specimens of the labours of the Etymological Society of Cambridge may not be without interest for the members of the Philological Society in general: and the office of drawing up such a memorandum of the Cambridge Society appears to devolve upon me more especially, inasmuch as the papers contributed by the members of that Society, except so far as they have been used for publication, remain in my hands."

On the last page of the letter Dr Whewell says, "A considerable portion of additional matter was printed from the MS. of the author of the paper on English Orthography, but has not yet been published." The author of this paper had been previously designated by the letters J. C. H.. these denote Julius Charles Hare. This additional matter was published in 1873 under the title of *Fragments of two Essays in English Philology,...Macmillan and Co.*, with an *Advertisement* by Professor J. E. B. Mayor.

Dr Whewell says near the end of the letter: "I might mention some others of the speculations of our Etymological Society; but though, as I have said, they were very instructive for us at that period, they have been superseded in a great degree by what has been done since by philologers, and especially by the members of the Philological Society. In particular we had a grand, but I fear hopeless, scheme of a new Etymological Dictionary of the English language; of which one main feature was to be that the three great divisions of our etymologies, Teutonic, Norman, and Latin, were to be ranged under separate alphabets." Some notes on the etymologies of words beginning with the letters A, D, E, F, H remain among Dr Whewell's collection of papers: these were doubtless drawn up by members of the Etymological Society for use in the proposed dictionary.

A letter to the Author of the Prolegomena Logica, by the Author of the History and Philosophy of the Inductive Sciences. This letter to Mr Mansel, afterwards Dean of St Paul's, is an octavo pamphlet of 18 pages, besides the title. It is dated Trinity College, Sept. 20, 1852. Perhaps it was printed for private circulation only. It is substantially reproduced in the *Philosophy of Discovery*, Chapter XXVIII. The parts not reproduced are the introductory note, and the first, second, and fourth paragraphs beginning on page 3 and ending on page 5. The note on page 334, and the first note on page 335 of the *Philosophy of Discovery* were not in the original letter. The most important part of the letter is that in which the writer tries to remove a misunderstanding of what he had said in his works : see the letter, page 10, or the *Philosophy of Discovery*, page 338.

The public opening of the New Grammar School at Lancaster took place on Sep. 10, 1852. Dr Whewell was present, and so also were his old friends and correspondents, Mr Morland and Dr Mackreth. Two speeches were delivered by Dr Whewell, of which a newspaper report has been preserved. In the introductory part of the first speech Dr Whewell said, "...I am, as the Mayor has stated, an old pupil of the Grammar School of Lancaster, and I owe to the instruction I there received, and to the kindness of my ever respected master, Mr Rowley, the commencement of a career of success in life, for which I cannot but be thankful."

Dr Whewell insisted strongly on the advantages of a classical education; he spoke also of mathematics and recommended early familiarity with Arithmetic, Practical Geometry, and Mensuration. He recorded the opinion that in the Universities the principle of examinations had been carried too far. He remarked that he was not "disposed to admire as a general class what are called self-taught classes." He continued, "A certain vigour there may be, but that is often far more than balanced by a corresponding quantity of self-opinion and conceit. There is an expression that appears to me equally acute and wise, on the part of Opie, the painter, when somebody was praised to him as a self-taught person: 'Depend upon it,' he said, 'a self-taught person is a person taught by a very ignorant man.'"

In his second speech Dr Whewell made an allusion to himself and his own works: "With regard to the occasion that calls us together on this day, most assuredly there is no one who can have more reason to think with kindness and gratitude of the Grammar School of Lancaster than I have. It was the first step in a career which has placed me, by the favour of my sovereign, in a situation which is the first situation the whole empire presents to a person fond of literature, and desirous of employing himself upon it. Upon literature and science I have undoubtedly employed a large portion of my time, and hope I may say, not with any personal egotism, but with the egotism of a Lancaster man, that I have so selected the subjects on which I have employed myself, and so treated them, that it may not be out of the way for the writers of future biographical dictionaries to enquire where the author of certain works was born. Certainly, as far as any one can accept the testimony of foreign countries of the present day, as representing the testimony of other ages, there has been something of that kind of evidence that such questions may be asked; to refer only to such events as have happened within the last few months; when no less an illustrious person than Humboldt, at his age of 83, came over from Potsdam to Berlin, in order to express as he said, the gratitude he felt for what I had done."

In 1850 Royal Commissions were appointed to enquire into the State, Discipline, Studies, and Revenues of the Universities of Oxford and Cambridge. The Cambridge Commissioners were the Bishop of Chester, Dean Peacock, Sir J. Herschel, Sir J. Romilly, and Professor Sedgwick; they issued in 1852 a *Report together with the Evidence and an Appendix*. Dr Whewell regarded the Commission as an unwarranted and undesirable intrusion into the affairs of the University; but he did not decline to answer various questions which were addressed to him in his official capacity. Accordingly communications from him will be found in the following pages of the *Evidence*, 99...102, 203...209, 271...274, 292, 414...419, 425. These relate to the Professorship of Moral Philosophy, to the regulations of Trinity College, and to the general subjects of Discipline, Expense, Private Tuition and Examinations; the opinions expressed are the same as Dr Whewell had published in his book on a Liberal Education.

An obituary notice of the Marquis of Northampton occurs in pages 117...120 of Vol. VI. of the *Abstracts of the Papers communicated to the Royal Society of London.* This was drawn up by Dr Whewell, and is an interesting memorial of one who occupied for ten years the office of President of the Royal Society, and gained universal respect by his attainments and character. The notice concludes thus: "It is difficult to abstain from dwelling longer on Lord Northampton's admirable gifts and accomplishments, and still more, on his virtues. He was full of kindness and benevolence for all who came under his notice, and seemed to be absolutely incapable of injustice or unfairness; and though a most clear-sighted judge of intellectual, scientific and artistical excellence, was with difficulty, if at all, moved to harshness towards shallow and petulant pretensions. He was zealous for the promotion of art as well as science in his native country; and even in the last days of his life his thoughts and his pen were engaged on a plan connected with that object."

In 1853 Dr Whewell published an edition of the work of Grotius *De Jure Belli et Pacis,* with notes and an abridged translation. The preparation of this work had occupied him during the months he spent in 1852 with Mrs Whewell at the baths of Kreuznach. It will be noticed together with the other contributions to Moral Science. An article on English Hexameters published in the *North British Review* will be noticed in the Chapter on that subject. The months spent at Kreuznach in 1853 were occupied in the composition of the *Plurality of Worlds,* which was published towards the end of the year; to this work the next Chapter will be devoted.

In 1853 Dr Whewell and Professor Sedgwick were requested to act as judges for awarding the valuable Burnett Prizes for two Essays on Natural Theology; they both however declined the office, and it was then undertaken by Professor Baden Powell and Mr Isaac Taylor.

One short paper has now to be mentioned. *Further Remarks on the proposal, That an ordinary Degree of B.A. shall be given on the ground of an examination in Classics, without requiring any other examination, except the Previous Examination: occasioned by*

"*A Reply*" *to previous Remarks on that proposal.* This consists of 5 small pages; it is signed W. W. April 26, 1853. It relates to some of the University legislation of the period with respect to Examinations, and is of no general interest. I have not seen the *previous Remarks,* nor the *Reply* to them.

CHAPTER XII.

TOWARDS the end of 1853 a book was published anonymously entitled, *Of the Plurality of Worlds: an Essay.* It is an octavo volume; the title and prefatory matter occupy viii pages, and the text 279 pages. The title is not very suggestive of the nature of the work, which is directed *against* the doctrine of the Plurality of Worlds: it will be convenient to speak of the book briefly as the *Essay.* In the early part of 1854 appeared a *Dialogue on the Plurality of Worlds, being a Supplement to the Essay on that subject;* this consists of answers to the arguments which had been urged in conversation, in writing, or in print, since the *Essay* was published. The *Dialogue* consists of 55 pages in small octavo, ending with an extract from a sermon by Professor Blunt. A second edition of the *Essay* was published in the summer of 1854; this has a new preface as well as the original preface: the *Dialogue* is reproduced with some additions, and is awkwardly placed first in the text; it retained this position in the next two editions. All the editions except the first are in small octavo size. The title and prefatory matter of the second edition occupy xvi pages, and the text 395 pages. The preface gives a good summary of the state of the question discussed in the *Essay.* The third edition of the *Essay* was published in the autumn of 1854; it consists of viii and 407 pages, and contains a new preface as well as those to the two preceding editions. The *Dialogue* is increased by about half a page, in which a supposed interlocutor designated as AC draws attention to a second article on the *Essay* which appeared in the *Westminster Review.* The fourth edition of the *Essay* was published early in 1855; it consists of xii and 411 pages, and contains a new preface as well as those to the three preceding editions. The fifth edition of the *Essay* was published in 1859. The title and prefatory matter occupy xii pages ; these include the preface to the first edition on four pages, and the preface to the fifth edition on a single page. The text of the *Essay* follows on 318 pages ; then we have a passage from Plato with a translation, the preface to the second edition dated June 23, 1854, the preface to the third edition dated Oct. 10, 1854, and the preface to the fourth edition dated Feb. 2, 1855 ; next comes the *Dialogue* on pages 337...403, and finally, on pages 404...416, an answer to an article in the *Edinburgh Review*, reprinted from the *Saturday Review* of Nov. 3, 1855. The passage from Plato with the translation, and the reply to the *Edinburgh Review*, do not occur in the previous editions. The *Dialogue* in the fifth edition is printed a little more closely than previously, so that some space is gained. All the prefaces, except the first, are curious, from the fact that the personal pronoun *I* is freely used, while no name is subscribed. I shall always cite the *fifth* edition when I have occasion to refer to a page of the *Essay* or *Dialogue*.

I shall begin with some remarks on the *Essay* and *Dialogue*, and then notice the principal reviews which they called forth ; next I shall advert to the opinions expressed on the subject by some of Dr Whewell's correspondents, and finally give an account of some unpublished matter relating to it.

At the date of the publication of the *Essay* the belief in the doctrine of the Plurality of Worlds was nearly universal. Men of science readily accepted a pleasing hypothesis which was not positively contradicted by evidence from their special studies. Some theologians may have hesitated ; but against them could be placed the high authority of Dr Chalmers, whose *Astronomical Discourses* had supported the popular persuasion with impressive declamation. A flippant remark in Coleridge's *Table Talk*, Vol. II, p. 293, is almost a solitary expression of a contrary opinion. "I never could feel any force in the arguments for a plurality of worlds, in the common acceptance of that term. A lady once asked me—'What then could be the intention in creating so many great bodies, so apparently useless to us ?' I said I did not know, except perhaps to make dirt cheap." The *Essay* then proposed to overthrow a common and well cherished belief. Although it was published anonymously it was soon ascribed to Dr Whewell; and his letters shew that he admitted the authorship to many of his friends. The period was little suited to literary enterprise owing to the Crimean war into which the country was drifting, yet the *Essay* excited considerable interest and attention almost immediately. It was reprinted in America with a preface by Professor Hitchcock.

In the margin of a copy of the *Dialogue* preserved among the papers of Dr Whewell the names of various persons whose opinions are represented are indicated by initials; in most of these cases the meaning of the initials is quite clear, and there can be now no impropriety in indicating the sources of the objections to which Dr Whewell replies. I will refer to the fifth edition of the work, and state the page and the line at which the initials are placed: Sir R. Murchison, page 339 line 8; Professor De Morgan, page 340 line 1; Sir James Stephen, page 343 line 3, page 380 line 30; Sir H. Holland, page 345 line 20, page 350 line 24, page 354 line 1, page 358 line 6, page 374 line 10; Sir J. Herschel, page 357 line 18, page 382 line 21; Professor J. D. Forbes, page 358 line 25, page 359 line 20, page 360 line 13 and line 26, page 361 line 25, page 362 line 13, page 377 line 26, page 385 line 2; Westminster Review, page 366 line 15, page 370 line 17, page 373 line 1. In all the preceding cases there is evidence which renders the meaning of the initials certain; in a few other cases there may be some doubt, namely Professor Sedgwick, page 353 line 13; Mr James Garth Marshall, page 357 line 7; Rev. Richard Jones, page 375 line 1. Also at page 357 line 1 is placed Mr H, and at page 383 line 22 is placed R. M. M.; the meaning of the former is not known, the latter denotes one who still survives but bears another name.

Of the ability shewn in the *Essay* there can be but one opinion; many persons have pronounced it the cleverest of the many works of the author; the style is vigorous and poetical, resembling that of the *Bridgewater Treatise* or of the most finished parts of the *History and Philosophy of the Inductive Sciences,* and contrasting strongly with that adopted in the *Elements of Morality.* The argument from geology is conducted with great skill, and although it appeared that some hints of the same kind had been previously thrown out by others, among whom Dr Whewell, in his *Dialogue*, names Hugh Miller and Professor Birks, yet he was not aware of this until after his own publication. But on a subject which cannot be settled by argument the analogies and conjectures which may be produced will influence various minds in various ways; and those who have not arrived at a decision for themselves will look with interest on the opinions recorded by able men. Perhaps no person was better qualified than Sir J. Herschel for appreciating the various considerations, physical, metaphysical, and theological, which were brought into discussion. He says in the *Proceedings of the Royal Society*, Vol. XVI, pages LX and LXI: "The essay on the 'Plurality of Worlds' (attributed to him, though published anonymously), can hardly be regarded as expressing his deliberate opinion, and should rather be considered in the light of a *jeu d'esprit*, or possibly, as has been suggested, as a lighter composition, on the principle of '*audi alteram partem*,' undertaken to divert his thoughts in a time of deep distress. Though it may have had the effect I have heard attributed to it, of 'preventing a doctrine from crystallizing into a dogma,' the argument it advances will hardly be allowed decisive preponderance against the general impression which the great facts of astronomy tend so naturally to produce." It will be seen from this quotation that Sir J. Herschel is rather against Dr Whewell than with him in the controversy; the suggestion however that the *Essay* can hardly be regarded as expressing its author's deliberate opinion seems to me inadmissible. We may refer to the work itself for evidence on this point, for instance on page 401 we read: "As to myself, the views which I have at length committed to paper have long been in my mind. The convictions which they involved grew gradually deeper...." There is a striking passage on page 281 which seems eminently sincere; and the Chapter on the *Argument from Design* is of the same character.

There is not much evidence as to the history of Dr Whewell's interest in the subject of the *Plurality of Worlds*, but I will state all that I have discovered. An old note-book records that he read Fontenelle's celebrated work at an early period, but adds no opinion respecting it. I have quoted on page 70 some sentences from the Bridgewater Treatise bearing on the subject; they seem to me in tone consistent with the doctrines of the *Essay*, but several reviewers appealed to them as indicating on the part of the author a complete change of opinion. A very interesting passage occurs in the *Second Memoir on the Fundamental Antithesis of Philosophy* read to the *Cambridge Philosophical Society*, Nov. 13, 1848; it is reproduced in the *Philosophy of Discovery*, pages 304 and 305. Dr Whewell there briefly contrasts two theories of the Universe, and while admitting that there are difficulties in both, considers that they are less in that theory which was in fact afterwards maintained in the *Essay*. Perhaps the opposition which this work encountered stimulated him to seek for arguments in its favour and increased his confidence in it.

Sir H. Holland in his *Recollections of Past Life*, pages 240 and 241, alludes to his intercourse with Dr Whewell; and makes some remarks on the *Essay;* the letter of criticism which he records that he wrote has not been found among Dr Whewell's papers. Sir H. Holland says, "It was easy to praise the great ability and ingenuity of the work while rejecting a conclusion defensible chiefly, if not solely, from the impossibility of ever *proving* the contrary." The venerable critic, speaking of the *Essay* a few weeks before his death, said, "it is the most brilliant of all Dr Whewell's writings, but the theory is false :" he also expressed the opinion that Dr Whewell was not convinced by his own arguments.

One question of interest is how far do the Astronomical researches of the last twenty years confirm or oppose the views of Dr Whewell. Probably the answer must be that the additional evidence is partly in one direction, and partly in the contrary. Such is in substance the opinion of Sir H. Holland given in a note on page 241 of his book. Thus on the one hand he allows that it seems established by spectroscopic observations that many of the nebulæ are really gaseous in their nature; this is curious because in the *Dialogue* the speaker who on page 345 objects to the "zeal for reducing nebulæ to a state of luminous vapour" is according to Dr Whewell's indication Sir H. Holland himself. On the other hand, as Sir H. Holland proceeds to say in his note, "recent discoveries, identifying many of the material constituents of the sun and stars with those of our globe, have much more direct bearing against his hypothesis." The current literature of the time supplies confirmations of both clauses of Sir H. Holland's judgment. Thus on the one hand with respect to nebulæ we are told that they are apparently nothing but floating masses of very rare nitrogen gas. *Academy*, Jan. 10, 1874, page 42. On the other hand we read, "It would be absurd to regard as a reasonable hypothesis *now*, either the theory of Whewell that Jupiter consists mainly of water, or the alternative suggestion of Brewster, that the substance of the giant-planets may be of the nature of pumice-stone." *Journal of Science*, Jan. 1874, page 18. The subject is discussed by Mr Proctor in his *Other Worlds than ours; the Plurality of Worlds studied under the light of Recent Scientific Researches.*

In the early editions of the *Essay* the critics detected a few mistakes and inconsistencies. As an example of the former it may be mentioned that the distance of the Moon from the Earth was put at a month's railway journey; this was afterwards extended to six months. As an example of the latter some statements as to the size of Mars compared with that of the Earth may be taken : page 7, " Mars...nearly as large ;" page 214, " his mass is so much smaller;" and page 215, " Mars is much smaller." In quoting the well-known speech to Jessica which occurs in the *Merchant of Venice* Dr Whewell adopted the reading *patterns* in the early editions; his attention was drawn to it in correspondence, and he changed it to the more usual reading *patines*.

In the *Dialogue* Dr Whewell quotes with strong approbation a passage from a sermon by Professor Blunt. Those who heard the preacher, so justly revered, still remember after the lapse of five and twenty years the great impression which the sermon produced. The first edition of the *Dialogue* finished with this extract ; it is on some account a matter of regret that owing to the addition of new matter this arrangement was altered in the subsequent editions, for the transition is unpleasant from the doctrine of the communion of saints, so earnestly set forth by the preacher, to the severe remarks made upon the *Essay* by Sir D. Brewster.

Rarely in recent times has a book received so much attention from reviewers as the *Essay*. Dr Whewell himself formed a large collection of the articles which were published about his book, though unfortunately in some cases the title of the periodical in which the article originally appeared has not been preserved. I shall give a brief notice of all the reviews which I have seen that bear on the subject, beginning with those which are decidedly opposed to Dr Whewell's views, and passing on to others which may be described as neutral or even favourable.

An article occupying 44 pages was published in the *North British Review* for May, 1854; it was written by Sir D. Brewster, and is a most impetuous attack on the *Essay*. The article is unworthy of the writer's great and just reputation; the science is not unfrequently at fault, and the rhetoric degenerates into extravagance. Even those who agree with his opinions must regret the tone in which they are expressed, and especially the depreciating terms in which he often speaks of his eminent antagonist. Passionately however as Sir D. Brewster loved the unseen inhabitants of the Sun, the planets, and the stars, there was one object, terrestrial indeed but perhaps equally visionary, which stood apparently still higher in his regard, and for which he had hoped and pleaded during a long career, namely, to use a current formula, *the endowment of original research*. He quotes a passage from the *Essay*, page 314, which may be regarded as favourable to this cherished scheme, and seems in consequence half disposed to forgive the author for depopulating the universe; he concludes his review thus, " ... we now part with him in better humour, as a distinguished philosopher—a profound thinker—an eloquent writer— a manly philanthropist—and a successful prophet." Dr Whewell replied to this review in pages 388...400 of the *Dialogue*. Sir D. Brewster expanded his review into a book entitled *More Worlds than One, the Creed of the Philosopher and the Hope of the Christian;* this passed through several editions. Longer consideration of the subject rendered the author more bold; for while in the Review he had hesitated to ascribe inhabitants to the Sun, in the book he has no misgivings, and suggests that they may find employment in watching for the opportunities of astronomical observation which will be afforded them by the occasional disruption of the bright covering of the Sun, when to earthly eyes spots are produced. Yet this is perhaps the most consistent form in which the doctrine of the Plurality of Worlds can be maintained; for a person who surrenders the Sun must in like manner give up the fixed stars, and has then at his disposal only the real planets, and the hypothetical planets which are believed to circulate round the fixed stars; he can therefore not advance the argument that it is intolerable to suppose so much room wasted that might be peopled, when he himself supposes the vastest globes to be desolate.

An article by M. Babinet entitled *De la Pluralité des Mondes* occurs on pages 365...385 of the *Revue des deux Mondes* for Jan. 15, 1855; it treats of the works of Dr Whewell and Sir D. Brewster. Dr Whewell refers to it in the preface to the fourth edition of the *Essay* as by a "writer whose authority is deservedly high on matters of physical science;" but he adds, "He has, however, in this instance chosen rather to play with the subject than to discuss it seriously, giving us to understand that such was the tone that suited the circles for which he writes." But I do not see in the article where M. Babinet gives us to understand this. According to the opinion of M. Babinet it is probable, and almost certain, that the planets which surround the Sun and the stars are inhabited, like the Earth, by all degrees of intelligence and all varieties of organisation, but we have no reason to suppose the Sun and the satellites to be inhabited. A statement which he makes as to Protestant sermons will probably not be confirmed by modern English experience. "Quoique M. Brewster ne soit pas, comme son antagoniste, un théologien de profession, les convenances religieuses n'y sont guère invoquées moins souvent, ce qui n'étonnera pas, lorsqu'on saura que dans leurs sermons les prédicateurs protestans ont l'habitude de développer beaucoup de thèses appartenant aux sciences d'observation; on cite dans ce genre un sermon du docteur Bentley, qui reçut de Newton lui-même les instructions nécessaires pour le composer."

An article by Professor H. J. S. Smith entitled *The Plurality of Worlds* occurs on pages 105...155 of the *Oxford Essays* for 1855;

it treats of the works of Dr Whewell and Sir D. Brewster. This
article is decidedly adverse to the views of the *Essay*, while acknowledging the literary excellencies and power of the work. It
is by far the ablest of the unfavourable reviews, and is especially
deserving attention because the writer was one of the few who
possessed scientific and philosophical knowledge at all commen-
surate with Dr Whewell's, and displayed the judicial calmness
and fairness which we naturally expect to find in an analysis
attested by the name of the author rather than in the irresponsible
effusions of anonymous critics. The writer makes a very ingenious
suggestion as to the choice of an umpire in the controversy—"One
is continually tempted to appeal from the judgment of the
Essayist, in his character of a victorious disputant, to his judgment
as a philosopher and a man of science." Dr Whewell was very
fond of discussion, and it is natural to conjecture that in his *Essay*
he was arguing rather than deciding; nothing however has been
found among his papers to suggest any wavering in his belief that
the Earth alone among stars and planets is the abode of intellectual, moral and religious creatures. Professor Smith gives up
one position as indefensible—"The Moon we are compelled to surrender at discretion; and we own ourselves, on this point, unable
to withstand the desolating rhetoric of the Essay." It is certainly
remarkable that the only celestial object which is near enough
for decisive examination by our telescopes should refuse so decidedly to countenance the Pluralists; for if analogy would have
placed rational inhabitants in any one of these objects it would
naturally be on the Moon, so near the Earth, and apparently in
circumstances of light and heat so similar to our own. Some
intrepid Pluralists indeed have undertaken to people the *unseen*
side of the Moon; it would seem that they hold the merit of
giving light to the Earth to be of sufficient importance to suspend
the operation of the general principle that where there is matter
to uphold them inhabitants must exist, while on the side of the
Moon which does not benefit the Earth the principle again revives
and prevails. Dr Whewell alludes to this point in the *Essay*,
page 412. The Reviewer observes with respect to the Dialogue:
" In the Dialogue at the beginning of the Essay, the earlier letters

of the alphabet, who appear as objectors, conduct themselves so much like simpletons that we wonder at their being thought worthy of so long an interview with the enlightened Z." But most of these earlier letters represent names of eminent distinction. In a slight sketch of the history of opinions on the subject the reviewer states that while men of science were usually in favour of Plurality theologians took the contrary side.

An article on pages 435...470 of the *Edinburgh Review* for October 1855 discusses the subject. Besides the works of Dr Whewell and Sir D. Brewster, two others are placed at the head of the article, namely *Essays on the Spirit of the Inductive Philosophy, the Unity of Worlds, and the Philosophy of Creation*, by Professor Baden Powell, and *A few more Words on the Plurality of Worlds*, by Captain Jacob, Astronomer to the East Indian Company. Passages are quoted by the reviewer from the latter two works against Dr Whewell's opinions. The spirit of the article is eminently one common to many of those which take the same view; the doctrine of the *Plurality of Worlds* is treated as if it were in *possession* of the field, and although it can shew no title-deeds whatever, yet it is to be upheld until the rival doctrine can exhibit incontrovertible proof of its claim. Dr Whewell replied to this review in an article in the *Saturday Review* of Nov. 3, 1855, which is reproduced in pages 404...416 of the *Essay*. One point to which Dr Whewell alludes on his page 409 is treated in so strange a manner by the Edinburgh reviewer that it deserves to be noticed. The *Essay* called attention to the very short time during which man had lived on the earth compared with other animals. The reviewer says, "Sir David Brewster has, however,...pointed out that the fact is not as stated.. ...For aught that is known, mankind may endure on the earth until the tables are turned upon its brute predecessors...." It is obvious that Sir D. Brewster cannot contradict the fact as at present stated; he can only *conjecture* that hereafter the circumstances *may* be different. The objection, such as it is, had been anticipated in the last section of Chapter XII. of the *Essay*.

I have not seen the publication by Captain Jacob which the Edinburgh Reviewer cites. One extract from it seems to me unsatisfactory. Dr Whewell is represented as beginning with the most distant, or at least the most obscure bodies, and "descending by degrees to the more and more distinct, he attempts to drag a little of the obscurity with him in his downward progress." But this quite overlooks the fact that Dr Whewell's treatment of some of the nearer bodies, the Moon for instance, is altogether independent of what he says respecting the nebulæ. To Captain Jacob is due a very inconclusive argument drawn from the theory of chances, of which Dr Whewell disposes on his page 407.

A notice of the *Essay* is given on pages 591...594 of the *Westminster Review* for April, 1854; it is quite adverse to Dr Whewell, and especially blames him for inaccuracy in the quotation of a passage which he ascribed to Professor Owen. Dr Whewell replied in his *Dialogue*. In pages 242...245 of the *Westminster Review* for July, 1854, the *Dialogue* is noticed, and the charge with respect to the inaccurate quotation is again urged. Dr Whewell replied to this in subsequent editions of the *Dialogue;* see pages 366...372. The *Westminster Review* of July, 1854, on pages 245 and 246 notices the book of Sir D. Brewster, and as unfavourably as the *Essay* and the *Dialogue.*

An article entitled *The Plurality of Worlds* occurs on pages 1...12 of *Fraser's Magazine* for March, 1854; it was written, I believe, by Professor J. D. Forbes. The reviewer obviously leans to the side of the Pluralists, though he does not explicitly go further than suggesting inhabitants for Mars and Venus; probably his strong regard and admiration for Dr Whewell placed some restraint on the exhibition of his own opinions. After quoting the description that Dr Whewell gave of the very unattractive population which he thought alone adapted for Jupiter, the reviewer says, "Alas! for the imagined seat of higher intelligences; alas! for the glories of the most majestic planet of our heavens, the stern will of the ruthless destroyer has dissipated with no sparing hand the threads on which we hung the network of our imagery. No unsentimental housemaid ever made with relentless broom a cleaner sweep of a geometrical cobweb!" The reviewer alludes to the case which is mentioned in the *Dialogue*, page 377, of a murderer who was held to be insane on the ground that he believed the Sun to be inhabited, and adds the extraordinary reflexion, "Now this curious history may be quoted as a strong proof of the *instinctive* belief of man in the diffusion, if not of his species, at least of his analogues." That is to say, the instinctive belief of *man* is inferred from the belief of one individual, though opposed to that of all his countrymen. It would be difficult to find a more striking example of a fault, traces of which may be seen even in Butler's *Analogy*—that of appealing to evidence on one side of a question and neglecting to observe that it is overborne by a vast preponderance of evidence on the other side. Moreover, the most ardent Pluralist might shrink from peopling the Sun with *analogues* of a madman or murderer. An interesting statement is made by the reviewer in a note, namely that the late Dr Chalmers was really the first who "happily reconciled the narrative of Moses with the demonstrated truths of science," by admitting the existence of a period of indefinite extent between the second and the third verses of the book of Genesis. The reviewer draws especial attention to the eleventh chapter of the *Essay*—that on the Argument from Design—as "probably the most interesting and original" in the book, and written with "great vigour and eloquence." This chapter was warmly praised by many of the reviewers of the *Essay*. It is stated in the beginning of the article, that "the preponderance of belief in all ages has been in favour of the Plurality of Worlds." Dr Whewell refers to this in his *Dialogue*, pages 377...380, where he maintains the contrary opinion.

The *Essay* is reviewed on pages 50...82 of the *Christian Remembrancer* for 1855. The article is written in a strain of banter which is very tedious; though it is decidedly adverse to Dr Whewell's opinions, it is also severe on Sir D. Brewster for the incompetent manner in which he has defended what it holds to be the right side. The review is almost unique in blaming the author of the *Essay* for "marvellous inelegance in the use of English," and for "heavy writing." The writer is an ultra Pluralist; for he says, "Had there been no such science as Astronomy, pious thought might have engendered the wish, the hope, the opinion, that there might be other worlds." On his page 73 he gives the following passage with the marks indicative of a quotation: "we are compelled by geological evidences to admit that a destitution of creatures who can know, obey, and worship God, has existed upon the earth..." In the margin of his copy of the review Dr Whewell has recorded against the words *destitution of creatures*, "no phrase of mine," and refers to his Chapter XI, section 21, for confirmation. The reviewer thinks this Chapter very striking and (as an effort of argument) worth all the rest of the book.

An article entitled *Modern Philosophy: The Plurality of Worlds*, occurs on pages 129...137 of the *Rambler* for August, 1854. This is a review of Sir D. Brewster's book, and strongly condemns the way in which the subject is there treated; though holding it probable that the planets are inhabited. The reviewer regrets the light esteem in which metaphysical subjects are regarded by the great majority of our learned men, and concludes thus: "Metaphysical science is conversant with the centre, where all the rays meet, physical science with the surface, where all are separate; human knowledge, in proportion as it is exclusively physical, is superficial: in proportion as any knowledge is superficial, is it absolutely powerless when it applies itself to things which do not lie on its own surface."

The *Essay* is reviewed in pages 407...425 of the *Christian Observer* for June, 1854. The reviewer, without expressing any strong belief in the doctrine that other worlds are inhabited, thinks that the arguments of the *Essay* against this are quite inconclusive. At the same time he bears witness to the excellencies of the book and draws a parallel between it and Warburton's *Divine Legation of Moses*.

An article entitled *The Plurality of Worlds* occurs on pages 22...28 of the *Evangelical Repository* for Sept. 1854; it reviews the *Essay* and the *Dialogue*. The writer of the article sent a copy to Dr Whewell with a note in which he gave his name and address, and stated that he had also reviewed Sir D. Brewster's reply in the same number of the *Evangelical Repository*. He holds that the Sun and the planets are inhabited, but gives up the planetoids.

A letter by Dr Croly addressed to the editor of the *Standard* occupies about two columns of the length of an octavo page; it maintains that the planets are inhabited. An allusion occurs in it to a former letter by the writer on the same subject.

An article entitled *More Worlds than One* occupies seven columns of the length of an octavo page; it is a review of Sir D. Brewster's book, and speaks of the author as "our great countryman." A singular remark is made with respect to Dr Whewell's statement that a man weighing 150 lbs. on the Earth's surface would weigh nearly 400 lbs. on the surface of Jupiter. The reviewer charges him with neglecting to notice the admirable contrivance by which the weight is kept within this limit, when if Jupiter had been smaller in size or more dense it would have been much greater. This is said to be "an example rather of special pleading than of sober and impartial investigation and inference." I am unable to name the periodical in which this article first appeared, as the copy preserved among Dr Whewell's papers has been separated from the pages with which it was originally connected; the same remark applies to various other short articles which I have to notice.

A series of four articles entitled *Are there more Worlds than One?* occupies about thirty-four columns of the length of an octavo page, and forms a review of the works of Dr Whewell and Sir D. Brewster; they were probably published in the *Mechanics' Magazine*. The writer is very diffuse, and it is difficult to say what is his main purpose. He seems to hold that Dr Whewell departs from the "spirit of the Inductive method," and that his reasoning is "inconsistent with every principle of a sound inductive logic." The judgment passed on Sir D. Brewster is equally unfavourable.

The Testimony of the Heavens to their Creator. A Lecture to the Enniskillen Young Men's Christian Association, delivered Feb. 3, 1857, by the Rev. Josiah Crampton, A.M. This is in 31 small octavo pages. Without alluding to any controversy the writer holds that the heavenly bodies are inhabited.

Thoughts, in verse, on the Plurality of Worlds. By the Rev. John Peat, M.A. 1856. This consists altogether of 15 octavo pages; it is dedicated to Sir D. Brewster, and leans to his opinions. I quote the last four lines, which are addressed to the systems of the heavens:

> Roll on, God's mighty clock-work! roll and chime,
> Ye tell the hours of Universal Time;
> God is your Mighty Mainspring—God your Soul,
> His word hath wound you up, and on ye roll!

Thoughts on the Controversy as to a Plurality of Worlds. By F. W. Cronhelm. This consists altogether of 24 octavo pages. The writer holds that the "earth is the *only seat of reproductive life,* and in this respect unique among the revolving globes created by the heavenly Architect." He thinks that the planetary bodies may be occupied by the angelic races, and by all the generations of mankind since Adam. He suggests that the "admirable, though somewhat austere, Milton" has given his sanction to a popular but erroneous creed about the angels. For on the visit of Raphael to Eden "Adam somewhat slily puts the question, whether he also had not an Eve, or something like her in the bowers above. To this the angel, as if taken by surprise, gives an evasive answer, but with

> 'A blush celestial rosy red'

by which Milton plainly intimates that our great ancestor was not far wrong in his surmise. This curious passage seems to have escaped the research of the numerous critics on the Paradise Lost." The precise words here given may well escape the research of the critics, as they are not Milton's; what he really wrote is probably as well known as any other passage of the poem.

I now proceed to those reviews which may be described as neutral in opinion, and then to some which are on the whole favourable; but the classification is necessarily not very exact.

An article entitled *The Plurality of Worlds* occupies 34 pages of the *New Englander* for Nov. 1854; it is by Professor Olmsted. It reviews the works of Dr Whewell and Sir D. Brewster in a grave temperate style. I will extract two sentences: "From the high character which the authors of these works justly sustain as men of profound science and accurate scholarship, we have felt some surprise at the number of erroneous statements in point of fact, or principle, met with in each essay." "After some reflection on the question of a Plurality of Worlds, (meaning thereby the question whether the planets are inhabited, and whether the stars are suns and centres of planetary worlds,) we are inclined to believe in the affirmative, although we do so with a full conviction that there is much to be said on the other side." It is stated that views agreeing with those of the *Essay* as to shooting stars and the zodiacal light were first proposed by the Professor of Astronomy in Yale College; and Dr Whewell in the preface to the fourth edition of his *Essay* recognised the priority of Professor Olmsted as to this matter.

An article entitled *The Plurality of Worlds* occupies six columns of the length of an octavo page. In its second sentence it praises the "good writing, ingenious thinking, and very extensive scientific knowledge" of the author of the *Essay;* and its last sentence concludes by recommending "the works both of Sir David Brewster and his antagonist, as two of the most remarkable pieces of controversy which the present age has produced."

An article entitled *Geology versus Astronomy* occupies about ten columns of the length of an octavo page. It may be safely attributed to Hugh Miller from the following circumstance. It quotes a passage of the *Dialogue* in which Dr Whewell says, "I now know that several years ago (in 1849), Hugh Miller, in his *First Impressions of England* (Chap. XVII) presented an argument from Geology, very much of the nature of that which I have employed;"...Then the article proceeds, "There is an error in the date given here: the argument to which the author of the Essay refers 'as much of the nature' of his own, was first published, not in 1849, but in October 1846, when it appeared in the columns of the *Witness,* as part of one of the chapters of 'First Impressions'—a work which was published in the collected form as a volume early in the following year." The article closes with an extract taken from the *Essay,* Chapter IX, sections 29...33, in which the claims of Geology are stated; the passage is pronounced to be "very admirable, both in form and substance."

In *Blackwood's Magazine* for Sept. and Oct. 1854 the works of Dr Whewell and Sir D. Brewster are reviewed in articles entitled

Speculators among the Stars. In the preface to the third edition of his Essay Dr Whewell speaks of these articles as "obviously the work of an acute and able judge." They were written by Mr Samuel Warren, and are reproduced in his *Miscellanies*, pages 437...494, with additions, which consist chiefly of more copious extracts; I shall cite the pages of this reprint. The review is one of the best which the controversy produced—giving a full account of the main parts of the subject in a calm judicial spirit; it is on the whole decidedly favourable, at least to the temper and ability, of the *Essay* and the *Dialogue*. Strong praise is expressed on page 442 in a passage beginning thus : "The author has an easy mastery of the English Language, and these pages abound in vigorous and beautifully exact expressions." On page 452 it is said, and quite truly, that the scope of contemporaneous criticism is hostile to Dr Whewell's views, and an extract is given from an unfavourable article which I have not seen ; no reference is given in the *Miscellanies*, but in *Blackwood's Magazine* the article is ascribed to an "accomplished diurnal London reviewer," to which a note, *Daily News*, is added. On page 454 Mr Warren alludes to the sensation which was produced by the delivery of the *Astronomical Discourses* of Dr Chalmers, one or two of which he himself heard. On pages 463 and 464 he strongly censures the tone adopted by Sir D. Brewster.

A review of the *Essay* occurs on pages 513...531 of the *Eclectic Review* of May, 1854; it may probably be attributed to Mr G. Gilfillan. This is a very interesting article, though more simplicity of style might have been desired. The writer speaks of himself as having been led more than ten years since to suspect that the evidence for the plurality of worlds might not be so strong as had been supposed. In proof of this he quotes a passage written in 1844 and published two years subsequently; it expresses opinions strikingly in harmony with those maintained in the *Essay*, though we may be sure it never came under the notice of Dr Whewell until he saw them in this article. The passage occurs substantially, though not in precisely the same words, on page 116 of Mr Gilfillan's *Gallery of Literary Portraits*, 1845. The reviewer says: "The book, particularly in the last four chapters, contains much that is as eloquent, powerful, and poetical in language, as it is piercingly acute and suggestive in thought." Dr Whewell notices this review on pages 400 and 401 of the *Dialogue*.

A review of Sir D. Brewster's book occurs on pages 35...62 of the *Christian Observer* for Jan. 1855. As we have seen, the *Essay* had already been noticed in the same periodical, but not by the writer of the present article. The review is very decidedly against Sir D. Brewster, and Dr Whewell alluded to it in the preface to the fourth edition of his *Essay*.

An article entitled *Unity of the World—Man's Place in the Creation*, occurs on pages 435...467 of the *New York Quarterly* for Oct. 1854. The writer, Mr Joshua Leavitt, sent a copy of it to Dr Whewell, together with a copy of the *Christian Review* for April, containing another article by him on the same subject; but the latter does not appear in Dr Whewell's collection of reviews. In the *New York Quarterly* the rival works are examined; the article deals chiefly with the religious side of the question, and with respect to this it strongly censures Sir D. Brewster. I will extract a few sentences from various parts of the article. "We find ample and exuberant stores of all that is needful, so as to be sure there is enough; and yet it is plain that they are exhaustible, and calculated for a world that is to be inhabited but for a limited period." "Evidently there must come a time, at however remote a period, when the earth will be worn out, and no longer fit for the abode of man." "The present knowledge in regard to the various [heavenly] bodies offers continually accumulating evidence that they are unfitted for the support of even animal life, and still less adapted to be the abodes of rational and accountable beings, capable of moral government and subjects of redemption. So that while all the analogies in the kingdom of morals are seen to be utterly irreconcilable with the theory, the supposed analogies in the material world are fading away in the clearer light of advancing science. We think Professor Whewell has rendered a good service both to science and religion, by the profound sagacity and the great learning which he has brought to bear on the question."

An article entitled *The Plurality of Worlds* occurs on pages

271...278 of the *Ecclesiastic* for June, 1854. The reviewer objects only to the Chapter on *The Future* in the *Essay;* and Dr Whewell noticed the objection in the *Dialogue,* page 401. Like others, he draws especial attention to Chapter XI of the *Essay;* he calls it a "masterpiece of close and logical reasoning," and extracts "one passage as noble in its thought as eloquent in its expression." The passage is Chapter XI, section 34, and following; this section 34 is also praised by Professor H. J. S. Smith, by Mr Gilfillan, and by Dr Lightfoot in his funeral sermon upon Dr Whewell: on the other hand, Sir D. Brewster could not find language to express the feeling with which he read it, and the reviewer in the *Christian Remembrancer* said it struck him with horror.

A review, occupying four columns of the length of an octavo page, speaks favourably of the *Essay,* and concludes by saying that it will "...show how entirely conjectural are the assumptions of such astronomers as delight to create fresh beings with endless profusion, and how the analogy of nature leans quite as much to the older theory of the unique character of the earth."

Astronomy and Geology as taught in the Holy Scriptures, Liverpool, 1855. This is an anonymous pamphlet in 23 octavo pages; the author gave his name in a copy which he sent with a letter to Dr Whewell. He refers to Dr Whewell by name, and adopts the view of the *Essay* that the planets cannot have inhabitants in any respect like man. He understands literally the Scriptural account of the Creation in six natural days.

In Mr De Morgan's *Budget of Paradoxes,* page 63, there is a slight notice of the *Essay* and *Dialogue.* When preparing this notice, Mr De Morgan wrote to Dr Whewell thus: "I have occasion for the date of your supplement-dialogue to the plurality of worlds; and no date whatever is given. Can you give it me? The omission is awful: for after all there may be more worlds, and how is ours to be the best of all possible worlds, as Leibnitz said, if books go without dates?" It is strange that so accurate a bibliographer as Mr De Morgan, after thus consulting the author of the *Dialogue,* should go wrong; his statement is "First found in the second edition, 1854; removed to the end in subsequent editions, and separate copies issued:" but there was a separate issue *before* that which occurs in the second edition of the *Essay*, and the *Dialogue* was not removed to the end before the *fifth* edition.

The collection of letters preserved by Dr Whewell contains many addressed to him on the subject of the Plurality of Worlds; they uniformly recognise the ability with which the argument of the *Essay* is conducted, and in some cases admit that it is quite convincing, though no eminent scientific name occurs among those who adopted his views. One correspondent offered a testimony which could not have been very highly appreciated; he attended to spirit-rapping—a fashionable folly of the hour—invoked the shade of Newton, and received satisfactory information respecting the nature of the sun and the comets. Short extracts from some of the letters may be published without impropriety.

Sir Benjamin Brodie suspected the authorship after reading a few pages; he says, " Whoever the author may be, I perceive that he is possessed of extensive and various knowledge combined with great sagacity in the application of it. He has shown very clearly that the other planets are not adapted for the maintenance of beings like ourselves." But the writer proceeds to observe that this leaves the possibility of the union of mind with other forms of matter. Mr Hallam, after giving up the moon, says: " This book carries the war much further, and it is impossible not to admire the scientific knowledge and the originality of the arguments against any plurality of inhabited worlds." " Upon the whole it is an original and very remarkable book, and will probably make an epoch in such speculations. I cannot deny that it leaves considerable difficulties, and will, I dare say, be unfavourably received by the majority. Chalmers's theories were well received, though they were very arbitrary." Sir J. Herschel, after alluding to the general opinion that Dr Whewell himself was the author of the *Essay*, says: " In common with other people I find myself obliged to admit that I should not have thought there was so much to be said on the non-plurality side of the question." " The book is full of striking things. The geological argument is very pointedly put.—The Magellanic Clouds are very availably brought into action.—Time and Space are duly and properly scorned and reduced to their true value." Lord Rosse says: " There seems to be strong evidence in support of your opinion that the nebulæ are not enormously distant in proportion to the Fixed stars, and in my last R. S. Address, which I enclose, I have taken that view. I confess, however, I believe the Fixed stars are bodies like our Sun, and that α Centauri is an example of them." General Sabine quotes a passage from the *Essay*, page 208, which conjectures that the atmosphere has probably solar and lunar tides, and states that the existence and amount of the lunar atmospheric tide has been ascertained at St Helena and Singapore.

A letter, of sixteen large pages in French, addressed to Dr Whewell, has been preserved. The writer says that he has published in France a little book in answer to the *Essay*, and with due apology he gives some account of his own views; they consist in fact in holding a form of the Nebular Theory. Each planet is slowly approaching the sun; thus each in turn comes within the range suitable to the existence of inhabitants, and is then peopled by the act of the Creator. The writer allows that he has never heard any one speak of the spiral movement thus attributed to the planets, but asks whether it is certain that the idea may not be received one day with a certain degree of favour. He also suggests that the planets may gradually contract so as to increase in density; and he asserts that when the density increases the orbit will diminish. He finally diverges into praise of the emperor Louis Napoleon, condemnation of Russia, and anticipation of unbounded benefits from the alliance of the English and French.

Humboldt in a letter dated Berlin, 21 February, 1854, says:

"Je ne saurois Vous exprimer assez vivement, Monsieur et illustre ami, combien j'ai été touché de votre aimable souvenir, en m'adressant, signé de votre main, l'important ouvrage sur la Pluralité des Mondes. Vous avez attaqué, avec cette puissance que donne une vaste étendue de connoissances et cette sagacité dans les vues dont brille votre admirable histoire des Sciences Inductives, un problème souvent dénaturé. On l'a rendu presque dangereux à traiter, en craignant d'empiéter sur la liberté d'une imagination toute poétique ou de mêler trop directement à la discussion des intérêts élevés de la religion positive. Je tiens beaucoup aux premières impressions d'une lecture rapide mais pourtant très instructive. Tout ce qui a rapport au Système Solaire et aux Nébuleuses est rempli d'idées grandioses et souvent très neuves; l'aperçu géologique, the theory of the Solar System, le 11me et surtout le 12me Chap^e, Unity of the World, m'ont paru pleins d'attraits. C'est un désir bien doux à remplir que de Vous offrir, Monsieur, le respectueux hommage de ma reconnoissance..."

The sheets of the *Essay* as they passed through the press were read by Sir James Stephen; and the correspondence which thus arose between the author and his accomplished friend and critic is very interesting. Partly in deference to the advice of Sir James Stephen Dr Whewell cancelled many pages of his work after they had been printed, as they seemed too metaphysical for English readers. He himself sent a set of the cancelled sheets to Professor J. D. Forbes, and requested his attention to them. A copy of the book in its original form has been preserved, so that the amount thus cancelled can be readily ascertained. In the first published edition the last three chapters are announced in the Table of contents thus:

In the original form we have:

Thus five Chapters of the original form occupying pages 249...332 are cancelled, and replaced by the Chapter on the *Unity of the World* on pages 248...264. There appears to be no other difference between the original form and the first published edition, except in the last section of the *Argument from Design*, and in section 21 of the *Future;* in the former case the original form is longer, and in the latter case shorter than the first published edition.

Portions of the cancelled sheets, amounting to about fourteen pages, were used in the Chapter on the *Unity of the World* in the published volume; and formed the whole of that Chapter except the first two sections, so that about seventy pages were printed and finally withdrawn. Thus sections 3 and 4 of the *Unity of the World* are the sections 20 to 23 of the *Argument from Law*; sections 5 and 6 are the sections 32 and 33 of the *Argument from Law*; section 7 is section 43 of the *Argument from Law*; sections 8 and 9 are sections 11 and 12 of the *Omnipresence of the Deity*; sections 10 to 15 are sections 5 to 10 of *Man's Moral Trial*; and sections 16 to 20 are sections 13 to 17 of the *Design of Animal Springs of Action*. The passages lose some of their interest by being thus dislocated from their original connection and recombined. The title *Unity of the World* is itself a witness to the awkwardness of the change of arrangement; for it seems to amount to giving to one Chapter the designation which may properly be assigned to the whole work. We may interpret it thus: in other Chapters it is argued that the stars and planets are not adapted for a rational population, and in the present Chapter it is maintained that the Earth and Man constitute an adequate field for the glory of the Creator. In consequence of their displacement from their original positions sections 3 to 6 of the *Unity of the World* seem not to bear upon the general subject; and section 7 begins by referring to "that argument," when nothing immediately precedes to which the phrase can apply. Instead of "that argument" we should read "the question of the Unity or Plurality of Worlds."

I proceed to give some account of the contents of the cancelled sheets.

The Chapter on the *Argument from Law* is a contribution to Natural Theology, throughout which the author seems to postpone his main subject until he comes to the last page; then he introduces a section coincident with the seventh of Chapter XII of the *Essay* which brings him back to the proposition that man is "a worthy object of all the vast magnificence of Creative Power."

A few detached passages from the Chapter may be extracted. "The argument from Design, and the argument from Law, in proof of the existence and activity of a Divine Creator and Governor, may indeed be stated briefly in similar modes. Those who urge the former argument, are accustomed to say, Design proves a Designer, and that Designer is God. In like manner, we may say, Law proves a Lawgiver, and that Lawgiver is God." "If man, the creature, the student of creation, the slow Learner, the mistaker, yet the Lover of Truth, must pursue his study, and acquire his truth, by the use of his geometry; must not God, the Creator, the Lawgiver of Creation, the Teacher, the Infallible, the Author of Truth, also have done his work, by introducing into it geometry?" "Why does the constitution of the world, as an object, correspond with the constitution of the mind, as a mind? First; because they are the works of the same Maker; but that is not the whole reason: further; because the constitution of the world is marked with the Thoughts of the Divine Mind, and the human mind is, in part, a sharer in the Thoughts of the Divine Mind." Speaking of Hegel's presumptuous criticisms on Newton, Dr Whewell says that "... they have been carefully and fairly considered in this country, and discussed, so far as they admit of discussion. And we do not see why persons who have thus weighed these speculations should not describe them plainly; and say that these, and many of the like speculations of other German philosophers, are ignorant, presumptuous, and illogical, to an extent to which no injustice is done by calling them childish." With respect to Hegel's speculations in other departments of human knowledge Dr Whewell says, "But that these parallels and contrasts, these antitheses and epigrams, these metaphors and phrases, are to pass for a Philosophy of History, or of Law, or of Taste, or of any large subject, shews, we may almost say, not only great presumption in him who delivers such a so-called *philosophy*, but a certain abjectness of spirit in those who receive it as a profound philosophy."

The question proposed in the Chapter on the *Omnipresence of the Deity* is, how are we to conceive that the Divine Ideas govern the Universe? The answer given and developed is that if we suppose "God, Infinite and Eternal, to be present in all parts of space, at all moments of time, and to carry into effect by his power the Laws of Nature, which he has established in his wisdom, many difficulties are removed, and many obscurities cleared." The discussion is connected with the main subject in the last two sections of the Chapter, which agree with the eighth and ninth of Chapter XII of the *Essay*.

The Chapter on *Man's Intellectual Task* supplies an important limitation of what had been said in the preceding Chapter. The question is, does God think and will *in us*, as He moves matter or developes plants? The answer is, that the events which take place in our *intellectual* world, the thoughts which we think, cannot be said to be God's thoughts or the direct result of His agency. Man's great business, considering him merely as an intellectual creature, consists "in controlling and subduing the influences within him, which may prevent or retard his discovery of truth, and in unfolding and perfecting the powers which he has in him tending to such discovery." The last section of the Chapter returns to the main topic—the possibility that the earth is the only seat of intelligent life, and concludes thus: "For how few of mankind realize the intellectual destiny thus placed before them! And if, by any act of the Divine Government, the number of those who rightly aspire to approach to the Divine Nature, were greatly increased upon earth, would not this be a far more suitable, and a far more intelligible method, of extending the Divine Kingdom of Intellect, than any multiplication of worlds could be, supposing the worlds to be such as this world of earth was, before that act of the Divine Government by which man was created, in his intellect at least, after the Image of God?"

The Chapter on *Man's Moral Trial* begins with some general remarks on *right* and *wrong*, and on the *springs of human action*, which are coincident with the author's views as expounded in his *Elements of Morality;* these form the first half of the Chapter, and the rest coincides with sections 10 to 15 of Chapter XII of the *Essay*.

In the Chapter on *The Design of Animal Springs of Action*, a doctrine is noticed which is often put forward—"that the happiness of man, and the enjoyment of all sentient creatures, is the pervading purpose of the Creation." Paley's well-known remark is quoted, "It is a happy world after all." Dr Whewell however does not admit that there is for the animal creation an obvious balance of pleasure: "But to many persons, the animal creation, and its fortunes, as they are most commonly seen, offer a far less cheerful spectacle than this. It appears to them, to be full of suffering and pain, unmerited, unredressed, uncompensated." Then some explanation is offered of the facts of animal life by the analogy of the general plan which can be traced through all the class of vertebrated animals. "And thus the springs of action in animals, may all be regarded as *rudiments* of the proper and universal springs of human action. And the general law of the animal creation, that such springs of action should exist, may be regarded as instituted, not for the sake of animals alone, but as a great feature of the animal creation, with man at its head, who is far more than a mere animal; and in whom, the animal springs of action assume a form and a purpose, which could not have been conjectured, from their operation in animals merely: just as the meaning of the rudimentary bones of the human hand could not have been conjectured, from merely considering the paw of the lion, or the foot of the crocodile." The sections 13 to 17 which finish this Chapter coincide with sections 16 to 20 of Chapter XII of the *Essay*. Here I conclude my account of the cancelled sheets.

Among Dr Whewell's papers are preserved some remarks on *Animal Instincts*, occupying six printed octavo pages, which were probably originally designed to form part of the *Essay*. The difference between *Instinct* and *Insight* is explained as in the *Philosophy of the Inductive Sciences*, Vol. II, p. 109. The doctrine maintained is that Instincts are the working of the Divine Mind. "At a certain period of philosophy, the doctrine was so familiar that the expression of it was almost proverbial. *Deus est anima brutorum*, 'God is the soul of brutes' was an accepted maxim." Grotius is cited as expressing the same opinion. The remarks conclude in the following manner: "And thus, the existence of Instincts in the animal world, even though they may be regarded as, in a peculiar sense, the operation of a Divine Mind, pervading such a world, does not make the existence of such a world a worthy scene of the Divine Care and Government, in comparison with the Divine Providence and Government which rules the affairs of this our human world; it does not do this, in any greater degree than the Divine Care and Government, by which the worlds of inanimate matter are directed and governed by the same Divine Mind. Whether the planets be uninhabited, or be inhabited by mere races of animals, which are impelled by animal impulses, and directed, by Instincts divinely implanted and maintained, in constant cycles of self-support, self-defense, and reproduction; still, such planets are stationary, unintelligent, unraised assemblages of being; and suggest no thought worthy of a Divine Author, compared with the humblest scene of human trial, discipline, sorrow, fear, hope, duty or virtue."

The following are the only manuscripts bearing on the subject of the Plurality of Worlds which occur among Dr Whewell's papers.

Astronomy and Religion, three Dialogues. The first dialogue is on 24 pages, and is dated June 13, 1850, K, that is, Kreuznach; it consists of a statement of the difficulties connected with the question of the Plurality of Worlds. The second dialogue is on 6 pages, and the third on 5; these are incomplete. On the whole we may say that there is nothing of importance in these papers which is not embodied in the *Essay*.

Philosophy and Theology. This consists of 74 pages, dated C. C. Aug. 27, 1851; the C. C. denotes Cliff Cottage, Lowestoft. The paper is interesting, though it contains nothing essentially new. It consists of the author's well-known doctrine about *Facts* and *Ideas,* and some discussion of the relation between the human mind and the divine, like that which is given in the cancelled pages of the *Essay,* and in Chapter XXX of the *Philosophy of Discovery.*

CHAPTER XIII.

1854...1866.

In the year 1854 the Archæological Institute met at Cambridge in July, and Dr Whewell was present. During the year he published two Inaugural Lectures delivered in London, and a pamphlet on the Oxford University Bill. These I shall now notice.

On the material Aids of Education: being an Inaugural Lecture on the occasion of the Educational Exhibition of 1854. *Delivered July* 10, 1854. This is a pamphlet in 39 small octavo pages. Education is defined to be " the process of making individual men participators in the best attainments of the human mind in general : namely, in that which is most rational, true, beautiful, and good." Some remarks are made under the four divisions here indicated. The lecture is well deserving of attention. A few brief extracts may be given.

After speaking of living and dead languages, Dr Whewell says : "you will find that the *living part of the English language is the dead languages ;* that is, the power of forming new words and new derivatives, is exercised by using, not Saxon, but Latin elements." After alluding to mechanical contrivances for making truths obvious to the senses, Dr Whewell says : "Let us by all means make our educational apparatus as complete as we can ; but let us always recollect that it is the thought involved in such apparatus which gives it its value ; and that it is the office of all such helps to teach men to think, not to save them the trouble of thinking." The great practical difficulty in education, at least in England, is frankly admitted by Dr Whewell in the brief notice which he takes of his fourth division : " The remaining one of

14—2

the four objects which I mentioned, forming the objects of a general education as generally understood,—the making men participators of that which the human mind has attained of *good*,—though in itself the highest and most important part of education, must on the present occasion be treated very imperfectly, and with great reserve. This must happen, unfortunately, I may say, in consequence of the very view which we are now taking of education, that it consists in making the individual a participator, in the best knowledge and belief, to which men in general have attained. For *what is* the best knowledge and belief, concerning good and evil, has unhappily been the subject of much controversy: so that men in general are not agreed as to what is the best and completest kind of such knowledge."

Dr Whewell's remarks on the English language were criticised in an article in the *Examiner*, reprinted on pages 508...510 of the *Life and Labours of Albany Fonblanque*. The critic condemns as a barbarism the word *prepaid* which Dr Whewell had used as an illustration, because the simple word *paid* expresses the sense completely. He proceeds to maintain a common proposition, that "The best styles are the freest from Latinisms; and it may be almost laid down as a rule that a good writer will never have recourse to a Latinism while a Saxon word will equally serve his purpose." The remarks of Dr Whewell, however, do not bear on this proposition; they relate to the fact that compound words are formed not from Saxon but from Latin elements.

On the influence of the History of Science upon intellectual education: a Lecture delivered at the Royal Institution of Great Britain, before H. R. H. Prince Albert. This lecture is the first in an octavo volume published in 1855, entitled *Lectures on Education, delivered at the Royal Institution of Great Britain.* Dr Whewell's lecture occupies 36 pages of the volume; it was delivered in 1854. Most of the lectures, including Dr Whewell's, were afterwards reproduced in a volume called *Modern Culture,* edited by Dr Youmans. The proposition which is supported by Dr Whewell is this: That every great advance in intellectual education has been the effect of some considerable scientific discovery, or group of discoveries. This is illustrated by the cases of Greek Geometry, Roman Law, and modern Inductive Philosophy. The lecture is written with great vigour, and presents in a compendious form the opinions of Dr Whewell on the important subject of education. Of Roman Law the lecturer speaks in high terms : " Cicero says, proudly, but not too proudly, that a single page of a Roman jurist contained more solid and exact matter than a whole library of Greek philosophers." The remark was adopted by Mr A. Elley Finch in a recent lecture, and came under the censure of a critic as an inexact rendering of Cicero's words. The lecturer replied that he adopted the passage from Dr Whewell, to whom he referred in a note, and added, " As a rhetorical paraphrase suited Dr Whewell's purpose (in an oral address) so it suited mine." Dr Whewell's lecture gave great delight to J. C. Hare, then drawing towards the close of his busy life.

Dr Whewell alludes in the lecture to the well-known story that Plato declined to receive among his hearers those who were ignorant of Geometry; but by inadvertently adopting a common error he represents the philosopher as disfiguring the gate of the Academy with an inscription in false Greek.

In the *Oxford Essays*, Vol. I, page 270, Mr Pattison observes in a note, " That improvements in education follow improvements in logical method has lately been insisted on by Whewell, *Lecture at the Royal Institution*, p. 7, thus very much modifying the opinions on the subject he had expressed in 1837, in his *Principles of English University Education*, p. 24, &c." I do not understand this; on page 7 of the lecture Dr Whewell merely adopts Mr Grote's view of the Sophists, and the phrase *logical method* does not occur in the lecture.

Notes on the Oxford University Bill in reference to the Colleges at Cambridge. This is an octavo pamphlet of 36 pages, which seems to have been printed for private circulation : it is dated Trinity College, May 1, 1854. The pamphlet disapproves very decidedly of all the changes which, according to the analogy of the Oxford Bill, might be proposed at Cambridge. A few sentences may be extracted. "If therefore it were proposed at Cambridge to provide a Fund for University purposes by suppressing some of the Fellowships of Trinity and St John's, and leaving the smaller Colleges untouched, I conceive that all persons interested in the improvement of the University would regard such a proceeding as enormously mischievous and perverse; since it would destroy or diminish the best parts of the system of things now existing at Cambridge, and encourage the worst parts. Trinity and St John's have been regarded as, in most respects, good examples of what Colleges ought to be; and those who blamed the effect of our College system have especially applied their condemnation to the smaller Colleges. Yet this part of the Oxford Bill, if applied to Cambridge, would leave the smaller Colleges as they are, and bring Trinity and St John's to resemble them more nearly."

In the beginning of the year 1855 Dr Whewell was requested by the Archbishop of Canterbury and the Bishops of London and Oxford to undertake in conjunction with Professor H. H. Wilson and Mr J. Muir the office of examiner of some essays on the Hindu Systems of Philosophy. I presume these essays had been written in competition for the prize mentioned on page 155, but I do not know whether Dr Whewell acceded to the request. In the course of the year he resigned the Professorship of Moral Philosophy, in which he was succeeded by the Rev. John Grote of Trinity College. In November he was appointed Vice-Chancellor of the University. In December Mrs Whewell died after a long illness. Three pamphlets connected with University Reform are all the publications of the year which I have to notice.

An octavo pamphlet of 8 pages marked *Private*, and dated April 19, 1855, relates to a point in the administration of the University. Objections were made by some persons to the power which the Heads of Colleges exercised with respect to the election to certain offices; and Dr Whewell proposes to answer the objectors. Since this date the matter has been practically settled in favour of the objectors, to whom it must be admitted that Dr Whewell himself belonged in his early days. The pamphlet seems scarcely consistent with itself, in treating the point in dispute as of small importance while strenuously upholding the existing condition.

Remarks on the proposed reform of the University of Cambridge, May 31, 1855. This is an octavo pamphlet of 25 pages, besides the title-page; it is marked *Private.* It is directed against two letters issued by four of the persons who had been engaged on the Royal Commission to enquire into the University of Cambridge, and their Secretary. The main point is the question of the position of the Heads of Colleges in the University Constitution. The progress of opinions and events seems to have decided against Dr Whewell. He claims great credit to himself as a reformer, but the changes which he advocated, or was willing to accept, were perhaps not of primary importance. There were many serious evils which his long experience must have brought under his notice—such as the undue pressure to take holy orders, the difficulty of securing efficient resident teachers, the inadequate stipends of professors, and the undesirable modes of election—and he seems to have made no attempt to remedy these. It is a matter of great regret that his power and influence were never used in favour of improvements, since almost universally recognized as desirable. But it would be ungrateful to forget that, both by his example and by his participation in University legislation, he did much to elevate the character of the University; especially by vigorously supporting the extension of the range of study, and by increasing the importance and the responsibility of the functions of the Professors.

An octavo pamphlet, entitled *Additional Remarks on the proposed Reform of the University of Cambridge,* consists of twelve pages; it is marked *Private* and dated June 16, 1855. This relates to the same matter as the pamphlet of April 19, 1855, and alludes by name to Dr Peacock, who was one of the objectors. A few words may be extracted. "Persons have spoken of putting down the tyranny of the Heads, after the manner of the Athenian hymn to Harmodius and Aristogiton. They have talked indignantly of the despotic character of Whitgift, the author, as they say, of the present University Constitution; though his character appears to have little more to do with the merits of the

Constitution for present use, than the character of the Barons who signed Magna Charta has with the present English Constitution."

In the year 1856, Dr Whewell seems to have attended an Archæological meeting at Edinburgh in July. He was present at the meeting of the British Association at Cheltenham, in August; and in the Transactions of the Sections he is reported to have made a communication *On the reasons for describing the Moon's Motion as a Motion about her axis.* From time to time the notion reappears among unscientific people that the moon does not rotate on her axis, and at this date Mr Jellinger Symons addressed a printed letter to Dr Whewell advocating the unsound opinion. Mr Symons was an Inspector of Schools and he complained that children were taught *most erroneously* that the Moon rotated on her axis. At the meeting of the British Association at Cheltenham he discussed the matter orally with Dr Whewell and others in the Mathematical Section. An account of the discussion will be found in the *Athenæum* for August 23, 1856, from which it appears, as might have been expected, that Mr Symons made no converts among the scientific people who were present. There is no record in the volume of the British Association of Dr Whewell's communication, but it probably coincided with a printed page on the subject in two columns, preserved among his papers, which agrees with the account in the Athenæum. The question is treated simply and conclusively. In December Dr Whewell visited Rome, furnished with letters of introduction from Bunsen.

The following are the publications of the year: a paper connected with the management of the Fitzwilliam Museum in the University, a sermon on the death of Mrs Whewell to be noticed hereafter, and Elegiacs on the same subject.

In January, 1856, a dispute arose between Dr Whewell, as Vice-Chancellor, and the Syndicate for managing the Fitzwilliam Museum. The Vice-Chancellor on his own responsibility altered the arrangement of the pictures in the Museum, and the Syndicate objected to this course as an unconstitutional exercise of authority. Various papers were circulated in the University on the subject, and among them one by the Vice-Chancellor in six quarto pages defending his proceedings.

Shortly after the death of his wife, Dr Whewell printed some *Elegiacs* which were circulated privately among his friends. They form a quarto pamphlet of 31 pages, besides the title-page and the list of contents. The matter consists of about 800 lines, and is disposed in 13 sections, which have the following titles: the first Sabbath after the burial; the second Sabbath, the Epiphany; the Picture; the Return home; recollections of the Burial Service; the Psalm; the Lesson; the Grave; the Thanksgiving; the Grave-stone; the Sermon; the Monument; Easter Day. The *Lesson* and a part of the *Grave* are in Hexameters, and the rest in Elegiacs. I will quote two passages as specimens.

The following is the conclusion of the Section on the *Monument:*

> Hope, when driven from earth, finds refuge in heaven, and beckons,
>   Beckons us onwards still: promises union there:
> Promises union there, in mansions of bliss and of gladness,
>   There in the presence of *Him* whom *she* so faithfully serv'd.
> So when the summer sunset gleam, on the top of Helvellyn,
>   Fades and passes away into the grey of the night,—
> Watch but a few short hours, and soon, through the melting of darkness,
>   Bright shines forth on his crest radiance of heavenly morn."

The following is from the section on the *Lesson:*

> One is the glory of earth: celestial glory another.
> One is the glory of Sun; of Moon the glory is other;
> Other the glory of stars; and stars, too, differ in glory.
> And even so man's body is changed to find resurrection.
> Sown is it in corruption, and raised in incorruption.
> Sown is it in dishonour, thereafter raised in glory.
> Sown it is in weakness, and raised in strength and in power.

Mr Longfellow in thanking Dr Whewell for a copy of the Elegiacs said: "I have read the poems with great interest and pleasure; and yet with a painful sympathy with you in your affliction, which has found expression in such musical numbers. I am delighted to see the Elegiac metre introduced into English verse; it is so beautiful, and satisfies the ear so completely. And to you we must give the honor of its introduction, if we pass over the few specimens of translation from the German." Mr Longfellow, as is well known, has himself cultivated English

Hexameters, and his *Evangeline*, written in that metre, was reviewed by Dr Whewell, as we shall see in a later chapter.

Early in the year 1857 Dr Whewell was elected a corresponding member of the French Institute in the place of Sir William Hamilton of Edinburgh; he mentions this in a letter to Professor J. D. Forbes as " a great and quite unexpected honour, and certainly not bestowed on account of the similarity of our philosophy." The Parisian metaphysicians could scarcely have given a more striking example of the eclecticism which was then their prevailing system than by the successive choice of such determined antagonists. In August Dr Whewell was in Dublin during the latter half of the time of the meeting of the British Association; but owing to illness he took very little part in the proceedings. There is only one publication of the year to notice, a review, to which I now proceed.

In the *Edinburgh Review* for October, 1857, pages 289...322, Dr Whewell reviewed the new edition of Bacon's works, then being published under the joint care of Mr James Spedding, Mr Robert Leslie Ellis, and Mr Douglas Denon Heath; at that time the first three volumes had appeared, containing the philosophical works. This is, I think, the only contribution of Dr Whewell to the Review, in which his own books had formerly been treated very harshly. Dr Whewell seems to have had the article reprinted for private circulation; the type is large, so that the reprint forms a pamphlet of 48 octavo pages. In the *Edinburgh Review* the article is headed *Spedding's complete edition of the works of Bacon;* in the reprint it is headed *Review of Spedding and Ellis's Edition of the Works of Francis Bacon;* and the list of books at the head of the article in the Review is omitted. The new edition of Bacon's works is justly praised; but it seems to me that the ability, the learning, and the disinterested labour of the editors deserved still warmer commendation from the official head of the great college to which they and their author alike belonged. One of the most striking Chapters in the *Philosophy of the Inductive Sciences* is that relating to Bacon; this Chapter, when reprinted in the *Philosophy of Discovery,* 1860, was followed by another consisting of *Additional Remarks on*

Francis Bacon : these two Chapters should be studied in connexion with the present review. Perhaps in the *Additional Remarks* we might have expected to find a reference to the very severe attack on Bacon which Professor Liebig had published not long before.

Dr Whewell in this article takes some notice of the hostile criticism on Bacon which appeared in a posthumous work of Joseph de Maistre ; he also contrasts Bacon with Comte, to the great disadvantage of the latter. A recent publication by M. Charles de Remusat on Bacon is praised; one by Kuno Fischer is also named at the head of the article, but there is no other allusion to it. The article seems to differ from the author's earlier notice of Bacon in two respects; on the whole the estimate of Bacon's scientific importance seems higher in the *Review* than in the *Philosophy of the Inductive Sciences,* and the practical applications of knowledge are here more fully considered. This matter of practical applications had been rather neglected, we might almost say depreciated, in Dr Whewell's earlier writings, but was made more prominent afterwards : see the *Philosophy of the Inductive Sciences,* Vol. II, page 246, and the *Novum Organon Renovatum,* pages 240...246.

There is probably no more difficult problem in the history of science than a correct estimate of the nature, the merit, and the value of Bacon's writings ; the criticisms and discussions connected with Bacon almost rival in bulk those which are associated with the names of Dante and Shakespeare, and the conclusions which have been deduced exhibit a bewildering discordance of opinion. Dr Whewell himself erected a statue of Bacon in the antechapel of his college; the thanks of the resident fellows were conveyed to him through Professor Sedgwick, and in reply Dr Whewell considered it fortunate that this communication had been made through one who knew and felt so well the purpose and the value of what Bacon did. But the venerable Professor seems never to have clearly revealed what he knew and felt so well. Dr Whewell himself discountenances the suggestion that Bacon's method consists in the accumulation of facts by ordinary observers, from which the elements of theoretical knowledge which they involve are afterwards to be extracted by intellects of a higher order. Some have fancied however that they do find in the functions of the *British Association* a near approach to the realisation of the Baconian scheme; they have regarded this scientific company as yielding a literal accomplishment of Bacon's favourite text: "Multi pertransibunt et augebitur Scientia." I see from an article in the *Revue des Deux Mondes* for March, 1875, that M. Charles de Remusat has again discussed the old problem of the special merits of Bacon.

Dr Whewell states in his review the points in which Bacon "has divined in a remarkable manner the characters of the true progress of science." The most important is the recommendation of what Bacon called the *Interpretation of Nature* in opposition to what he called the *Anticipation of Nature*. It is illustrated in two ways. "Then only may we have hope of the sciences, when we ascend by a true ladder..." "The sciences are like pyramids, experiment and history being their indispensable basis." But these images of a ladder and of a pyramid seem far less appropriate to the course of science, than that of a stream gradually increasing from an insignificant source till it becomes an ample flood. Stripped of metaphor Bacon's essential precept amounts to an injunction to abstain from guesses and theories— a fatal discouragement to the efforts of the highest genius if it could be really enforced. But fortunately the great discoverers in science have never submitted to the fetters of even Baconian rules. Dr Whewell exemplifies his commendation of Bacon by considering the sciences of astronomy and magnetism. It may be said however that the triumphs of astronomy were gained quite independently of Bacon; his name indeed seems never to have occurred to Newton. And magnetism consists of a mass of facts and observations which as yet have led to nothing like an *Interpretation of Nature;* the only theory that exists consists of some exquisite mathematical investigations which Bacon would have scorned in ignorance.

The works of Bacon consist altogether of such a large mass, that it may be well to remember that so far as science and the history of science are concerned we may restrict ourselves to the *Novum Organum.* The *Essays,* famous as they are, have no importance in this respect; as Bacon himself says of the clipt juniper trees of his time, "they be for children." The fascination of the *Novum Organum* is irresistible ; a student may not unfrequently think that he will explore some other parts of the author's writings, and yet soon find himself drawn back to that which seems so incomparably the greatest of them all. The latest and best of Bacon's editors expresses the opinion that few persons will care for Baconian Latin when they might read Baconian English. I am aware that I shall appear presumptuous in confessing to be among the *few ;* to me the English of Bacon presents no peculiar attractions compared with that of his contemporaries, while the Latin seems to be especially adjusted to the striking thoughts which it preserves. It must be admitted, however, that Hobbes, who is said to have assisted Bacon in the composition of some of his works, enjoys a just reputation for the vigour of his English style, while his Latin writings are not conspicuously eminent. On the whole Bacon may be likened to an unknown divinity, and no interpreter has yet revealed to us the true character of the object of our ignorant worship.

I extract a passage from Dr Whewell's article which applies in some measure to his own manuscripts : "There is a question of prudence and propriety which must often be very perplexing to editors who have in their hands a great number of successive drafts of an author's work, or of its parts. How far are they bound by deference to the author so as to acquiesce in what he thought and said after all his fluctuations of judgment ? How far are they to indulge the curiosity of the reader, by shewing him the trials which the master made before he succeeded to his own satisfaction ? Perhaps we may venture to say that in such cases the completed work more truly represents the author's mind ; whilst at the same time it is difficult for his ordinary admirers not to wish to catch him in the unaccomplished effort. The real reverence for the teacher is shewn by listening to him when he begins to teach, not by overhearing his soliloquies. In some kinds of composition, poetry for instance, nothing can be more adverse to a pure poetical enjoyment of a fine passage, than to compare a rough draft, full of false rhymes, flat lines, superfluous epithets and broken constructions, with the 'full majestic march and energy divine,' which the same passage finally acquires, and which it then seems to have had from the moment that it sprang from the poet's mind. We know manuscript drafts of portions in Milton, which it seems almost profanation to read, or when they have been read, to recollect."

In July, 1858, Dr Whewell married Lady Affleck, widow of Sir Gilbert Affleck and sister of Robert Leslie Ellis of Trinity College. He travelled in Scotland during the summer, and spent a few days with Professor J. D. Forbes at Pitlochry in Perthshire; he attended the meeting of the British Association at Leeds in September. He was present at the ceremony connected with the completion of a statue to Newton at Grantham, on which occasion Lord Brougham presided and delivered an address. I pass to the publications of the year, namely an address at Leeds, a translation of Bürger's *Lenore, The History of Scientific Ideas, The Novum Organon Renovatum,* and two academical pamphlets.

At the meeting of the British Association at Leeds in September, 1858, Dr Whewell was President of the Section of Mathematics and Physics. His introductory Address to the Section is given on pages 1 and 2 of the Transactions of the Sections in the *Report* for the year. The Address alludes to the Report by Professor Powell on Luminous Meteors, and to the Report by Mr Cayley on the Progress of Theoretical Dynamics. It concludes thus : "And now, having explained that we must often be necessarily difficult to follow in this Section, I must ask the ladies and gentlemen here present, as the *Spectator* asks his readers, to believe that, if at any time we are very dull, we have a design in it."

In 1858 a little book appeared, entitled *Two Verse Translations of Bürger's Lenore;* it was published at Cambridge by Macmillan and Co. It consists of 39 pages. In the preface the first paragraph of the preface to the *Verse Translations...* of 1847 is repeated. The original German is given with two translations; that on the left-hand pages is the same as in the *Verse Translations,* and is therefore by Dr Whewell; I do not know

the author of the other translation. Various English translations
of the piece had appeared before Dr Whewell's; and since then,
Sir John Herschel has published another; see his *Essays*..., 1857,
pages 712...731. If I may venture to offer an opinion on the
subject, I should say that the original scarcely deserves the great
attention which it has received. The ghastly catastrophe rather
shocks than instructs; and though the moral which the writer
intends to draw is the wickedness of rebelling against Providence,
yet it seems more natural to grieve over the permitted existence
of such evils as dreadful battles than to blame the unhappy girl
for her want of resignation to their results. It is difficult to
understand the fascination which has made so many eminent men
linger over this depressing ballad, to which Dr Johnson's fine
remark is especially applicable: "Pleasure and terror are indeed
the genuine sources of poetry; but poetical pleasure must be such
as human imagination can at least conceive, and poetical terror
such as human strength and fortitude may combat." An article in
Fraser's Magazine for May, 1858, entitled *Bürger and his Translators*, treats of eight published translations and one in manuscript; Dr Whewell's is included among the eight.

    In the year 1858 Dr Whewell issued his *History of Scientific
Ideas* in two volumes, and his *Novum Organon Renovatum* in one
volume. The volumes are in small octavo, to range with the
third edition of the *History of the Inductive Sciences*. The *History
of Scientific Ideas* is substantially a reproduction of the first
volume of the second edition of the *Philosophy of the Inductive
Sciences;* there are a few additions which are indicated by being
enclosed within square brackets: also each of the ten Books
of which the work consists is now furnished with a passage which
may serve as a motto. A discussion of the question whether
Cause and Effect are successive or simultaneous, which was one
of the Essays at the end of the second volume of the *Philosophy*,
is transferred to these volumes. The *Novum Organon Renovatum*
is a reproduction of part of the second volume of the second
edition of the *Philosophy of the Inductive Sciences*, namely of
Books XI and XIII and the *Aphorisms*. Bacon's well-known work
is called *Novum Organum:* it is not obvious why Dr Whewell

preferred the Greek form *Organon*. The title-page has the words *with large additions;* these are principally given in two parts of the book. Pages 237...246 are almost all new; they relate to the application of inductive truths to the explanation of natural phenomena, and to the improvement of the processes of the Arts. Pages 346...370 are entirely new; they are entitled Further Illustrations of the Aphorisms on Scientific Language, from the recent course of Sciences: the illustrations are derived from Botany and Comparative Anatomy. The remainder of the original *Philosophy of the Inductive Sciences* not comprised in these two works of 1858 was reproduced in the *Philosophy of Discovery,* published in 1860.

A valuable suggestion occurs in a letter to Dr Whewell from the late Sir G. C. Lewis, dated Nov. 3, 1859, which may find an appropriate place here. " Will you allow me to suggest that your third part on the Philosophy of Induction would be much enriched by a full exposition of the theory of *Experiment.* When I wrote my work on Political Method, I was unable to find, either in your writings or elsewhere, a satisfactory explanation of the nature and limits of experiment. It is commonly assumed that all the physical sciences are experimental, whereas it is only a few of them which admit of experiment. The majority of them—as Astronomy, Geology, Natural History, Physiology— are purely sciences of observation. Even the Political and Moral Sciences admit of experiment to a greater extent than many, if not most, of the physical sciences."

Suggestions respectfully offered to the Cambridge University Commissioners. This is an octavo pamphlet of 13 pages; it is marked *For the Commissioners only.* Dr Whewell states his objections to various principles which the Commissioners recommended ; such as the opening of the Fellowships to general competition, the limitation of the tenure to a fixed number of years, and the contributions of the Colleges to the funds of the University.

An octavo pamphlet of 11 pages; is marked *For the Fellows only.* It invites the Fellows of Trinity College to record their dissent from certain propositions of the University Parliamentary Commissioners. One of these propositions was, that clerical

Fellows like others, were to vacate their fellowships at the end of ten years after the standing of M.A.; but to retain for ten years more the right of succession to a college living. Dr Whewell says: "To any such Rule as this, there is an obvious objection in its extreme cruelty to the clerical Fellows. A clerical Fellow is put out of his Fellowship at ten years' standing from M.A.: and is allowed to take a College living till twenty years' standing: but what is he to live on in the intervening ten years? His profession we know, will not necessarily support him..." But it is difficult to see how the clerical Fellow is worse off than the lay Fellow. In the case of the lay Fellow the question is not how he is to live for *ten years*, but how he is to live *permanently;* his profession we know will not necessarily support him. Moreover, there is one great advantage in the clerical profession, that it requires no long and costly training like that of a barrister or a physician; and the numerous valuable educational appointments which are practically if not necessarily limited to clergymen, give them opportunities which are denied to laymen.

In the year 1859, Robert Leslie Ellis, brother of Dr Whewell's second wife, died after a long and painful illness. There has probably been no member of the University in recent times to whom could be more justly applied the words ascribed to Newton with respect to the early death of Cotes—if he had lived we should have known something. He had consulted Dr Whewell with respect to his very acute paper on the Theory of Probability, and on the other hand he contributed valuable aid in revising the Chapters on Roman Law in the *Elements of Morality.* In the course of the year Dr Whewell delivered a lecture on Plato at Leeds, which appears to have been printed, though I have not seen a copy of it; probably it was embodied in the *Platonic Dialogues,* the first volume of which appeared in 1859. The publications of the year consist of the volume just named, which will be noticed in connexion with the works on Moral Science, an account of Barrow and his Academical Times, and a prefatory notice to the Literary Remains of Mr Jones.

The Rev. Alexander Napier, of Trinity College, edited the works of Barrow for the Cambridge University Press. The editor

W.   15

was naturally desirous of the opportunity of connecting the names of two illustrious Masters of the College, and at his earnest request Dr Whewell contributed to the work a paper entitled *Barrow and his Academical Times as illustrated in his Latin Works*. This paper is published in Vol. IX of the edition, 1859; it occupies pages I...LV. The illustrations are drawn chiefly from a volume of Barrow's Latin writings, published by his father after his death. The paper is very interesting in connexion with Barrow and with the history of the University during his time. The passages which are translated give striking evidence of Barrow's copiousness, not to say redundancy—and this although Dr Whewell says that he very rudely reduces the dimensions of Barrow's rhetorical sentences. Dr Whewell adduces evidence on a point which always interested him, namely, that the Cartesian system of vortices never made converts at Cambridge: see page VI. Dr Whewell adopts a much-quoted saying, though he does not fall into the popular error of attributing it to Buffon: "In Barrow we have, so far as we can judge, a striking exemplification of the saying, that the written style is the living man:" see page XLVIII.

Barrow while holding a fellowship was allowed by Trinity College to travel, and in his letters to the Master and Fellows he strongly commends the companions of his academical life. Dr Whewell draws attention to these praises on his page XIV; he has, however, printed on his page LIV the formal permission given to Barrow for travelling, from which it would seem that he was officially bound to render some such tribute for the indulgence received: "provided also that the said I. B. during the term of his absence shall every year write letters to the Master and Senior Fellows, in which he shall acknowledge the benefit..."

Dr Whewell shews that Barrow quoted from Bacon, and to this Professor De Morgan alludes in a letter dated Oct. 10, 1859: "Many thanks for the Bacon which you found in the Barrow. It all amounts to wondrous little, if, as you say, Bacon was known to the Cambridge men generally. How could Bacon be so little quoted. The conceits of which that age was fond were taken out of puerility by him—and made into wit and covered with taste.

And yet they know nothing of him—to speak of. Newton's silence is emphatic. When I have time and opportunity I intend to work out the thesis—that Newton was more indebted to the schoolmen than to Bacon, and probably better acquainted with them."

The Rev. Richard Jones was one of the earliest friends of Dr Whewell, and perhaps on the whole the most intimate; he was Professor of Political Economy in the East India College, Hailey-bury, but best known first as one of the Tithe Commissioners, and afterwards as one of the Charity Commissioners. He died in 1855, and his widow entrusted his papers to Dr Whewell with the view of publishing such as seemed suitable. After some time Mr Cazenove, a warm friend and admirer of Mr Jones, wrote to Mrs Jones to enquire about the state of the proposed publication. Mrs Jones sent the letter to Dr Whewell, who invited Mr Cazenove to come to Trinity Lodge and assist in a joint investigation of the papers. Accordingly a volume was published in 1859, entitled, *Literary Remains, consisting of Lectures and Tracts on Political Economy by the late Rev. Richard Jones... Edited with a Prefatory Notice by the Rev. William Whewell, D.D....* The volume is in octavo, containing xl and 620 pages; Dr Whewell's prefatory notice occupies thirty-two pages. The notice is very interesting; it is remarkable, however, to see how little trace there is in it of the long and stedfast connexion which subsisted between the writer and the subject of the memoir. There is no reference to the zealous interest with which Jones's work on *Rent*—the only important publication of his life-time—was almost forced into existence by Dr Whewell himself, who extracted the manuscript piecemeal from the procrastinating author, procured assistance from the University Press towards the expenses of publication, and corrected the proofs. For a long time too Dr Whewell persisted in urging Mr Jones to continue his labours by bringing out his promised work on *Wages;* but this was in vain. At the end of his notice Dr Whewell declares that he shall rejoice if Mr Jones's friends derive any satisfaction from what he has said, and the phrase might almost suggest that he had been speaking about a comparative stranger. In the notice Dr Whewell uses the old talismanic formulæ about induction and deduction; and Mr Jones is highly commended as an inductive philosopher. On page xiii more praise is assigned to Mr Ricardo than Dr Whewell was accustomed to allow to him either in print or in his letters. On page xvii a note is aimed at J. S. Mill without naming him; he is charged with disparaging Mr Jones, while he adopted the classification and the facts of the work on *Rent*. The University friends of Jones are named on page xxi, and the list is interesting, because they were doubtless for the most part friends also of Dr Whewell himself. At the end of the notice Dr Whewell mentions Mr Cazenove, and says that to him properly belongs any credit which the editor of the volume may deserve. Mr Cazenove still lives and retains his esteem for Mr Jones and for his fellow labourer in the publication of the *Remains*.

On Jan. 29, 1860, Dr Whewell preached in St Paul's Cathedral at the request of the Bishop of London, now the Archbishop of Canterbury; the sermon will be noticed hereafter. He was present at the meeting of the British Association at Oxford in June; and afterwards went to Spain to observe the total eclipse of the Sun on July 18. The following are the publications of the year: Platonic Dialogues, Vol. II, Translation of a passage of Homer into English Hexameters, a paper relative to St Mary's Church at Cambridge, a paper submitted to the Board of Visitors of Greenwich Observatory, an edition of the Mathematical Works of Barrow, and the Philosophy of Discovery; the first two of these will be noticed hereafter, and I now proceed to the others.

On the proposed Alterations in Great St Mary's Church; this is on three octavo pages, dated May 9, 1860. A plan was at that time under discussion, and afterwards carried into effect, for removing the Eastern Gallery and making other great alterations in the internal arrangements of the Church used by the University; Dr Whewell thought the alterations undesirable, and issued this paper in reply to one which advocated them.

Dr Whewell printed a paper on four octavo pages, which is headed *Private. For the Board of Visitors of Greenwich Observatory;* it belongs, I think, to the year 1860. The paper supports the merits of Hansen's Researches in the Lunar Theory and Tables against some objections which had been urged: there is a discussion of the same subject in Sir J. W. Lubbock's *Theory of the Moon*, Part x, 1860. It may be fairly presumed that Dr Whewell's opinions were formed in reliance on the authority of some friends whom he consulted rather than on his own investigations. One sentence runs thus: "Sir J. Lubbock expresses his opinion, in which he conceives MM. Plana and Pontécoulant to concur with him, that Mr Hansen's is rather a retrograde than a progressive step in the Lunar Theory." I quote these words, which Dr Whewell conceives to express an erroneous opinion, in order to support them by the judgement of the present Astronomer Royal, given in the *Monthly Notices of the Royal Astronomical Society*, Jan. 1874. While great praise is awarded to Hansen for his discoveries and his Tables, the general form of his theory is characterised as "a retrograde step."

I have next to notice *The Mathematical Works of Isaac Barrow, D.D., Master of Trinity College, Cambridge. Edited for Trinity College, by W. Whewell, D.D., Master of the College*, 1860. This is an octavo volume of about 800 pages, besides numerous plates of diagrams. Dr Whewell contributed a preface of ten pages. In the main the work consists of three sets of lectures, the Lectiones Mathematicæ, the Lectiones Opticæ, and the Lectiones Geometricæ; in the case of the first Dr Whewell annexes "at the foot of the page, a brief summary of their contents, which may enable the English reader to trace the general course of the argument." The first paragraph of the preface explains the occasion of the publication. "It was thought right, by Barrow's College, that a new edition of his Mathematical Works should accompany the edition of his Theological Works, lately published by the University, under the editorial care of Mr Napier; and I have willingly undertaken to superintend the printing of the edition thus agreed upon. I have already, in the Preface to Vol. ix of Mr Napier's edition, given an account of Barrow's mathematical writings; but for the sake of convenience I will here resume the subject."

The *Philosophy of Discovery* appeared in 1860 in one volume small octavo, ranging with the last edition of the *History of the Inductive Sciences*. The text consists of a republication of Book XII of the *Philosophy of the Inductive Sciences* with additional matter nearly equal in bulk; the Table of Contents prefixed to the volume indicates the additional matter, and points out the portions of it which had already appeared in print. Besides various shorter passages the pages 238...399 consist of Chapters not contained in the original *Philosophy;* they treat of the following subjects: Mr Mill's Logic, Political Economy as an Inductive Science, Modern German Philosophy, The Fundamental Antithesis as it exists in the Moral World, Of the Philosophy of the Infinite, Sir William Hamilton on Inertia and Weight, Influence of German Systems of Philosophy in Britain, Necessary Truth is progressive, The Theological bearing of the Philosophy of Discovery, Man's Knowledge of God, Analogies of Physical and Religious Philosophy. An Appendix is formed of ten Essays which had all originally been published in the Memoirs of the Cambridge Philosophical Society, except the Remarks on Sir J. Herschel's Review of the *History* and the *Philosophy.* There were six Essays at the end of the second edition of the *Philosophy of the Inductive Sciences;* one as we have seen appears in the *History of Scientific Ideas*, three are among the ten of the *Philosophy of Discovery;* and the other two are omitted, namely one *On the Nature of the Truth of the Laws of Motion*, and one *On Mathematical Reasoning and on the Logic of Induction:* the last however remained in print in the *Mechanical Euclid.*

It seems to me a matter of regret that the original *Philosophy of the Inductive Sciences* was broken up into the three works which appeared in 1858 and 1860; it becomes somewhat difficult to discover where the *Philosophy* really resides. The *whole work* as published in 1840 and republished in 1847 bears the title of the *Philosophy of the Inductive Sciences;* on the other hand in the Introduction to the *History of Scientific Ideas*, page 17, we are told that the Philosophy of the Inductive Sciences " contains these two parts, *The History of Scientific Ideas*, and the *Novum Organon Renovatum.*" Thus Book XII of the original work seems now not to be considered as part of such Philosophy. In the *Novum Organon Renovatum*, page 142, Dr Whewell says, "I would present to the reader the Philosophy, and if possible, the Art of Discovery:" thus according to this statement the *Philosophy of Discovery* is the subject of the latter part of the *Novum Organon Renovatum*. But the volume which is now called the *Philosophy of Discovery* is something different from this *Novum Organon Renovatum*, and only half of it appeared in the original *Philosophy of the Inductive Sciences*: moreover in the preface to the volume of 1860 we are told respecting the original work "the Philosophy at which I aimed was not the Philosophy of Induction, but the *Philosophy of Discovery*." It seems difficult to harmonise these contradictions into a distinct and consistent statement of the design of the original work and of the separate pieces into which it was finally decomposed. In particular the relation between the *Novum Organon Renovatum* and the *Philosophy of Discovery* is hard to define.

But setting aside the uncertainty as to the precise place which it occupies in Dr Whewell's system, the *Philosophy of Discovery* may be pronounced one of the most interesting of his works: much of it is critical, and as a critic his ability is conspicuous. The Chapter which reviews Mill's Logic is especially deserving of study; it is always striking even when not convincing. On some of the points discussed I venture however to side with Mr Mill; thus I agree with him in considering Bacon wrong in enunciating "as a universal rule that induction should proceed from the lowest to the middle principles, and from these to the highest, never reversing that order...." And I go far with Mr Mill in his estimate of the relative importance of Deduction and Induction, which Dr Whewell condemns in his page 282...284; many persons perhaps will be content to be classed, with Mr Mill, among the "common minds" of Dr Whewell's page 267. On page 243 Dr Whewell speaks of the different ways in which he and Mr Mill understand the term *Induction:* it would have been more convenient if each of the writers had invented a word for himself, and used it in his own sense, instead of disputing as to the proper application of a word involving so much controversy.

The discussion on Comte which this volume contains in common with the *Philosophy of the Inductive Sciences* may be supplemented by the Article on the subject which Dr Whewell published in *Macmillan's Magazine* in 1866; and the discussion on Plato by the three volumes entitled *Platonic Dialogues for English Readers.*

In the last three Chapters of the *Philosophy of Discovery,* the titles of which I have given above, Dr Whewell brings his doctrines to bear on religious philosophy. He says on his page 354, "I have hitherto abstained in a great measure from discussing religious doctrines; but such a reserve carried too far must deprive our philosophy of all completeness. No philosophy of science can be complete which is not also a philosophy of the universe; and no philosophy of the universe can satisfy thoughtful men, which does not include a reference to the power by which the universe came to be what it is." He then proceeds to remarks on the Divine Mind similar to those which are briefly suggested in Chapter XII, Section 5 of the *Essay* on the Plurality of Worlds. In the course of these remarks there seem indications of the necessity which the author felt of limiting the range of his favourite doctrine of Fundamental Ideas more closely than he had formerly supposed. Thus after mentioning the sciences of Geometry, Arithmetic, Chemistry, Classification and Physiology as involving such Ideas, he says on page 372: "To these we might have added a few others; as the sciences which deal with Light, Heat, Polarities; Geology and the other Palætiological Sciences; and there our enumeration at present must stop. For we can hardly as yet claim to have Sciences, in the rigorous sense in which we use the term, about the Vital Powers of man, his Mental Powers, his historical attributes, as Language, Society, Arts, Law, and the like. On these subjects few philosophers will pretend to exhibit to us Ideas of universal validity, prevailing through all the range of observation." Again, on page 386, "Only a few sciences have made much progress; none are complete; most have advanced only a step or two. In none have we reduced all the Facts to Ideas. In all or almost all the unreduced Facts are far more numerous and extensive than those which have been reduced. The general mass of the facts of the universe are mere facts, unsubdued to the rule of science."

The strictly scientific writings of Dr Whewell end with the *Philosophy of Discovery*, and in no unworthy manner; the last paragraphs draw an argument for the Immortality of the Soul from the intellectual and moral progress of which it is susceptible. A few sentences may be extracted though their effect is diminished by separation from the context. " And when we contemplate a human character which has, through a long course of years, and through many trials and conflicts, made a large progress in this career of melioration, and is still capable, if time be given, of further progress towards moral perfection, is it not reasonable to suppose that He who formed man capable of such progress, and who, as we must needs believe, looks with approval on such progress where made, will not allow this progress to stop when it has gone on to the end of man's short earthly life ? Is it not rather reasonable to suppose that the pure and elevated and all-embracing affection, extinguishing all vices and including all virtues, to which the good man thus tends, shall continue to prevail in him as a permanent and ever-during condition, in a life after this ?" Similar remarks occur in a very early manuscript written by Dr Whewell, suggested by the last sentence in the fourth paragraph of Sir J. Herschel's *Discourse on Natural Philosophy*.

In 1861, Dr Whewell travelled in Switzerland in the summer. He published the third volume of his *Platonic Dialogues* in this year.

In 1862 he was present at the meeting of the British Association at Cambridge. He was very strongly in favour of the North in the civil war which broke out in the American States; his old antagonist Mr Mill was delighted to find that they agreed in opinion on this important question. It appears from a letter addressed to him by his friend Mr Everett, formerly the United States ambassador to England, that Dr Whewell thought the arrest of Messrs Mason and Slidell warranted by the doctrines of belligerent rights and the duties of neutrals as hitherto understood and enforced by Great Britain. The writings of the

year consist of two articles relating to English Hexameters to
be noticed hereafter, a note on Bacon, and Lectures on Political
Economy.

A note from Dr Whewell to W. Hepworth Dixon occupies
about half a column of page 661 of the *Athenæum* for May 17,
1862. The first two sentences explain the purport of the note.
"I have again to thank you for your kindness in sending me
your 'Story of Lord Bacon's Life. I have read it through, and
cannot understand how it can fail to convince readers of the
absurd injustice of the representations of Bacon's character and
history given by Macaulay and Campbell." The note is followed
by a letter of three columns addressed to Mrs Anne Sadler a
daughter of Lord Coke, which Dr Whewell caused to be transcribed from the original in the Library of Trinity College.

*Six Lectures on Political Economy delivered at Cambridge
in Michaelmas Term,* 1861...*Cambridge: printed at the University
Press,* 1862. This is an octavo volume; it consists of a Half Title,
the Title Preface and Contents on viii pages, and the text on
102 pages. The volume was not published, but Dr Whewell
distributed copies among his friends. The lectures were delivered to
the Prince of Wales and a few other students; the preface well
deserves to be reproduced here: "The following Lectures were
delivered at the request of one of the wisest and best fathers
who have ever lived, for the instruction of a son on whose
education he bestowed much careful thought: and indeed that
education was a matter of national as well as of family concern.
He kindly judged that I could deliver a short course of Lectures
on Political Economy which might forward his purpose: and
I willingly undertook the task, rendered acceptable by the prospect
of submitting to him afterwards the purport of what I had said
in the Lectures; and of having the gratification and the advantage of hearing his remarks, instructive and interesting as they
always were. This satisfaction I was not permitted to enjoy.
The Allwise Disposer did not allow the father to see with his
earthly eyes the completion of his well-devised plans for his
son's education: though I doubt not that those plans will bear
their fruit in national blessings.

"The scheme of the following Lectures will, I hope, carry with it its own excuse. It seemed to me unwise on such an occasion to aim at any originality beyond that which selects the best passages of writers of acknowledged authority and weighs them against one another. And though the extracts given in the earlier of the following Lectures may by some be thought lightly of, as being common-places, they are common-places which young men of rank, such as those to whom these Lectures were addressed, ought to know, and which they were not likely to learn unless they were brought before them in some such way as this. The later extracts and the reflexions which I have added to them will be found, I think, to contain views of which the importance is now only beginning to be duly felt."

The plan of selecting what Dr Whewell calls *standard and classical passages* from the best writers on Political Economy has the merit of leading to a very interesting volume; but considering the elementary purpose which the Lectures were designed to promote, seems liable to one objection. Writers on Political Economy do not perfectly agree as to the meaning of some of the important terms of the science; now if we study the whole of any author attentively we shall be able to determine precisely the sense he attaches to his technical language: but a beginner who has before him only short extracts from various works will scarcely obtain evidence sufficient for certainty as to the exact force of propositions which discordant theorists enunciate, each in his peculiar dialect.

The part of the work in which the author expresses most decisively his own opinions is where he treats of the cause which has produced the rise of rents in England in modern times; he undertakes to prove that this has *not* resulted from the cause which Mr Ricardo assigned, namely, the rise of prices in consequence of the increased difficulty of production. For if such were the cause the *rent would necessarily become a larger fraction of the produce*—which is contrary to observed facts. He establishes the proposition just enunciated by the aid of numerical examples, but the process is rather strange.

Rent is defined as the produce of the land diminished by the cost of cultivation. Suppose that the cost of cultivation is £10 per acre; and that we have land A of the best quality of which the produce is worth £12 an acre, land B of the next quality of which the produce is worth £10 an acre, and land C of the worst quality of which the produce is worth £8 an acre. In this case the land C would remain uncultivated, since the expense would be greater than the produce. The whole produce would be $12 + 10$ that is 22; the whole cost would be $10 + 10$ that is 20. Hence the rent is 2, and it is $\frac{2}{22}$, that is $\frac{1}{11}$, of the whole produce.

Now let there be a rise in the value of produce, say to the extent of one-fourth of the original price. The produce of A becomes $12 + \frac{12}{4}$, that is 15; the produce of B becomes $10 + \frac{10}{4}$, that is $12\frac{1}{2}$; the produce of C becomes $8 + \frac{8}{4}$, that is 10. In this case C would come into cultivation, though it would yield no rent. The whole produce would now be $15 + 12\frac{1}{2} + 10$, that is $37\frac{1}{2}$; the whole cost would be $10 + 10 + 10$ that is 30. Hence the rent is $7\frac{1}{2}$, and it is $\frac{7\frac{1}{2}}{37\frac{1}{2}}$, that is $\frac{1}{5}$ of the whole produce.

Then Dr Whewell says:

"The numbers in this case result from the supposition that the quantities of each of the kinds of land A, B, C, are equal.

"But the general result, that the rent is a greater portion of the produce at the latter stage than at the former, does not depend on any supposition as to quantities. It will be the same, whatever be the quantities of different kinds of land.

"For each kind of land will have the rent a greater fraction at the latter stage than at the former; and therefore all the fractions at the latter stage, multiplied by their quantities, and added together, must be greater than at the former stage."

It is with respect to the last paragraph that I wish to make a remark. The *result* which Dr Whewell states will be obvious in a moment to a person who is acquainted with the elements of Algebra, whatever be the quantity of each kind of land, and whether the expense of cultivation is the same or not for different kinds of land, assuming only that it is constant for each kind: but the *method* which Dr Whewell suggests for obtaining the result is unsatisfactory. In the first place, if it is obvious to any person that for each kind of land the rent is a greater fraction at the latter stage than the former, then to such a person the whole theorem is obvious; whatever consideration satisfies him with respect to each kind of land will be found to apply to the aggregate. In the second place, the argument which Dr Whewell offers in the latter clause of the paragraph is useless, for it will not warrant the conclusion that the *whole rent is a greater fraction of the whole produce* in the second case than in the first.

Dr Whewell speaks very highly of the work on *Rent* published by Mr Jones; and on the last page seems to anticipate with confident satisfaction the ultimate disappearance of the various kinds of Rent described in that work except that called *Farmers' Rents*. I do not know what are the *reflexions* to which allusion is made at the close of the preface, except it be to what is contained in the last page of the text. By mistake Dr Whewell gives the name *Pons Asinorum* to the *fourth* Proposition of Euclid's Elements: see his page 61.

There are some interesting letters preserved among Dr Whewell's papers from friends to whom he sent copies of his *Lectures*. From one of them I extract the following passage: "During the late French war Rents were very high in England; partly, no doubt, arising from the depreciation of the currency, and partly from the fact that we did not grow enough for our own consumption, and the difficulty of importation. But Ricardo was so imbued with the notion that the main cause of the rise was owing to the necessity, he supposed we were under, of cultivating inferior soils, that he sold out a large sum of government stock and invested the amount in land. If he had waited a few years longer, he would have sold his stock 30 per cent. higher, and bought his land 30 per cent. cheaper, which he afterwards found to his cost." A brilliant but paradoxical writer, well-known for his works connected with art, gave to it probably the highest praise he could in saying, "Like all other books I ever opened, from Adam Smith

downwards, written by clever men on this subject, it fills me
with wonder... You know (I suppose by your sending me the
book) that I am entirely opposed to all the modern views on this
subject." In *Fraser's Magazine* for April 1863 there is an article
on Political Economy, and a note refers to the opinions on *Interest*
collected by Dr Whewell on page 41 of his *Lectures;* the writer
says, " ...it never seeming to occur to the mind of the compiler,
any more than to the writers whom he quotes, that it is quite
possible, and even (according to Jewish proverb) prudent, for men
to hoard, as ants and mice, for use, not usury, and lay by something for winter nights, in the expectation of rather sharing than
lending the scrapings."

In 1863 Mr J. Evelyn Denison requested Dr Whewell to contribute to a work which he had suggested, and which is called
after him the *Speaker's Commentary on the Bible.* Mr Denison
wished to have a "statement of the present relations between
science and revelation;" but I presume Dr Whewell did not
accede to the request.

A paper entitled *Of certain Analogies between Architecture
and the other Fine Arts*, by Dr Whewell, was read at the ordinary
General Meeting of the Royal Institute of British Architects, on
March 9th, 1863, and occupies pages 175...181 of the volume
containing the published Papers of the Institute of that date.
The most curious part of the paper is the analogy pointed out
between Architecture and Music. I quote a few words: "I do
not know whether this comparison of exterior architecture to
melody, and of interior architecture to harmony, will seem to my
hearers too fanciful to be tolerated. To me it seems to be something very real; and I think I may venture to say, that if any
lover of architecture will, in looking at edifices, give up his mind
to it, he will find that it may serve to bind together many floating
thoughts that, however vague and fugitive, are not without their
value as elements of the enjoyments of art."

A Blue-Book was issued in 1864, entitled *Report of Her
Majesty's Commissioners appointed to inquire into the Revenues
and Management of certain Colleges and Schools...* Vol. II. This
contains on page 43 a letter from Dr Whewell on School Education.

He gives a summary of the opinions he had expressed on some leading points of the subject in his work *Of a Liberal Education*. The letter concludes thus: "I had also at an earlier period (I believe in 1837) written a little book, 'On the principles of English University Education;' but this treats the subject in a more general manner, and I do not mention it as at present worthy of the notice of the Commissioners. I send herewith a copy of the book to which I have referred." Large extracts from the *Liberal Education* are printed on pages 43...47 of the Blue-Book.

Three papers relating to University matters belong to the year 1864.

Reasons for voting against the purchase of Colonel Leake's Collection of Coins. This consists of three quarto pages, dated Feb. 22, 1864. It was proposed to buy the coins for £5000 out of the funds belonging to the Fitzwilliam Museum. Dr Whewell thought that the money could be better spent in finishing the Fitzwilliam Building, and that coins were not very suitable objects for the Museum. A reply to the paper was issued by Mr Churchill Babington, now Professor of Archæology in the University, and the coins were purchased and placed in the Museum.

The Professor's Certificate. This consists of four quarto pages, dated Trinity Lodge, March 8, 1864. Its object is to uphold a regulation which had been established for about fourteen years in the University, enforcing the attendance of Candidates for the Ordinary Degree on a course of Professorial Lectures on some part of Moral or Natural Science.

The last of the three papers consists of three quarto pages, addressed by Dr Whewell to the Fellows of his College, urging them to maintain the institution of Fellow-Commoners which it was proposed to abolish.

Dr Whewell's second wife, Lady Affleck, died on April 1, 1865.

On Feb. 24, 1866, Dr Whewell met with an accident while riding on horseback, and died on March 6. The last book which he had read before his accident was Sir E. B. Lytton's *Tales of*

Miletus; on this he made some remarks to the author, who acknowledged himself much gratified by the kind reception given to his little volume.

Two Magazine articles belong to the year 1866, one of them published after Dr Whewell's death.

Comte and Positivism. This article was published in *Macmillan's Magazine*, March, 1866, pages 353...362. The article is founded on two relating to Comte, one by J. S. Mill, first published in the *Westminster Review*, and one by G. H. Lewes in the *Fortnightly Review.* Dr Whewell's article is very interesting, and we may notice especially the high praise which he awards to his old antagonist, then recently elected Member of Parliament for Westminster. Dr Whewell says: "We have especially the great authority of Mr J. S. Mill calling upon us to give again our attention to M. Comte and his philosophy. No authority of our own time can be greater than this. Beside Mr Mill's profound philosophical thought and wide sphere of knowledge, the dignity of his position naturally makes us look where he points. His love of truth and fearlessness of consequences have given him an eminence which all must rejoice to see generally acknowledged. It is no small glory of our times, that one of our most popular constituencies has fully and practically adopted the great Platonic maxim, that it will never go well with the world till our rulers are philosophers, or our philosophers rulers." The unfavourable opinion which Dr Whewell in his *Philosophy of the Inductive Sciences* had expressed of Comte is maintained in this article. It is said that "his pretensions to discoveries are, as Sir John Herschel has shown, absurdly fallacious:" see also the *Philosophy of Discovery*, page 275. The reference to Sir John Herschel seems to apply to a passage in the address which he delivered as President of the British Association in 1845, and reprinted on page 666 of his *Essays.*

Dr Whewell makes some interesting remarks on the followers of Saint Simon: "The Saint-Simonians formed a very striking epoch in French speculation. I think M. Comte's admirers have not done them justice. There are, perhaps, not many Englishmen who now recollect to have read their writings when they were published (about 1820 and after); but those who do must regard them as very striking works. Most readers at that time were deeply impressed by the largeness, subtlety, and ingenuity of their views of society. Their doctrine of the alternation of *critical periods* and *organic periods* was really a startling theory, bringing together in a general view many historical facts." Dr Whewell quotes part of the same passage about the eye as in the last edition of his *Bridgewater Treatise*, and comments on it in a similar manner: see page 71. The comic lines at the end of the article, "Thus did the youth...," occur in a manuscript by the late Mr Drinkwater Bethune, Feb. 7, 1848; I do not know if they are in print elsewhere.

Grote's Plato. This article was published in *Fraser's Magazine* for April, 1866, pages 412...423; it was not quite finished by the author, and was put into a fit state for printing by the Rev. H. R. Luard, the Registrary of the University. A short notice respecting Dr Whewell and his works is prefixed to the article.

In the article Mr Grote is praised "for having rejected an established system of vilifying and misrepresenting Plato's opponents, the Sophists, and ascribing to them in every thing that they say, sophistry in its modern English meaning." On the other hand it is suggested that "Mr Grote has not thoroughly purged himself of an established system of seeing everywhere a profound meaning and a solid philosophy in the Platonic *Dialogues*, or at least steps towards such a philosophy." This is the main point in the article; it is illustrated at some length by an examination of the *Lysis*. To an English reader we may exemplify the difference of opinion by saying it amounts to this; Mr Grote might class the *Lysis* with Berkeley's *Three Dialogues between Hylas and Philonous*, while Dr Whewell might class it with Mrs Barbauld's *Evenings at Home*, or Mrs Marcet's *Conversations*. Dr Whewell also combats Mr Grote's notion that none of the Platonic Dialogues were published before the death of Socrates. The article contains no substantial addition to the matter which we find in the author's *Platonic Dialogues for English Readers*, but is interesting as putting his special opinion in a prominent form.

w.   16

CHAPTER XIV.

In the present Chapter I shall notice the writings of Dr Whewell connected with Moral Philosophy; they are as follows:

Preface to Mackintosh's Dissertation, 1835.

On the Foundations of Morals. Four Sermons, 1837.

Two introductory Lectures to two courses of Lectures, 1841.

Elements of Morality including Polity, 1845.

Lectures on Systematic Morality, 1846.

Butler's Three Sermons on Human Nature, 1848.

Butler's Six Sermons on Moral Subjects, 1849.

Sanderson's Lectures, De Obligatione Conscientiæ, 1851.

Lectures on the History of Moral Philosophy in England, 1852.

Hugonis Grotii de Jure Belli et Pacis Libri Tres, 1853.

Platonic Dialogues for English Readers, 1859...1861.

Sir James Mackintosh contributed to the Seventh edition of the *Encyclopædia Britannica* a *Dissertation exhibiting a general view of the Progress of Ethical Philosophy, chiefly during the seventeenth and eighteenth centuries.* Dr Whewell thought it would be convenient for students to have this Dissertation apart from the Encyclopædia, and he seems in 1835 to have urged the publishers to issue it in a distinct volume, offering to contribute a preface. The publishers accordingly accepted the proposal. Dr Whewell submitted the manuscript of his preface to his friends Mr Jones and Dr Holland; the latter had discussed the original Dissertation with Mackintosh before its publication. See Sir H. Holland's *Recollections,* page 239.

The Dissertation with Dr Whewell's Preface first appeared in Dec. 1835; it has been reprinted at various times. The earliest edition I have seen is dated 1837. In 1852 both appeared in the eighth edition of the *Encyclopædia Britannica;* here a few brief notes are added to the Preface, dated 1852. The publishers in 1861 asked Dr Whewell for any notes he might have to be used in a new edition; I do not know whether he gave any : the last edition which I have seen is dated 1872. I have read the preface in the issue of 1852. I may remark that the text of the Dissertation seems to have been inadequately revised. On page 294 for *addition* we must, I presume, read *edition;* on page 424 the statement that Brougham wrote the *Pursuit of Knowledge under Difficulties* should not be allowed to remain; on page 442 for *proscribed* we must, I presume, read *prescribed.*

The Preface is intended to give an account of Mackintosh's own views and reasonings; they are delivered in various parts of the Dissertation, and Dr Whewell renders good service to students by collecting them, arranging them, and supplying an appropriate commentary. Thus, in the Preface, Mackintosh is contemplated not as a critic and historian, but as an independent contributor to Ethical Philosophy. Notwithstanding these original portions, however, the *Dissertation* is mainly a review of the works of other writers, and it would have been profitable to invite attention to one of its most distinct characteristics. From native kindness of disposition, and from an eclectic habit of mind, Mackintosh was strongly disposed to take the most favourable view of every author and every system that came before his tribunal : it has been well said of him that " he was tolerant even of mediocrity" (Sir H. Holland's *Recollections,* page 239). Hence there is naturally some want of discrimination in the praise which he delights to award; and a reader who wishes to detect his more valuable remarks, must watch carefully for the slight ripples of dissatisfaction which break the stately flow of his panegyric. A simple example will suffice, taken from his account of Butler. It might at first appear that this account is extremely favourable, but a careful reader will see that Mackintosh was much dissatisfied with what is usually held to be the essential element of the Morality of Butler, namely, the so-called *supremacy of Conscience.* Mackintosh says, page 362 : " He makes no attempt to determine in what state of mind the action of conscience consists. He does not venture steadily to denote it by a name. He fluctuates between different appellations, and multiplies the metaphors of authority and command, without a simple exposition of that mental operation which these metaphors should only have illustrated. It commands other principles. But the question recurs, why or how?" His opinion is still more clearly seen in a familiar letter, page 294 : " ...I have endeavoured to slip in a foundation under Butler's doctrine of the supremacy of conscience, which he left baseless." Dr Whewell says in one of the notes to the Preface, dated 1852 : " The use of the term *Supremacy*, as ascribed to the conscience, is liable to convey an erroneous and dangerous doctrine, that every one's conscience is for him the supreme rule of right and wrong. The word *Authority*, which is all that Butler and Mackintosh's argument requires, avoids this inconvenience." This is followed by references to Whewell's edition of *Butler's Three Sermons*, and to Whewell's *Elements of Morality*.

The notice of the Life of Dr Whewell which is contained in the *Monthly Notices of the Royal Astronomical Society*, Vol. XXVII, mentions among his works " his *Metaphysical Introductions* to the Encyclopædias." The Preface to Mackintosh is the only foundation for this absurd phrase.

Mr Hallam in acknowledging the receipt of the book alluded to difficulties which he felt as to the theory maintained in the *Dissertation* and the *Preface ;* the old objections he thought were not removed against the existence of a moral faculty, as a standard of right and wrong, drawn from the great diversity of emotions connected with actions which we find among mankind. Nearly twenty years later, on looking at the book again, he was surprised to find some hasty and incorrect expressions due to Mackintosh which had been left unnoticed. " Is it not strange he should charge all moral philosophers with confounding the theory of moral sentiments with the criterion of virtue, when Adam Smith has so fully distinguished them at great length, by pointing out the several theories with respect to each?" Dr Pye Smith drew the attention of Dr Whewell to Wardlaw's *Christian*

Ethics, saying that the "fourth edition contains valuable additions in reference to these topics and to yourself, Chalmers, Mackintosh."

I have next to notice the publication entitled *On the Foundations of Morals. Four Sermons preached before the University of Cambridge, November* 1837. These sermons were printed very soon after their delivery; a second edition appeared in 1839; this contains, in addition to the original publication, a preface to the second edition, and a Syllabus of the Sermons. The second edition consists of xx and 76 octavo pages. The Sermons were reprinted in America, as appears from a letter written to Dr Whewell by C. S. Henry of New York, April 20, 1839. The Sermons are dedicated to J. C. Hare, Rector of Hurstmonceaux: and an allusion to a well-known work by Hare and his brother is contained in the words: "Yet to you I may say, without any doubt of receiving your assent, that this employment of Guessing at Truth, is both in itself praiseworthy, and if carried on with a humble trust in the Divine Giver of all Truth, is full of deep and wide sources of practical blessing."

The Four Sermons bear the following titles: Conscience is the evidence of God's righteousness; Conscience requires cultivation; Conscience is in accordance with God's government; Moral Good is superior in kind to Pleasure. The last sermon contains some strong censure of two of the principles of Paley's Moral Philosophy. The Preface alludes to this, and might suggest the notion that it is the main drift of the Sermons: but though the first three Sermons in treating on Conscience may implicitly be in opposition to Paley, yet it is only in the fourth that he is explicitly attacked. In the original Preface Dr Whewell alludes to the construction of a detailed system of Ethics on the foundation laid in the Sermons; he says: "It appears to me that such an undertaking is both possible and highly interesting; but, even if I felt myself prepared for such a task, other avocations and objects with which I am already engaged, would probably long prevent my making the attempt." The principal work then on his hands was the *Philosophy of the Inductive Sciences* to succeed the *History* which had recently been issued; subsequently he published a system of Ethics, namely his *Elements of Morality*.

At the end of the original Preface Dr Whewell announces his intention of editing a selection of Butler's Sermons, with a few illustrations. Shortly afterwards he was appointed Professor of Moral Philosophy; and he says at the end of the Preface to the second edition: "And as it has, since that time, become my business to give a more continued attention to the subject of Moral Philosophy, I have laid aside, at east for the present, this temporary project." The intention was, however, executed in 1848 and 1849.

With respect to the Four Sermons, t may be remarked that from their nature they are exhortations rather than discussions; the work of a preacher rather than of a professor: and as such they have high merit, being earnest and persuasive. But the agreement does not seem very close between these addresses and the systematic treatise which Dr Whewell afterwards published: the authority of Conscience is not made so prominent in the treatise as might have been anticipated from the importance which is here assigned to it. In a rhetorical passage towards the end of the first sermon, the preacher gives his sanction to a common, but probably inadmissible sense, of a well-known passage, "When day shall cease to utter forth to day and night to night; when the angel shall swear that there shall be time no longer...."

Mr Hare was much gratified by the dedication of the Sermons, and a few extracts may be given from his letter of thanks, dated Hurstmonceux, Jan. 4, 1838: "The honour itself is one in which I should feel great pride, if pride were not overpowered by stronger and far more delightful feelings. Yet a strange wonder will come over me now and then, that to me in my littleness such honours should have befallen as to have books dedicated to me by Niebuhr and by you." "And it is a very justifiable piece of irony, that you will not allow the man, who pretends to correct the orthography of the nation, to know even how to spell the name of his own parish." Dr Whewell in the second edition of his Four Sermons changed his spelling to *Herstmonceax*; while *Hurstmonceux* is the form which now seems to be adopted. Mr Hare after saying that he agrees with the preacher's main object in heart and mind, alludes to "secondary or tertiary points" in which there might be some difference. In the first text, Romans i. 20, he did not think that St Paul is speaking of God's moral being; he says, "The θειότης is indeed a most puzzling word: but I should be inclined to suppose that the meaning of this one passage is merely that the visible world shews that there is an Almighty Maker and *Ruler*. Besides, I am not sure whether you do not strain the passages from St Paul, in which the word *conscience* occurs, too far at the beginning of your second sermon. Sedgwick, I thought, certainly did. You rest less on them. In several of them at least *conscience* is nothing more than *consciousness*, and compatible with any view as to the origin and nature of our notions of right and wrong."

Two introductory lectures to two courses of lectures on Moral Philosophy, delivered in 1839 *and* 1841. This is an octavo pamphlet; the title and dedication occupy iv pages, and the text 52 pages. Part of the dedication to the Rev. Thomas Worsley, M.A., Master of Downing College, may be with propriety reproduced : "My dear Worsley, I think you already know that, in my opinion, one of the greatest pleasures which the writing of books brings is, that we may make them memorials of our most cherished friendships. The few pages now before you, I have peculiar and paramount reasons for dedicating to you. It is through your kindness that I hold the Professorship in virtue of which these Lectures were delivered; for the encouragement which you gave me when the vacancy occurred, you being, as Vice-Chancellor, one of the electors, principally induced me to offer myself as a candidate for the office. Yet the subject which the Professorship embraces was one which had occupied your own thoughts so much, that the vacancy might readily have suggested other wishes to a person who thought more of himself and less of others..."

The first of the two lectures was delivered April 22, 1839; this is reproduced in Dr Whewell's *Lectures on the History of Moral Philosophy* : it occupies pages 18...39 of the second edition of that work. A paragraph on page 13 of the original, which relates to the proposed series of lectures, is omitted in

the reprint. The second of the two lectures was delivered
Feb. 9, 1841; this does not seem to have been reproduced. It
may be said to consist in tracing an analogy between those
Inductive Sciences, to the History and Philosophy of which the
author had devoted so many years, and the subjects now occupying the sphere of his professorial activity. He admits, however,
that, "To many persons, it will, I doubt not, appear (at least
at first sight) a forced and fantastical notion, that any analogy
should exist between the material sciences and moral doctrines,
in regard to the nature of the truths which they contain, and
the mode of arriving at our knowledge." The lecture is interesting, but rather vague, and occasionally a little too rhetorical.
A few sentences may be extracted: "The whole world is to
the moralist one vast school, in which he sees a body of pupils
engaged in their various exercises. They repeat and dissert,
they dispute and wrangle, they wrestle and strike; while he
looks on, and listens, and determines in what each is right, in
what wrong; how far he has rightly caught the meaning of
his lesson, how far he is still ignorant and dull. It is a noisy
school; for men's emotions and passions, their hearts and their
swords, happiness and ruin, glory and infamy, courts and prisons,
tribunals and armies,—are the weapons with which their disputes
are carried on. Fierce wars and faithful loves, and all the fierce
and all the loving affections of their nature, moralize their
unceasing outcries. And the moralist must often needs look at
the turmoil with sadness, and mourn over the cost of his own
studies; for the weak and innocent is thrust down, and he
cannot interfere to prevent it. Virtue shivers, and he can
praise, but may not warm her..." "But allow me, before dismissing this subject, to dwell for a moment longer upon the
image which I have presented to you. I will do so only for
the sake of supplying what many of you will probably think a
strange omission. The moralist looks upon this strange school of
the world, full of pupils of such varied characters and employ-
ments; but where then is THE MASTER?..."

Mr Hare wrote thus to Dr Whewell on March 18, 1841:
"The first lecture on casuistry is interesting, but is scarcely

complete enough to be substantive, and would have come much better, I think, as an introduction to the course on the history of moral philosophy in England, of which I heard several lectures last year. By the by, there were some good remarks on casuistry, if I remember right by De Quincey, in an article on Duelling, in the February number of Tait. In your second Lecture, as I read it hastily before I left home, there seemed to me several things that I could not fully go along with, and that appeared to me at least to need the being workt out more in detail. However, they both hold out a promise that much will be done for Moral Philosophy in Cambridge, and that we shall take a wiser course than......at Oxford."

We now arrive at the largest contribution made by Dr Whewell to Moral Philosophy. He published in 1845 his *Elements of Morality including Polity* in two octavo volumes: the pages in the first volume are xxix and 374, and in the second xxiii and 401. The second edition of the work appeared in 1848 in two volumes small octavo; the pages in the first volume are liii and 388, and in the second xxi and 286. The original preface is not reproduced, but there is a new preface which defends the general scheme of the book, and points out the changes and the omissions which had been made in the second edition. The third edition appeared in 1854 in two volumes small octavo; the pages in the first volume are xxxv and 414, and in the second xxiv and 338. The third edition is a reproduction almost page for page of the second, with the addition of a new preface and a Supplement. The fourth edition appeared in 1864, in one octavo volume containing xl and 611 pages; this seems to coincide with the third edition, with the addition of half a page at the end. The preface to the fourth edition is almost identical with that to the third. Thus the work assumed in the second edition the form which it permanently retained. It will be sufficient to cite it by the numbers of the *Articles*, for this suits any edition after the first. Dr Whewell regretted that the American booksellers stereotyped his first edition, and so would not adopt the improvements of his later editions. The first edition is dedicated to the poet Wordsworth, and the date affixed is April 14, 1845; the dedication is omitted in the second edition, it is reproduced in the third edition, but with the date unaccountably changed to April 5, 1841; finally it is omitted in the fourth edition. The poet, in a letter dated May 12, 1845, returned his acknowledgment of the compliment paid to him: "I regretted very much missing you when you were so good as to call on me in Dover Street, also that I could not thank you immediately for your two Vols. on the Elements of Morality, and express my sense of the honor I have received by your dedicating them to me, and *that* in terms which all my friends have read with the greatest pleasure; and indeed you must allow me to say that both for delicate warmth of feeling and importance of the points touched upon, I have rarely seen the dedication of a work which was more honourable to both parties."

The work is divided into six Books, with the following Titles: I. Introduction. Elementary Notions and Definitions. II. Morality. Of Virtues and Duties. III. Religion. Of Divine Laws, and their Sanction. IV. Jus. Of rights and Obligations. V. Polity. The Duties of the State. VI. International Jus. Rights and Obligations between States. Speaking generally it may be said that Books I and II contain all that is specially characteristic of the author as a Moral Philosopher. Book III is the work of a commentator on the Bible. Book IV consists of a sketch of the Laws of ancient Rome and modern England. Book V treats much of the English Constitution. Book VI is a brief outline of the relations between nations in War and Peace.

The difference between the first and the second edition consists mainly in the following particulars: The Book which is called IV in the list just given was originally Book II. Dr Whewell says, in the preface to the second edition: "I have made this transposition for this among other reasons, that the parts of the work in their former order might possibly suggest an erroneous view of the grounds of Morality; as if *Jus*, or Positive Law, were the foundation of moral truths, instead of being merely a condition of the application of moral results to actual cases." In the first edition the Book on Religion contained discussions on Forms of Prayer, the Christian Sabbath, and other ordinances. In the second edition these were withdrawn. Dr Whewell says, in the preface to the second edition: "In the present edition I have excluded all that relates to Christian Ordinances; perceiving that the discussion of such matters is not a part of the Moralist's proper province, and involves him in various unnecessary difficulties. I trust that none of my readers will regret the absence of these discussions." There are some omissions from the Book on Polity. Dr Whewell says, in the preface to the second edition: "... I have in the present edition omitted the general statements formerly given respecting the Relation of Church and State, and have given to the few remarks on this subject which I have now introduced, an historical turn."

It will be convenient to give an outline of Dr Whewell's main principles. He enumerates five classes of Springs of Action, namely: the Appetites or Bodily Desires; the Affections; the Mental Desires; the Moral Sentiments; and the Reflex Sentiments Art. 25. The Springs of Action are distinguished by the nature of their objects. The Appetites have for their objects, Things; the Affections, Persons; the Mental Desires have Abstractions; the Moral Sentiments, Actions; and the Reflex Sentiments have for their objects the thoughts of other persons, or our own, about ourselves, Art. 61. The primary and universal Rights of men are five: the Right of Personal Security; the Right of Property; the Right of Contract; Family Rights; and the Rights of Government, Art. 80. There are five Cardinal Virtues, namely, Benevolence, Justice, Truth, Purity and Order. The old list of four Cardinal Virtues, namely, Temperance, Fortitude, Justice, and Wisdom is "too unphilosophical a division to be employed with any advantage in Morality," Art. 128. We are told that, "in a general manner, without pretending to any great precision," the fivefold divisions of Virtue depends on five elements of our nature: Love, Mental Desires, Speech, Bodily Appetites, and Reason, Art. 123. Finally, the Rights and the Virtues are connected; for there is a relation of approximate parallelism between the five classes of Rights and the five Cardinal Virtues:

see Art. 124. These divisions naturally encountered adverse criticism; Dr Whewell defends them in a section of his Supplement, which commences thus: "We enumerate five Desires or Needs, as the leading Springs of human action; and five Primary kinds of Rights. But why, it is asked, do these Rights so exactly correspond to the five Desires? Why, as it has been put in a lively way, are they so exactly face to face with the Springs of Action;—so many Policemen, watching so many Thieves?"

In his original Preface Dr Whewell made some remarks on the analogy which he traced between his subject and Geometry. He said, "Morality, and the Philosophy of Morality, differ in the same manner and in the same degree as Geometry, and the Philosophy of Geometry. Of these two subjects, Geometry consists of a series of positive and definite Propositions, deduced one from another, in succession, by rigorous reasoning, and all resting upon certain Definitions and Self-evident Axioms. The Philosophy of Geometry is quite a different subject; it includes such Inquiries as these:—Whence is the Cogency of Geometrical proof? What is the Evidence of the Axioms and Definitions? What are the Faculties by which we become aware of their truth? and the like." He proceeds then to assign Geometry to the Mathematicians, and the Philosophy of Geometry to the Metaphysicians. He wished the *Elements of Morality* to be considered analogous to Geometry, and held that the construction of such a system should "precede any attempt to settle the disputed and doubtful questions which are regarded as belonging to the Philosophy of Morality." Perhaps Dr Whewell did not keep very closely to the design which he thus indicates; he introduced matter that seems to belong rather to the metaphysical than to the systematical part of the subject. The main principles of his scheme were accepted by few of his contemporaries. Some of the most eminent of his correspondents recorded their hesitation or dissent in letters to the author. His friend and relative, the late Mr Myers, sent him a long criticism on the work; he praised much of it highly, but objected altogether to the analogy with Geometry, and would not allow that Morality could claim higher rank as a science than those which involve merely observation and classification. Dr Whewell's own letters in reply appear to shew that he was not satisfied with the result of his labours.

Morality seems indeed to have but feeble claims to the character of an independent science; it is made up of fragments of Theology, Law, and Political Economy; and thus resembles some modern kingdoms which have risen to importance by the spoliation of their neighbours. In particular, for one who thoroughly accepts the Christian Revelation ethical systems as such are superfluous; the moralist is superseded by the theologian.

A few remarks may be offered on special points of Dr Whewell's work. The most difficult matter to be treated by those who belong to what Dr Whewell calls the independent school of moralists is probably the *conscience*. Much of what he delivers on this head seems very good. He puts very strongly in the preface to the second edition the argument which is urged against the morality of conscience as a system, and says that it "has been of efficacy enough to prevent the morality of conscience from being generally adopted as a *System*." Similar remarks are made by him in his Arts. 262...276, where he treats of conscience; thus he says in Art. 272: "Since Conscience is thus a subordinate and fallible Rule, it appears, that for a man to act according to his conscience, is not necessarily to act rightly." But then on the other hand he asserts in Art. 267, that "He who acts *against his Conscience* is always wrong." So again in Art. 275, "To disobey the commands and prohibitions of Conscience, under any circumstances, is utterly immoral; it is the very essence of immorality." This seems to me extravagant. The foulest crimes in History—the burning of Latimer and Ridley—the massacre of St Bartholomew—the Gunpowder Plot, —may be palliated or even justified by the assertion that the perpetrators followed the command of their consciences, which it would have been the essence of immorality to disobey.

The Chapter on Cases of Necessity is very interesting; but we must remember that they are defined to be "those in which a man is impelled to violate Common Duties and Common Rules by the pressure of extreme danger or fear." According to such a definition, a certain amount of opprobrium, or at least of suspicion, attaches to the whole class, and we ought not to refer any case to the class which may justly demand a more favourable consideration. For example, in Art. 328, Dr Whewell speaks of the Duty of Obedience to Government being sometimes put aside under the pressure of Necessity. But such a case would frequently be more properly treated as one of Conflict of Duties. On the one hand there may indeed be the ordinary obligation of obedience to government; but on the other hand there may be the patriotism which enjoins resistance to a wicked tyrant, and the self-devotion which incurs personal risk for the sake of the liberty and happiness of others. On Dr Whewell's principles the noblest examples of political and religious martyrdom might be included in the harsh definition of Cases of Necessity.

The Chapter *Of Things Allowable* contains some striking remarks; especially we may notice part of Art. 340, "The more entirely a man's whole being is governed and directed by Moral Principles, the more does the circle of Things Indifferent narrow and dwindle. As the moral light grows stronger, every thing assumes a colour of good or bad, between which he has to choose. Everything, however trivial or mean, affords aliment and occasions to virtue. And as all things thus become good or bad, nothing is merely allowable. If it be allowable it is right; and is what must be done because it is right, not what may be done because it is allowable."

The references to the relations of the Sexes and to Marriage are frequent, and they indicate invariably the purity and tenderness characteristic of the author. I do not know, however, whether the enthusiastic advocates of the rights of women will admit that there is any permanent and universal truth in a few sentences of Art. 228: "The Desires and Affections of both Sexes lead to the Conjugal Union: but according to the natural feelings of most persons, and the practice of most communities, the man proposes and urges the union, before it takes place; the woman yields and consents. The man is impelled by a love which he proclaims to the object of it; and he asks for a return in which he has the character of a conqueror. The woman is led to consent, not only by affection, but by the hope of a life filled with those family affections, and family enjoyments, for which, as her heart whispers to her, she was made."

The subject of slavery is treated in a manner which became a member of the University that had the honour of counting Paley and Clarkson and Wilberforce among her sons.

The third Book is entitled *Religion. Of Divine Laws and their Sanction.* Here the author presents himself rather as the commentator on Scripture than as the Professor of Moral Philosophy. A brief sketch is given of Natural Religion; this it is said presents to us the Ideas of the Moral Government of God, and his Providence. "But we cannot reasonably be satisfied with a mere *Idea* of the course of this World. We must attend to the *Fact* also, that is, to the History of the World: and thus we are led to Revealed Religion," Art. 474. Next we have chapters on Christian Revelation and Christian Morality; these are followed by five chapters containing Christian Precepts, arranged under the five heads already adopted, namely. Benevolence, Justice, Truth, Purity, and Order. After these we have the following subjects: the Christian Rule of Conscience, Natural Piety, Christian Piety, Religious Belief, Christian Edification, Oaths, and Oaths of Christians.

One or two difficulties in Scripture are noticed, but too briefly for complete satisfaction. Thus we read in Art. 504: "The reasons which in these precepts are connected with the injunction, must be accepted in several cases as imperfectly expressing the Christian ground of the duty;" this is illustrated by a quotation from the Sermon on the Mount. Again, in Art. 518, we find "Perhaps to some readers, justice in matters of property may seem to be made light of, in the parable of the unjust steward whom *the lord* (that is, his lord) commended (Luke xvi. 8), and of the unjust judge (Luke xviii. 6) of whom Christ said, *Hear what the unjust judge saith;*" this is followed by some explanatory remarks. In Arts. 543...545 we have a consideration of the apparently conflicting duties of obedience to the powers that be and of altering the government and even deposing the

Sovereign in extreme cases. We are told, "In a constitutional form of government, in which the whole or a large part of the citizens possess more or less political power, the Constitution, as much as the person or family of the Sovereign, may be considered as *the ordinance of man*, to which all are commanded to submit themselves. And every citizen, who thus possesses by Law a share of political power, is one of *the powers that be.*"

The analogy and the difference between the course of truth and knowledge in Science and in Religion are very well treated: see Art. 593.

The fourth Book is entitled *Jus. Of Rights and Obligations.* This gives an outline of the actual Laws of ancient Rome and modern England. As it was revised for the first edition by Professor Empson, and for the second by Mr Robert Leslie Ellis, there is sufficient guarantee that the facts are accurately exhibited.

The fifth Book is entitled *Polity. The Duties of the State.* Some of Dr Whewell's friends were unable to follow him in so strongly insisting on the moral character of the State; they held that there was nothing in this apart from the moral character of the individuals comprising the State: Dr Whewell himself objected to the phrase, *the Conscience of the State*, in his Art. 808. The question of a National Church is discussed in this Book, and Dr Whewell seems to have found it difficult to settle it in a satisfactory manner; we have seen that he made some changes with respect to it in his second edition: I shall return to the question in noticing his next work. He is strongly against the pretensions of the Popes, as became a successor of Barrow and of Bentley: see his Art. 1023.

The sixth Book is entitled *International Jus. Rights and Obligations between States.* The subject is one which had a peculiar interest for the author, who had a strong desire to elevate the standard by which the mutual duties of states should be regulated, and especially to diminish and mitigate the horrors of war.

The *Supplement* was first published in 1854 in the third edition of the work. It consists of four Chapters: Of Enumerations and Classifications of the Springs of Human Action;

Objections considered; Of Paley's Moral Philosophy; Of Mitigations of the Laws of War. In the preface to the third edition there is a Note, not reproduced in the last edition, which refers to the last Chapter of the Supplement. "In doing this I have availed myself of the speech made by Sir William Molesworth in the House of Commons on July 4, 1854."

The style of Dr Whewell's *Elements of Morality* is frequently impressive from its dignified seriousness, but it wants the animation which distinguishes the *History* and the *Philosophy of the Inductive Sciences*, and which was again conspicuous in the *Essay* on the *Plurality of Worlds*. This is remarkable, as the work was written in the most fortunate period of his career, and in the maturity of his powers; while the Essay did not appear until the long illness of his wife and the course of academical events had thrown some gloomy shadows over him. Perhaps he did not allow himself sufficient rest after the exhausting labour connected with his two great scientific publications before he undertook the new task of constructing a system of Morality. But it would appear from a passage which will be quoted presently that he deliberately adopted what he considered a suitable style for a systematic work.

In 1846 Dr Whewell published a work entitled *Lectures on Systematic Morality delivered in Lent Term*, 1846. This is an octavo volume of 205 pages, besides the Title, Preface and Contents. The preface is unusually short, occupying less than a page. The first two sentences may be quoted: "I have stated at the outset of the following Lectures that though I hope they may have an independent interest for some readers, they contain a kind of commentary on some parts of the two volumes on *The Elements of Morality* which I lately published. I gladly take advantage of this opportunity of offering explanations on some of the points treated of in the former volumes; for a further attention to the subject has made me aware of very serious defects which are to be found in the work." This admission of *very serious defects* in the former work does not seem quite consistent with the tone of the preface to the second edition of that work in 1848; for there, while allowing that a few superfluities have been withdrawn, he offers in substance a resolute defence of the whole system.

The lectures are eight in number, and on many accounts well deserve study.   They present a clear and lively account of the author's views respecting a System of Morality, and thus possess the interest which must attach to the exposition by a writer of great ability of the conclusions he had obtained by long investigation of an important subject.   The volume is also a good example of what professorial lecturing may accomplish.   Many persons maintain that the day for this species of teaching is gone by,—at least for all the sciences except those which are called experimental,—and that the press is now the natural and appropriate channel of communicating knowledge.   Even in those Universities where professorial instruction is more predominant than it is in Cambridge, the success of the process is far from decisive. The following testimony for example has recently been recorded by an earnest student respecting some lectures.   "As to Wilson's political economy, I regret to say he had neglected to get up the subject; and certainly upon the whole cut but a poor figure, often coming before us quite unprepared."   The case of Dr Whewell in the present course might have seemed very unpromising at first, for he had just published a large work on the subject, which was thus apparently exhausted; and yet this volume shews that he could deliver animated and eloquent addresses, well calculated to charm his hearers and to impress their memories, whether they had studied his published system or not.

A few miscellaneous points may be noticed.   The first four pages give an account of the course of professorial lectures which Dr Whewell had delivered from his appointment in 1838 up to the current date.   From this date until his resignation in 1855 I think he was mainly occupied with Plato in his lectures.   On page 4 he seems to imply that the dry style which characterises his *Elements of Morality* was deliberately adopted; he says: "...the barenness [bareness?] and conciseness of language which appeared to me to be required in a printed systematic work, leave great room for explanation and illustration...."   Speaking of the *Elements* he says, on page 5 : "I have no doubt the book has many and grievous defects and blemishes..." This however does not seem to go so far as the words which I have quoted from the preface; it is not so much a confession that faults really exist as a suggestion that they may be expected in any human work, with a sort of challenge to critical opponents to find them. On page 19 we have a concise formula on an important subject: "the true guide of man is Conscience, only so long as the guide of Conscience is Reason." On page 67 we have a lively passage containing the answer which a student of systems may be supposed to give to some practical man who depreciates them. "If any man comes to us, and says, "Why do you sit here discussing about virtue ? Get up Go out into the world, and *be* virtuous. We reply, That is our object: but we are reasonable beings; we cannot help wishing and trying to understand what we do, as well as to do something......If you do not choose to join with us in such deliberations, go you forth and be virtuous; and may all good omens attend you! Go if you will, and be virtuous, without knowing what it is to be virtuous..." And yet the practical man thus bantered might have had some reply—to the clergyman at least if not to the professor,—by quoting St John vii. 17. The lecturer himself says in a later work "For morality is, as we have seen, eminently and peculiarly a practical science; and its truths must be acted first and contemplated afterwards." (*Lectures on Hist. of Moral Phil.* 1862, *Additional Lectures* page 63.) On pages 142...147 we have some remarks on Butler as a Moralist; their tendency may be summed up in the opinion that by Conscience Butler meant Reason, or at least ought to have meant Reason. "This is Conscience,—the Law written on the Heart. The Law is written there, but to read the writing is a matter, often at least, requiring a careful and continued study. And this being so, Conscience is no permanent body of Rules or ultimate Tribunal." On page 165 Puffendorf is introduced, and is held to be far inferior to Grotius. On page 179 the author's favourite subject of international law is handled in a very eloquent passage. On page 204 the views of citizens and magistrats as to Church Establishment are thrown into the form of an address beginning thus: "We desire," they say, "that this country should be divided into parishes..." The printer however has omitted to inform us by the aid of inverted commas where this address ends and the author resumes his own course.

The last lecture is entitled The Relation of Church and State; it occupies pages 180...205. Towards the beginning of the lecture Dr Whewell mentions his *Elements of Morality* and says "I take the liberty of referring particularly to what I have published on this head; because what I have said in this part of the work, more than in any other, appears to me to require some additional explanation, to prevent its being misunderstood..." The intricate nature of the subject is well shewn by the difficulty which Dr Whewell experienced in forming or communicating his opinions respecting it. Instead of the published eighth Lecture Dr Whewell seems to have delivered two, namely an eighth entitled, The Establishment of the Church ideally and historically considered, and a ninth entitled Establishments, and the Separation of Church and State. The whole of the former was set up in proof, occupying pages 180...206; part of the latter was set up in proof, occupying pages 207...224; the manuscript of the remainder of the ninth lecture, as delivered, has not been found among the papers. The proof agrees with the published volume up to about the middle of page 191; the remainder of the published volume is fresh matter except the passage on pages 197 and 198; "This is, as Chevalier Bunsen calls it,......which could befall the nation:" this passage formed part of the ninth Lecture.

A few points in these proofs may be noted: The author maintains the principle of an established Church, but says we must not hold that every thing which occurs in the English system is necessary to the principle: "It must therefore be considered as a mistake, if any writer should represent any particular scheme of provisions of this kind, as an essential part of the Establishment of a Church;—that is, of the Church being by the State conjoined with itself for the purposes of Education. It must, I say, be reckoned an errour, if any of these details of polity should be put forwards as a necessary portion of such a conjunction of Church and State:—if, for instance, any one should assert the admission of Bishops to the National Councils, and the exclusion of Dissenters from civil offices, to be the necessary conditions of the establishment of the Church." He thinks that Warburton made this mistake, and that some of the remarks in his own *Elements* might appear to involve it. To illustrate the variety which may occur consistently with the general principle of an Established Church an account is given of the ideal system proposed by Bunsen for Prussia in his *Church of the Future*. The Polity of the United States of North America with regard to Religion is spoken of rather more favourably than in some other parts of the author's writings: " That the present Polity of the North American States with regard to Religion is suited to their present condition; and that it would neither be right nor wise to attempt to introduce in that country an Established Church in the form in which it exists in any of the nations of Europe, is a matter of which no thoughtful person can entertain any doubt....Again : when the adoption of religious doctrines as a leading part of the political principles of the rulers of States has led to the most glaring and oppressive intolerance, far exceeding all that was ever dreamt of as a security for an establishment, wise legislators may fitly seek a remedy for this evil, in excluding religious doctrines altogether from the consideration of the civil governor...."

A lecture on *Toleration* which occurs among Dr Whewell's papers seems to have formed part of the series delivered or intended to be delivered ; it is however incomplete. Dr Whewell adopts, as he says, Locke's arguments in favour of Toleration, though taking a different view of the office of the State. The following sentence occurs : " These doctrines of Toleration are necessary consequences of our principles, and I the more willingly state them now in a distinct form, inasmuch as I have omitted to do so in my book on Morality, and have only considered Toleration in a collateral and transient manner. The importance of the subject merits a different treatment in a Treatise on Morality and Polity."

In 1848 we have *Butler's Three Sermons on Human Nature*

and Dissertation on Virtue, edited by **Dr** Whewell, with a
Preface and a Syllabus of the Work. This is in small octavo,
the preface occupying about 30 pages; the publication went
through various editions, the last being the fourth, in 1865.
The preface is interesting; it contrasts Butler and Paley as
Moralists. Dr Whewell allows that Butler's language about the
natural *supremacy* of conscience is liable to be misunderstood,
but considers that Butler himself was free from error: his estimate of Butler in this respect is more favourable than that of
Mackintosh which has been formerly noticed.

An example may be given of a difficulty which sometimes
occurs in Dr Whewell's writings and to which attention has
been drawn in the account of his *History of the Inductive
Sciences.* He is speaking of the phrase *Moral Sense* suggested
by Butler, and makes the following statements: "It may be
doubted whether such a crude and physical notion of a Moral
Sense was ever entertained by any thoughtful moralist..." "It
is plain, at least, as I have already said, that Butler never
dreamt of asserting a Moral Sense in any such use of the term
as this." "As I have said, Butler does not assert a Moral
Sense to exist in any technical or distinct form..." Here the
as I have already said of the second passage must allude to
the first passage, because no other has preceded it which can be
meant; but it is plain that the words are not very strictly applicable. So again the *As I have said* in the third passage
must refer to the first or second passage; but the words are
not very strictly applicable: something *similar* has indeed been
given, but not the precise statement.

After some criticisms on Paley the following high praise is
awarded: "If the work had been entitled *Morality as derived
from General Utility,* and if the Principle had been taken for
granted, instead of being supported by the proofs which Paley
offers, the work might have been received with unmingled
gratitude; and the excellent sense and temper which for the
most part it shews in the application of rules, might have produced their beneficial effect without any drawback." I do not
know exactly how much of Paley's work is to be omitted before

we can thus receive it with *unmingled gratitude*, but even if we make a liberal estimate of the quantity, it seems that the language is scarcely consistent with the strong condemnation of Paley pronounced in the Sermons on the *Foundation of Morals.* The praise is reproduced in the *Lectures on the History of Moral Philosophy*, 2nd edition, page 168.

In the *Dissertation on Virtue* a passage is thus given by Dr Whewell: "... whether considered as a perception of the understanding, or as a sentiment of the heart...;" and the following note is added. "The editions have 'a sentiment of the understanding or a perception of the heart;' but I think it cannot be doubtful that Butler intended to write it as I have printed it. W." A writer in the Dublin University Magazine for May, 1857, draws attention to the passage, and prefers the old reading.

At the end of the preface in the later editions Dr Whewell adverts to the University Regulation made in 1855, by which an acquaintance with Paley's work was no longer required from the students.

The late Robert Leslie Ellis addressed a letter to Dr Whewell on Jan. 22, 1848, in which he gave his opinion with respect to the use of Butler's *Three Sermons* for examination purposes; this well deserves to be preserved.

"I make no apology for communicating to you the impression which I was led to form at the college examination in December with respect to the two subjects of examination in moral philosophy.

The examination in your work was on the whole very satisfactory: that in the three sermons on human nature was certainly much less so.

In the first place, the shortness of the subject gives great facility in getting it up, as the phrase is, by rote.

Secondly, the sermons are essentially a controversial argumentation against those who in Butler's day denied the reality of moral distinctions. Their scope and purpose are therefore but imperfectly apprehended by those who are not acquainted with the history of the subject. In the first stage of his progress the student can only be perplexed by a refutation of errors which he had perhaps never heard of.

Thirdly, the sermons, perhaps from their controversial character, contain statements which may fairly be called unguarded, and which I cannot doubt have misled many of those whom I examined. The language they made use of, when they spoke of vice and sin as if actually they were rare and exceptional phenomena, or when they asserted that there is a *natural* tendency in man to act rightly, could not, I think, be recognised as orthodox by any body of Christians, the unitarians being of course set aside. I do not say that Butler's teaching is heterodox, but that from the end he had in view in the sermons in question, it easily suggests to unwary readers errors opposite to those which he refutes.

Fourthly, the dramatic and metaphorical character of Butler's illustrations is also a serious objection. Conscience comes forward, steps in, sits in judgment and the like, and by this kind of anthropomorphic psychology the minds of many students are not a little bewildered.

Lastly, they learn from Butler to look on conscience as a faculty which decides, without the necessity of any analysis, on the right and wrong of any proposed course of action: a view which would make moral philosophy, or rather morality, an otiose study. They thus lose sight of that which you so distinctly point out, that our moral judgments are to be *derived* from our primary moral conceptions."

In 1849 we have Butler's *Six Sermons on Moral Subjects. A Sequel to the Three Sermons on Human Nature*, edited by Dr Whewell, with a *Preface and a Syllabus of the Work*. This is in small octavo, the preface occupying about 10 pages; the publication reached a third edition. The preface is very brief and seems to be less successful than the author's usual efforts of this nature. For an example take the following ascribed to Butler as part of "the distinction between our self-love and our other affections:" "...that self-love produces *interested* actions, the affections, if unrestrained, produce *passionate* actions...." Butler, however, really says, "passionate, ambitious, friendly, revengeful, or any other, from the particular appetite or affection from which it proceeds." Thus, in Butler, *passionate* actions form but one division, whereas in the preface they form the whole class. Again, Butler says that Hatred and Revenge are *disinterested;* the Preface replies to this in a long sentence, which amounts to saying that if we *suppose* and *imagine* we have a pleasure and interest in the gratification of hatred and revenge, then these feelings are *interested.*

I cannot agree with the high estimate given by Dr Whewell and others of Butler's Sermons; they seem to me to shine with a pale light entirely borrowed from the splendour of the *Analogy,* nor would they, I think, have received so much attention if they had proceeded from an author not otherwise famous. The answers to Hobbes have long since lost what interest they may once have had; it is difficult to suppose that Hobbes himself could have seriously believed in his own audacious paradoxes, and it is not likely that any person would now revive them.

I have next to notice a publication with the following title: *De Obligatione Conscientiæ Prælectiones Decem Oxonii in schola theologica habitæ Anno Domini M.DC.XLVII a Roberto Sandersono S. Theologiæ ibidem Professore Regio postea Episcopo Lincolniensi. With English Notes, including an abridged translation, by William Whewell, D.D., Master of Trinity College, and Professor of Moral Philosophy in the University of Cambridge. Edited for the Syndics of the University Press, Cambridge,* 1851. This is an octavo volume containing xix and 331 pages. The editor's preface occupies three pages: it acknowledges the aid of Mr Holden, who verified and completed the references and compiled the indexes. The English notes not unfrequently refer to points as to which the Editor differs from the Author.

Towards the end of the preface the Editor, after praising the Author generally, adds: " In saying this, we of course except the cases in which Sanderson maintains that doctrine, of the divine right of kings, which was then common in England, and which is now obsolete; but which does not vitiate other parts of his reasoning." This statement, however, seems likely to shake the confidence of a reader; he may suspect that if an important part of the author's teaching is to be dismissed as *obsolete,* then much of the remainder is likely to be precarious, and that the whole work is rather a collection of opinions than an effort of reasoning. We find that in the course of the seventh Lecture the Editor has often to protest against the Author; thus on page 212, after Sanderson has made some extravagant remarks in support of the royal authority, the Editor says, " This was not a just representation of the English constitution, even under the Stuarts." So also on page 281. There is a very good note at the end of the eighth Lecture; this Lecture treats the question, How far Human Laws are binding upon the conscience: the Editor plainly admits that there is little value in what the Author says. The whole of the tenth Lecture and part of the ninth are given to a subject which seems of very subordinate importance in a system of morality, namely, the meaning of the aphorism *Salus populi suprema Lex.* If such an aphorism had any *authority,* derived from the Bible, or from the law of the land, it would be of course of great moment to ascertain its exact significance: but to discuss the meaning of the words in a phrase of Cicero's drawn from the old Roman law, to which no person in modern times was bound to pay regard, was altogether a waste of time.

Sanderson finds frequent opportunity for striking blows at the Pope and the Papists: see pages 86, 99, 111, 125, 181, 234. He holds that the laws of a ruler *de facto* are to be obeyed: see page 149 with the Editor's note. Sanderson's language seems occasionally hard to understand exactly, though the general drift is clear. An example may be given from page 22 which is obscure from the metaphorical allusions. " Lapsu enim amissa est *Conscientiæ rectitudo* et integritas; non *Natura.* Et cum renascimur, non infunditur de novo, quæ defuit ante, *Conscientia,* sed quæ prius inerat, fœda peccatis atque *impura,* respersa jam sanguine Christi, fide *purificatur,* ita ut placeat Deo. Etiam in *infantibus* admodum parvulis, *ut in lactentibus* magnificetur Dei nomen, emicat cum primis e veteris incendii ruderibus, et quamvis admodum debiliter, spargit tamen utcumque igniculos suos rationis hæc *scintillula;...*" Here the word *infunditur* suggests the pouring in of a fluid, but the *respersa* does not fit so well with a fluid as with a solid body. Then the clause *emicat...scintillula* is perplexing: Dr Whewell makes a *spark emerge* from the ruins of that ancient conflagration, which is an awkward metaphor. Sanderson appears to speak of the fall as an ancient conflagration, and of conscience as a little spark sending forth little flames: that is, a very evil thing and a very good thing are both treated as fires, the difference being in size alone.

As an example of obscurity arising from brevity we may refer to page 282. Sanderson is discussing whether a People may change their government from an elective monarchy to an hereditary monarchy; he decides that they may, but only when the throne is vacant, *for otherwise injury would be done to the reigning king.* It is hard to conceive how a person who has been elected king can be injured by a law that in future the crown shall be hereditary. The editor seems to find a difficulty; but the precise force of his addition is not apparent, namely "He overlooks the case in which monarchy is made hereditary in the reigning king's family."

Sanderson's Lectures though delivered in 1647 were not published until 1660. But it appears that manuscript copies were in circulation; two were brought to a London bookseller with a request that he would procure the publication of them: the bookseller very properly declined to do any thing without the author's sanction, and wrote to him. Sanderson's remarks in the Preface, though complimentary to the particular person, seem to indicate an unfavourable opinion of the booksellers of the time. "Laudavi, imo amavi, in homine, et ante id tempus mihi penitus ignoto, animi candorem; et ex eo genere, quibus fere unius lucri studium est, æqui reverentiam."

On the whole the edition of Sanderson seems a work scarcely deserving of the time and attention of Dr Whewell; had it been advisable to disturb a slumber of two centuries the task might have been left for some junior member of the University at which the lectures were originally delivered. In the *Lectures on the History of Moral Philosophy in England,* Dr Whewell has compressed into less than a page all that he has found it necessary to say about the three works of Sanderson connected with Moral Philosophy.

Lectures on the History of Moral Philosophy in England. This was published in 1852, in one octavo volume, containing xxxii and 265 pages. The preface is short, and I extract two sentences from it : " The following Lectures have, for the most part, been repeatedly delivered, in substance at least, in the Courses which I have read as Professor of Moral Philosophy." " Being written for oral delivery, they will be found to contain repetitions, and certain irregularities of style which if I had composed them for the general reader, it would have been my business to avoid." The second edition appeared in 1862, in small octavo, ranging with the last edition of the *History of the Inductive Sciences.* Besides the lectures of the first edition there are others called *Additional Lectures on the History of Moral Philosophy;* these are very inconveniently paged separately : the whole volume now contains xvi, 280, and 130 pages. Unfortunately there is no Index to the volume. I shall refer to the second edition in citing the volume. The original preface is not formally reproduced, but it is substantially embodied in that to the second edition ; this however is chiefly occupied with a reply to some remarks on Dr Whewell's opinions made by Mr Mark Pattison in the well known *Essays and Reviews.*

In the first part of the volume we have an Introductory Lecture, then Lectures I to XII which treat of all the principal English Moralists down to the epoch of Paley inclusive; and finally Lectures XIII to XVIII which are devoted to Bentham. The preface begins thus : "The first Twelve Lectures of the following Collection were delivered, nearly as they are here printed, by me in the first year of my tenure of the office of Professor of Moral Philosophy, to which I had been appointed in 1838." This sentence refers to the Lectures I to XII; but the Introductory Lecture, which bears the second title of *The Point of View,* must belong to a later period, when the Professor had constructed his own system of Morality, of which this Lecture is an epitome. The lectures are very interesting, for the author always excels as an expositor; but they seem much inferior in execution to the historical and critical parts of the *History and the Philosophy of the Inductive Sciences.* They might with advantage have been reconstructed, and presented in some more connected order, so as to form a complete critical and biographical narrative of English Moral Philosophy. In fact, they illustrate an evil of which there are other examples in academical literature; a professor fresh to his subject writes a course of lectures amidst the pressure of his first year of office, and then continues to repeat them annually and finally to publish them, without giving to his students or to the public an adequate share in his more mature knowledge.

The Chapters on Bentham are remarkable. Far more space is given to him than to any other writer, and the criticism is here especially animated and severe. It is instructive to compare this with what Mackintosh in his *Dissertation* says under the same head. Mackintosh, as usual with him, is tolerant even to the verge of indecision, and seems to shrink from open condemnations. He offers general observations on the utilitarian system, which are not specially applicable to the writer whom he proposes to discuss; and such censure as he pronounces is directed against Mr James Mill a disciple of Bentham's, rather than against the Master himself. The two critics seem to have exchanged their professions. Dr Whewell exhibits the tenacious grasp, the impetuous spirit, and the incisive sarcasm of a lawyer; while in Mackintosh we have the serene temper, the tranquil hope, and the persuasive eloquence of a clergyman.

As I remarked with respect to the *History and the Philosophy of the Inductive Sciences* so I may repeat with respect to the present volume, that the second edition ought to have exhibited signs of a more stringent revision. I will mention a few examples: page 53, the second Greek word in the title of one of Hobbes's works is wrong; page 150 line 4, for *the author of the thoughts* read *the author of the Notes;* page 165 line 7 from the foot, the word *resemblance* yields no sense, it should be *restriction.* The phrase *as I have said* or something similar is frequent, and it is not always easy to determine the reference.

The second part of the volume purports to consist of fourteen lectures; but it is obvious that these are not in all cases complete lectures, as their average length is only nine pages. Thus in many cases we may conclude that what is printed is only a portion of what was delivered; and in consequence the treatment of some of the subjects seems too brief. For instance Lecture XI, of about three pages is devoted to Peter Lombard, and Lecture XII, of about six pages to Thomas Aquinas. In each case some illustrative extracts might have been given with advantage, especially as the following high praise is awarded to the Schoolmen. "If we were to invest with modern forms many of their dissertations, it would be found impossible to give either a better analysis of the question, or more solid arguments, or a more judicious decision." The last lecture is on the distinction Mr Coleridge wished to introduce between the Reason and the Understanding, which Dr Whewell here and elsewhere altogether rejects; see his *Philosophy of Discovery* page 445 and *Platonic Dialogues* Vol. III page 296. He says here "...if we were to adopt Mr Coleridge's account of the distinction of the substantives, we should have to assert that *by the Understanding we reason, and by the Reason we understand.* But, as I have ventured to say elsewhere, this is neither good English nor good philosophy."

In 1853 Dr Whewell's edition of the great work of Grotius appeared under the title, *Hugonis Grotii de Jure Belli et Pacis Libri Tres accompanied by an abridged translation by William Whewell, D.D. Master of Trinity College and Professor of Moral Philosophy in the University of Cambridge with the Notes of the Author, Barbeyrac, and others. Edited for the Syndics of the University Press.* The work consists of three octavo volumes, containing on the whole about 1400 pages. Dr Whewell's preface occupies fourteen pages; it acknowledges the aid of Mr Holden who verified the references. There are a few notes by Dr Whewell himself, but they are not extensive; and he occasionally gives references to his *Elements of Morality* for his own opinion on the matters discussed by Grotius. He rarely records his dissent from the positions of his author; there is however an example of such a kind with respect to the election of the Roman emperors, where he finds that Grotius has set up *five fictions;* see Vol. II pages 13 and 14. It would be superfluous to make any remarks on the famous work of Grotius; the editor received the thanks of many distinguished men for his labours, which seemed well timed with respect to the condition of Europe.

A few words may be quoted from a letter by Mr Hallam. " I have again to thank you for a literary present—your leisure is nobly employed, and your chair of Moral Philosophy no sinecure in your vigorous hands. But I have additional reason to thank you for the very handsome manner in which you have mentioned my own little attempts to place the fame of Grotius, in his great work, in the light which it appeared to deserve. You have taken infinite pains to do much more than I could do; and the English reader will have the advantage, in your edition, of reading a well printed text or of economizing his time by an abridged translation."

The following is from one eminent both as a statesman and a scholar. " Pray accept my best thanks for the excellent edition of Grotius which you have had the kindness to send me. I have read the preface, and examined the plan of the work, and feel satisfied that you have rendered a most useful service to all students of International law and politics by your labours. Grotius was unfairly run down by the writers of the eighteenth century—he was considered heavy and pedantic—but a fairer estimate of his immortal work now prevails, and your edition will contribute to restore it to its proper estimation.

Your remarks on the project of Perpetual Peace are very just—the extravagances of a few fanatics ought not to be allowed to cover this lofty ideal with ridicule.

I have observed in reading Livy lately that there are several passages in which *jus gentium* is applied by him to rules of international law—and if the passages were taken singly, the phrase might seem to be used in these phrases in its modern sense. In fact, however, it is used in these places in its ancient sense of *law generally received among nations;* but it is used to comprehend the prevailing rules of international law as well as those of civil law. For instance, a Roman might have said that the slaughter of ambassadors was contra jus gentium—but then he would also have said that it was contra jus gentium for a man to have several wives, or for a master not to have power of life and death over his slave.

In Vol. I p. 40 of your edition, the proverbial Greek line expressive of the jus talionis is incorrectly cited by Grotius. Its correct form may be seen in Aristot. Eth. Nic. v. 8.

Barbeyrac is an excellent commentator. His notes to Puffendorf are full of information, as well as his notes to Grotius.

Wheaton's treatise is a meagre performance. It is remarkable that no English writer has produced any work of authority on International law. We have nothing but Lord Stowell's judgments.

I wish some University man, who understands law and modern history would undertake a treatise on the subject. In the present state of the science, an extensive knowledge of positive law would be indispensable."

Dr Whewell's works on Moral Science encountered various hostile reviews; to these he sometimes replied without naming the publication in which they appeared. He excused himself for this omission of reference on the ground that "in all subjects, the more *impersonal* our controversies can be made, the better they will answer all good ends; and certainly controversies on Morality are most likely in this way to be really moral :" see the Preface to the *Lectures on Systematic Morality,* and the Preface to the third edition of the *Elements of Morality.* But whatever may be the advantages of this course, there are obvious reasons why critical students should be able to turn to the original statements of objections; and, therefore, I shall now advert to the principal of these.

I first notice three articles in the *Prospective Review,* all obviously by the same author, and written with great vigour and ability: there is a long unpublished reply to them among Dr Whewell's manuscripts.

We have a review of the *Elements of Morality* on pages 577...610 of the *Prospective Review* for 1845. Among many complaints one is especially prominent, namely, that the book deduces morality from positive law: "All virtue is made to grow out of judges' decrees, and the Will of Heaven is reached through the Institutes. And so insecure seem the steps of this strange ascent from the Basilica and Westminster Hall to the throne of God, that we are astonished at the intrepidity that trusts to them." The analogy between Morality and Geometry, which Dr Whewell had suggested, is pronounced to be unfortunate and fallacious. His leaning to the highest doctrines of authority; both in Church and State, is briefly but emphatically condemned. " Not Wolsey himself could find more magnificent pleas for state prerogatives; and scarcely Innocent, had he lived now, make grander claims for an exclusive church." The only words of commendation are contained in an allusion to "some better things, equally earnest and hearty, especially the indignant severity with which Slavery is everywhere treated."

A review of the *Lectures on Systematic Morality* is given on pages 400...427 of the *Prospective Review* for 1846. The *Lectures* are considered, and justly, as mainly a defence of the *Elements of Morality* from certain objections advanced against the book. " The first half of the defence is directed against some critic imbued with Mr Carlyle's mode of thought, and sympathising with his aversion to all systematic definitions of human duty. The remainder is a manifest reply to our review; unless indeed the very same series of strictures has been repeated in some other quarters unknown to us." The reviewer does not allow the propriety of Dr Whewell's course in omitting an exact reference to the criticisms he answered, but does not impute to him " the slightest degree of conscious injustice." The reviewer anticipates me in a remark which I had made before reading the article— that Dr Whewell treats *Cases of Conscience* " as arising from the obscurity of a single precept, instead of from the collision of two." The reviewer in order to test the doctrine—which he ascribes to Dr Whewell—that Morality is dependent upon Law—suggests two examples, in which he implies that Morality and Law would pronounce different verdicts in the same transaction. " A female captive from Dacia is given to a lady of fashion about Trajan's Court, as her *ornatrix*, or lady's maid. The lady is passionate, and particular about her head-dress : and day by day the poor maid is submitted to the thong for the imperfection of a braid, or hung

W. 18

up by the hair to be lashed for the scratching of a comb. The
humanity of a Christian neighbour is excited by her cries: and
he secures her escape and restores her to her country." " Again,
if a captive girl is sent into the harem of an oriental tyrant, and
a noble-minded youth, knowing something of her history, and
regarding her with pity and affection, rescues her and marries
her, is the Moralist to accept the legal determination that she
is another's wife, and to pronounce the young man guilty of
adultery ?"

We have a review of the *Lectures on the History of Moral
Philosophy* on pages 545...565 of the *Prospective Review* for 1852.
This is not so severe in its condemnation as the other two articles;
but still it is very unfavourable.   In particular, the account given
of Dr Samuel Clarke is held to be extremely inaccurate; and in
fact Dr Whewell corrected this in his second edition.   In this
article a question is asked, to which a reply is given in the *Supplement* to the third edition of the *Elements of Morality*, namely,
"Why are they [the Rights] five? why so exactly face to face,
one by one, with the Springs of action—so many policemen
watching so many thieves ?"

An article in the *Westminster Review* of October, 1852, by
J. S. Mill, reviews the *Lectures on the History of Moral Philosophy*,
and also the *Elements of Morality*.   This is partly a defence of
Bentham against Dr Whewell's objections, and partly a criticism of
Dr Whewell's own system: to the latter part of the article a
reply is given in the *Supplement* to the third edition of the
Elements of Morality.

An article which uses as a text the *Lectures on the History
of Moral Philosophy* appeared in the *Dublin University Magazine*
of May, 1857: the writer, however, is mainly occupied with giving
his own account of Grotius, Sanderson, and Hobbes.   The chief
fault found with Dr Whewell's Morality is that it is too systematic.
"Surely Dr Whewell's *pentagonal* morality— (his five elements
of our nature, five classes of rights, five virtues, five branches of
the general trunk of morality, and we know not what other fives),
postulates its own failure.   It is a sort of Philistine superstition,
reproducing the five golden emerods, the five golden mice, the

five lords, the five cities." A passage which occurs towards the end of the fourth Lecture is censured as too rhetorical; but the reviewer himself was certainly not justified in rebuking any other person for the obtrusive display of literary ornament. Dr Whewell in his second edition added a note to the passage which he said had been criticised as rhetorical and unmeaning, but he gives no reference to the place where the criticism is to be found.

An article in the *British Quarterly Review* for 1863 discusses several of Dr Whewell's Moral Works and pronounces an adverse judgment on them all.

In the *North American Review* for July, 1846, there is an article on the *Elements of Morality*. Some faults are pointed out, but on the whole a very favourable opinion is expressed. " But we have read the work through with growing gratitude to the author for the distinctness of his definitions, for the transparency of his statements, for his accuracy in the use of terms, and for the minuteness and thoroughness of his analysis of moral ideas and conceptions."

Shorter notices of the works on Moral Science are given in various periodicals. From one of them I extract a sentence relating to the *Lectures on the History of Moral Philosophy*. "Moreover, since it would be in most people a shameful waste of time to read the works of all the men whose names we chronicled just now, it will be found convenient to have an abstract of their chief opinions, arranged (as it is arranged here) by a master hand."

The critics of Dr Whewell's works are almost unanimous in condemning the derivation of Morality from positive Law which his system seemed to involve; and in spite of his justification of his course it is difficult to acquiesce in its propriety. Thus, to take an example, he replaces the old four *Cardinal Virtues* by five, namely, Benevolence, Justice, Truth, Purity, and Order; and with respect to *Purity* he often makes substantially the following remark : "Morality says, You shall not desire her who is another's wife : Law determines whether she be his wife." It is scarcely possible to imagine a frailer foundation on which to rest one of the most important of the five pillars of his Moral Edifice. A duty which is to be incumbent on both sexes, and on every condition of life, is connected with the formalities which the law enjoins as to the marriage ceremony—formalities which have for their main purpose the regulation of the succession to property, and which are independent of such important distinctions as monogamy and polygamy, permanence and divorce. In fact, history shews that the laws of marriage lend only a precarious support to the demands of morality. The powers of dispensation and dissolution exercised so freely during the middle ages, promoted the rapacity of priests and the licentiousness of kings rather than purity of heart and conduct. The facility of divorce among the ancient Romans must almost have destroyed the obligation of ties which could be so easily cancelled. The followers of Mahomet found in his laws rather a charter of indulgence than a warning and restraint. Nay even the Mosaic institutions were condemned by the Divine Master as a lapse from primeval innocence—a concession to a hard-hearted race.

There are numerous manuscripts relating to Moral Philosophy among Dr Whewell's papers, and preparatory to a brief notice of them it will be convenient to mention the courses of Lectures which he delivered as Professor.

In 1839, 1840, and 1841 he traced the history of Moral Philosophy in England, and his Lectures were embodied in the volume published in 1852. In 1842 and 1844 he lectured on the Difficulties of Constructing a System of Morality; in 1843, being Vice-Chancellor of the University he did not lecture. In 1845 his lectures consisted of portions of his *Elements of Morality*, then about to be published. In 1846 he delivered a course of lectures on Systematic Morality, which was published. The engagements of subsequent years are not distinctly recorded; they consisted partly of historical criticism, which was published in the second edition of the *Lectures on the History of Moral Philosophy*, and partly of lectures on Plato, which were embodied in the *Platonic Dialogues for English Readers*.

The manuscripts include many of the lectures in whole or in part; the most complete are those of the courses of 1842 and

1844. They contain little which is not substantially involved in the *Elements of Morality*, though the matter is generally presented in a more lively and rhetorical form in the lectures than in the printed book. The manuscripts also include the first drafts of large portions of the *Elements of Morality*; and these differ considerably from the forms finally assumed.

There are also various papers which may be properly described as unpublished, of which the following are the principal:

(1) A reply to the criticisms in the *Prospective Review;* This consists of 139 quarto pages. Among other things the two illustrative examples suggested in the volume of the Review for 1846 are here discussed; namely, that of the Dacian captive, and that of the girl removed from the harem of an oriental tyrant. (2) A manuscript of 54 pages entitled *Cases of Conscience*. The case of fugitive slaves already suggested by the examples of the preceding manuscript is here thrown into the form of a fictitious narrative : the matter has lost much of its practical interest since the abolition of slavery in the United States. (3) An account of the work of Mr Austin on Jurisprudence. (4) A criticism of Coleridge's doctrine of the Reason and the Understanding, much longer than that which was published in the second edition of the *Lectures on the History of Moral Philosophy*. A small volume might perhaps be hereafter published, consisting of these papers, together with a selection of passages from the Lectures.

With Dr Whewell's works relating to Moral Philosophy we may properly connect his *Platonic Dialogues*, for this work doubtless derived its origin from various courses of lectures which he delivered as Professor of Moral Philosophy. The *Platonic Dialogues for English Readers* appeared in three volumes in small octavo; the first volume was published in 1859, and reached a second edition in 1860, the second volume was published in 1860, and the third volume in 1861: the three volumes together contain about 1200 pages. The work may be described as a condensed translation of a large portion of the Platonic Dialogues, but it must be observed that in this condensation the form of a *dialogue* is often abandoned. Dr Whewell himself justifies this on the ground of the gain of clearness even with the loss of vivacity: see for example his Vol. III, page 174. He says in the Preface to the first volume: "I have given both the matter and the manner with all fidelity, except in so far as I have abridged several parts, in order to avoid prolix and obscure passages. And I can venture to say that my task (including translations of most of the other Platonic Dialogues as well as of those given in this volume) has not been lightly executed. It has been a labour of many years; each part has been gone over again and again;...." The labour of many years refers most probably to the lectures which Dr Whewell had delivered on Plato, not only at Cambridge, but elsewhere, as at Leeds, and at the College for Ladies in London.

Dr Whewell himself had drawn the attention of English scholars, and especially of Cambridge scholars, to the advantage which would follow from good translations of the ancient classics, particularly of the philosophical works; see his *Liberal Education*, Arts. 93 and 326. His *Platonic Dialogues*, however, are of a less elaborate kind than he probably contemplated, in the passages just cited, being adapted for the general reader rather than for a laborious student. The design and execution of the work received the approbation of some very eminent scholars, and were in general favourably noticed by the reviewers: an exception must be made, however, in the case of a sharp criticism in the *Saturday Review* on the second volume.

The first volume contains Dialogues of the Socratic School, and Dialogues referring to the Trial and Death of Socrates; in all thirteen dialogues. Dr Whewell does not lavish the undiscriminating eulogy on his author which we find in some commentators on Plato; he does not hesitate occasionally to record his opinion that, instead of profound philosophy, we have only school-boy puzzles. He defends the genuineness of several of the dialogues against the attacks of Schleiermacher and Ast; and especially dissents from the dogmatic utterances of the former critic. The section of the first volume which relates to the trial and the death of Socrates, is very impressive; the Apology especially is rendered in a most attractive manner. At the end of the first volume are some interesting remarks on the subject of the Phædo, namely, the Immortality of the Soul.

The second volume of the *Platonic Dialogues* is devoted to the Antisophist Dialogues; there are nine in all, the most important being the Protagoras and the Gorgias. Perhaps the volume is less pleasing than the first; and indeed the closing scenes of the life of Socrates have a fascination for the ordinary reader beyond that of any other thing which Plato wrote. It may be too that in dialogues where the metaphysical element is more prevalent, it is less easy to present the argument to modern readers, than in those which are almost entirely ethical. Moreover Dr Whewell would naturally choose for his first volume those parts of Plato which he felt most suited to his mode of treatment.

A few miscellaneous points in this volume may be noticed: In the preface Dr Whewell praises and adopts the opinions expressed by Mr Grote as to the so-called Sophists. Dr Whewell regards the Phædrus and the Menexenus as intended among other things to shew that Plato "could, if he chose, produce compositions of the same kind of excellence which was admired in the productions of the rhetoricians, whom we may regard as his rivals." A curious remark is made on page 206. In a passage of the Gorgias a sarcastic description of Plato and his school is put into the mouth of Callicles; and Dr Whewell says: "Plato so far acknowledges the sting of this sarcasm that he makes Socrates, at a later period of the Dialogue, after he has silenced Callicles, say that he would have liked to answer the attack." But this seems to neglect the fact that Plato himself sets his characters in action and speech; we might as plausibly maintain on the contrary, that Plato, by reproducing the sarcasm shews that he cared nothing about it, and contents himself with saying at a later period that he should have liked to answer it. Here in fact Dr Whewell seems to look on the Dialogue as if it were a *report* instead of an *imaginary scene;* and he leans in the same direction elsewhere when he remarks on the *dates* of the Dialogues. But on page 306 he is obliged to take the contrary direction in reference to an "anachronism which makes Socrates speak of events which happened long after his death." It is obvious that remarks and criticisms will bear a very different aspect according as the Dialogues are regarded as reported occurrences or as imaginary conversations; and yet this simple consideration seems sometimes forgotten. Thus Dr Whewell observes on page 174 that "Plato knew too well Gorgias's talents and skill, and also his reputation, to think it prudent to represent him as a contemptible adversary or an easy conquest." This is surely a natural and convenient remark; the Dialogue is regarded as a work of fiction, and it is suggested that even a fiction will be displeasing if it represents a man of known ability and learning as foolish and ignorant. But a hostile critic on Dr Whewell's book says: "The insinuation is, that Plato had no regard for truth, and was quite dishonest enough to misrepresent if he had not been deterred by prudential considerations." This criticism assumes that the Dialogues aspire to be treated as histories and not as inventions.

The third volume of the *Platonic Dialogues* is devoted to the *Republic* and the *Timæus*, the former occupying by far the greater space. The preface is very interesting, as is generally the case in Dr Whewell's works; he passes briefly in view the whole series of Dialogues which he has expounded, and considers that they became more striking by being taken in the order in which he has presented them. In the treatment of the *Republic* the main peculiarity is the division of the whole into five parts and five digressions, which are supposed to have been written at various times; see pages 1...5, and 146. In pages 103...112 Dr Whewell objects to the awkward phrase *the spirited element* which had been adopted by some English writers, and appeals to Bishop Butler and St Paul as admitting *anger* to be a necessary element of human virtue. The *passionate element* of a recent exposition is certainly much better English than the *spirited element*. There is not much ethical criticism on the *Republic;* more might certainly have been expected from one who had been for many years Professor of Moral Philosophy. The well known doctrine on the condition of women elicits only the epithet *strange* twice, and *monstrous* once. A recent more advanced Platonist, though at first shocked, seems finally half inclined to approve.

With respect to the *Timæus* Dr Whewell says in his Preface that in many parts he gives a mere abridgement. He refers to the *Études sur le Timée* of M. Théodore Henri Martin; a work which I remember he once strongly recommended to me. The Timæus is in fact very obscure, and does not present a favourable subject for an English version.

Since the publication of Dr Whewell's *Platonic Dialogues* English students have received two valuable expositions in the great works of Mr Grote and Professor Jowett; the subject may be thus studied with double advantage, as it may be approached respectively with the spirit of Comte and of Hegel. The beautiful introductions which Professor Jowett has prefixed to the several Dialogues cannot fail to delight a reader who is attached to ancient or modern philosophy. So far as Plato is concerned, the task of presenting the ancient classics to English readers, which Dr Whewell recommended to the attention of scholars, may be said to be accomplished; for in addition to these two works named above various single Dialogues have been translated and explained by skilful hands. But notwithstanding the far more elaborate treatment which the subject has received in the learned works to which I have referred, the three volumes published by Dr Whewell may retain their value as a popular and elementary introduction. One at least of Plato's accomplishments was a profound acquaintance with mathematics, and it cannot be without interest to listen to a commentator who in that respect can sympathise with his author, and who combines in himself a variety of erudition which may rival that of Leibnitz.

Among Dr Whewell's manuscripts there are drafts, more or less complete, of numerous lectures on Plato, which he delivered in his courses as Professor of Moral Philosophy. Many of them are very interesting, especially those which treat on general subjects, such as the Socrates of Xenophon and Plato, and the Connexion and Series of the Platonic Dialogues; but it would be undesirable to print these because the matter which they contain is substantially reproduced in the *Platonic Dialogues for English Readers*, though in a different order and connexion, by which the impressiveness is somewhat diminished. Dr Whewell seems to have intended to write an *Introduction* to his volumes on Plato, and part of this is executed with so much spirit that it is to be regretted the design was not carried into effect. Interesting letters from Lord Macaulay and other scholars on the subject of the *Platonic Dialogues for English Readers* are preserved among Dr Whewell's papers.

CHAPTER XV.

THE subject of English Hexameters engaged so much of Dr Whewell's attention that it will be convenient to give a connected account of his publications relating to it. The general term *hexameters* may be used, though in some cases the verses which will have to be noticed are composed of hexameters and pentameters alternately, which are technically called *elegiacs*. To save repetition I may state here that the articles which I shall notice in this Chapter, though published anonymously, are known to be by Dr Whewell, from allusions in his correspondence; there is one exception which I mention in its proper place.

In 1837 in the *History of the Inductive Sciences* a passage of eight lines from Göthe is given with a translation into English hexameters; it occurs on page 360 of Vol. III of the third edition of the work. Sir J. Herschel quoted the translation with praise in his review of the *History and Philosophy*.

In 1839 a translation of Herman and Dorothea was printed, I think for private circulation only. It is in oblong form, containing on iv pages the title and introduction, and the text on 139 pages. It was reprinted, as we shall see, in 1847.

In 1840 the *Isle of the Sirens* was printed in 7 quarto pages. This consists of an adaptation of some passages of Carlyle's Chartism published in 1840. It is dedicated to a friend, not named, but easily recognisable as the present Lord Houghton, in these words:

> Senator! Poet! who long, driven on in course Odyssean,
> Seekest thy home, the dwelling of Truth, through scenes of enchantment,
> Seek'st in the Dark and the Strange and the New for counsel and wisdom,
> List, from the lyre of a friend, the lay of THE ISLE OF THE SIRENS,
> And shun thou, as thou bidst, "base fear and servile contentment."

Mr H. Crabb Robinson, so well known for his acquaintance with German literature, was much pleased with this piece. He said in a letter to Dr Whewell "You must be content to hear that the author of the Isle has beaten the translator of the German Epos out of the field as an Hexameterist." The *Isle of the Sirens* was reprinted in 1847 in the volume entitled *Sunday Thoughts and other Verses.*

In 1846 three *Letters on English Hexameters* were published in *Blackwood's Magazine;* these are signed M. L., the final letters of the name William Whewell.

The first letter is on pages 19...21 of the number of the Magazine for July. It consists mainly of remarks on a version of the first book of the Iliad, published with the signature of N. N. T.; these are the final letters of the name John Gibson Lockhart: the authorship was disclosed by the publisher to Dr Whewell with Mr Lockhart's permission.

The second letter is on pages 327...333 of the number of the Magazine for September. In this Dr Whewell insists on his main principle, that in English hexameters we are not to attend to long and short syllables, but to strong and weak syllables; that is we are to regard accent not quantity. Thus in fact an English *hexameter* is to be something very different from what is known under that name in Latin and Greek. A further great difference according to Dr Whewell, is that, to use technical terms, we must have trochees where the ancients had spondees; that is to say where according to the classical hexameter we should have two strong syllables we are to have a strong syllable followed by a weak one. This objection to spondees is an essential doctrine with Dr Whewell, and is enforced in all the three letters. Immediately after the second letter we have translations of two pieces from Schiller entitled *Columbus* and *Odysseus :* these are reprinted in the volume noticed immediately after the three letters.

The third letter is on pages 477...480 of the number of the Magazine for October. It begins thus: "Dear Mr Editor, I hope you will be of opinion, that I have, in my two preceding letters, proved the hexameter to be a good genuine English verse, fitted to please the unlearned as well as the learned ear; and hitherto prevented from having fair play among our readers of poetry, mainly by the classical affectations of our hexameter writers—by their trying to make a distinction of long and short syllables, according to Latin rules of quantity; and by their hankering after spondees, which the common ear rejects as inconsistent with our native versification." This is a very good summary of Dr Whewell's opinions on the subject. He holds that "Klopstock's *Messiah* did a great deal to give the hexameters a firm hold on the German popular ear," and that "if Pollok's *Course of Time* had been written in hexameters, its popularity would have been little less than it is, and the hexameter would have been by this time in a great degree familiarised in our language." Dr Whewell quotes with some satisfaction a translation by William Taylor of Norwich of a passage from Klopstock; to some persons the translation will recall Coleridge's opinion that Klopstock was a *very* German Milton. However the reader can judge for himself, and perhaps more accurately by having the passage printed continuously as he can then exercise himself in resolving it into hexameters; it must in fairness be stated that the passage does not begin with the beginning of a line. "While spake the eternals, thrill'd through nature an awful earthquake. Souls that had never known the dawning of thought, now started, and felt for the first time. Shudders and trembling of heart assail'd each seraph; his bright orb hush'd as the earth when tempests are nigh, before him was pausing. But in the souls of future Christians vibrated transports, sweet foretastes of immortal existence. Foolish against God, aught to have plann'd or done, and alone yet alive to despondence, fell from thrones in the fiery abyss the spirits of evil, rocks broke loose from the smouldering caverns, and fell on the falling: howlings of woe, far-thundering crashes, resounded through hell's vaults." Towards the end of the letter Dr Whewell speaks of *pentameters;* he says, "The alternate hexameter and pentameter are, for most purposes, a more agreeable measure than the hexameter by itself." Most persons appear to acquiesce in this opinion; but Lord Lindsay, a very strenuous advocate of English hexameters, is against the pentameter : see a note on page 4 of his *Theory of the English Hexameter*, 1862. Dr Whewell's third letter is followed by a translation of the *Dance* from Schiller into Elegiacs; this was reproduced in the volume I have next to notice.

English Hexameter Translations from Schiller, Göthe, Homer, Callinus, and Meleager. London: John Murray, 1847. This book is in oblong form; it contains viii pages of introductory matter, and 277 of text. The volume was edited by Dr Whewell, who started the scheme; it contains pieces by himself, Sir J. Herschel, J. C. Hare, J. G. Lockhart, and E. C. Hawtrey. The profits of the publication were given to the Royal Literary Fund. Much of the volume consists of matter which had already been printed, at least for private circulation.

Dr Whewell's principal contribution is the translation of Göthe's Herman and Dorothea, which he had printed in 1839. Among Dr Whewell's papers there is one in the handwriting, I believe, of Sir J. Herschel, which consists of a series of suggested corrections of the first edition of the translation of *Herman and Dorothea*. Dr Whewell adopted none of them, except supplying the lines occasionally marked as omitted. Probably when he received the paper he filled up the omissions in his copy of the first edition, which is preserved, and then put the paper aside and never consulted it again. Dr Whewell's pieces *Columbus, Odysseus,* and the *Dance,* which were published in *Blackwood's Magazine* in 1846, are reproduced in the present volume; besides these he gave the *Sexes* from Schiller, and the *Metamorphosis of Plants* from Göthe, which do not seem to have been previously published. The quatrains which occur on pages viii, 1, 59, 241 of the volume are also, I believe, due to Dr Whewell; but Sir J. Herschel gave some aid for that on page 1. The quatrain on page 271 is by Dr Hawtrey.

Sir J. Herschel's contribution to the volume consists of a translation of Schiller's *Walk;* this had been printed in 1842.

Mr Hare's principal contributions are the *Poetical Epistles* and *Alexis and Dora,* all from Göthe; these had been published in the *Athenæum* about eighteen years previously, and were now reprinted with corrections. Besides these, Hare furnished translations of some epigrams from Schiller, which he had executed long before.

There are four pieces by Dr Hawtrey, two from Homer, one from Callinus, and one from Meleager; these I think had all been written some time before, and perhaps printed for private circulation.

Mr Lockhart contributed one piece, a translation from Homer, which he wrote specially for the volume.

Letters from the contributors are preserved among the papers of Dr Whewell. It may be inferred from them that the editor himself and Sir John Herschel were firm believers in English Hexameters; Mr Hare was, at least at this date, much less confident, while Dr Hawtrey and Mr Lockhart were little better than sceptics. Sir J. Herschel shewed his faith some years later by translating the whole of the Iliad. It would be unreasonable to complain of the relaxation which such a laborious student and investigator allowed himself, but many who have read his famous *Discourse* will regret that *hexameters* instead of *science* absorbed so much of his precious time. It may be observed that Sir J. Herschel spontaneously admitted that a certain line of his translation has nothing to correspond to it in the original; it is line 4 of page 11. "What shall coerce the strong, when at the lovely it spurns?" But he liked the line, and so let it stand. Dr Whewell himself had specially complained of this line when the translation was first printed. Mr Hare and Dr Hawtrey speak in terms of high praise of Sir J. Herschel's translation; I am sorry to see however that the two great scholars were unable to spell correctly the name of so eminent a philosopher. Dr Hawtrey commends for its rhythm Dr Whewell's translation of the *Metamorphosis of Plants;* but his most emphatic praise is awarded to Mr Lockhart's piece, which seems to me the most pleasing in the volume. Perhaps the reason may be that while the other translators followed elaborate theories as to accent and quantity, Mr Lockhart wrote his version straight off at a country house without any books to consult. It has, I think, been stated by Mr J. C. Wright that

Dr Hawtrey subsequently lost his admiration for English hexameters; but I cannot give an exact reference.

Dr Whewell was a strong advocate for the cultivation of English hexameters, as we know from his various publications on the subject; for my part I must confess that his arguments do not convince me, and the examples do not please me. It has been commonly said that hexameters have been produced with success in Germany, and that there is no reason why they should be rejected in England. Now the question of the difference between the English and the German languages has been frequently discussed, and we need not stay to point out the advantage which the Germans may have over ourselves in this respect, though it is obvious that in the formation of compounds and in the arrangement of words they have resources which we do not possess. It may be doubted however whether the success of hexameters in Germany has been very triumphant; the subject treated by Klopstock may have ensured his popularity more than the form into which he cast his work, and the interest of the story may have assisted *Herman and Dorothea.* But even admitting that hexameters have obtained favour in Germany two reasons may be assigned for the absence of a corresponding result in England.

In the first place the greater attention which is paid in England to Latin poetry, and especially to the composition of Latin verse, is a strong obstacle to the reception of the so-called English hexameter. The ears of scholars are trained to admire the classical hexameters, and thus they cannot endure what is presented to them under the same name, but with a different nature, in English. Dr Whewell himself thought that scholars were far more hostile than ordinary English readers to the hexameters. In the next place the main hindrance to the toleration of English hexameters is the extraordinary extent and beauty of the poetry we possess in the ordinary metres. It is remarkable that a nation which is assumed to be preeminently practical should enjoy a poetical literature which for amount and excellence probably surpasses that of all other modern nations put together. The causes of this result need not be examined here; they depend partly on the circumstances of national history. It is however the fact that from the days of Chaucer, to use Dr Whewell's favourite emblem, the torch of poetry has been transmitted from hand to hand in England, down to our own days; and this cannot be said with respect to any other country. In Germany the Thirty Years War, the wars of Frederic the Great, and the wars of the French Revolution, have successively for a time almost arrested the course of literature. In France the religious wars and the political convulsions have often wasted the energies of the nation; in Spain the inquisition seems to have burnt them up. In Italy the ease of the language, and the licence taken by the early writers, have so much alleviated the task of verse composition that the poetical powers seem to be impaired for want of due exercise. In England the efforts to conquer the difficulties of a rugged language have been sustained through many continuous centuries, and the result is the acquisition of a matchless wealth of poetry. Thus a writer in a new or forgotten metre, like the hexameter, has implicitly to contend against all the range of English poets; the minds and memories of readers supply an inexhaustible stock of passages of the highest excellence, and the new aspirant is at a hopeless disadvantage. We are reminded of Voltaire's remark, that Thomas Corneille would have been a poet if it had not been for his brother. Southey said that Hexameters made no way with the public because instead of a large array of several thousand lines only a few straggling hundreds were brought into the field; but, to carry on this military allusion, it should not be forgotten that the position to be won is guarded by the poetry of Shakespeare and Milton, Dryden and Pope, Byron and Wordsworth. In the *Dialogues on English Hexameters* 1847, and in the review of *Dart's Homer* in 1862, Dr Whewell seemed to find an argument for English Hexameters in the fact that women and children can write them; but the ease of any literary achievement is a very untrustworthy evidence of its permanent value.

Something of the cold reception awarded to English Hexameters may be due to the want of high merit in the pieces

w.   19

selected for translation in this form. The admiration which Dr Whewell often expressed for Schiller seems to me extravagant, and though Göthe may occupy a more assured place, yet his hexameters are not his most famous efforts. It may be said that the hexameter will secure a position in England when it is adopted by some eminent poet, and that its failure hitherto has resulted from the fact that it has been mainly used by persons who were only accomplished versifiers. Dr Hawtrey ascribed real poetic power to Lockhart, and assigned this as the reason for the excellence of his piece; but, on the other hand, we must remember that a poet so distinguished as Southey tried hexameters and failed most decidedly. Still, if I may may venture to offer an opinion, I think the cause of hexameters will be served more by a writer who can trust his ear and his instinct than by one who works in conscious reference to technical rules. To give an example; in the introduction to the first edition of his *Herman and Dorothea*, Dr Whewell wrote a line thus:

"And the tumult of day-life, softening, sinks to its slumber."

Mr J. C. Hare said the line "seems to want a *cæsura*. A line that divides thus into two halves appears to me objectionable." Accordingly the line was altered to,

"And the tumult of day sinks softening into its slumber."

But it may be doubted whether the line is really improved by the *cæsura*.

In *Fraser's Magazine* for December 1847, pages 665...670, we have a *Dialogue on English Hexameters*. Here *Marcus* condemns and *Ernest* praises English Hexameters. *Ernest* quotes with approbation various passages from the *English Hexameter Translations*. Göthe's *Herman and Dorothea* is said to be consummate in its narrative interest and dramatic truth.

In *Fraser's Magazine* for March 1848, pages 295...298, we have a review of Longfellow's *Evangeline;* there are several quotations, and the work is highly commended. No special remarks are offered on the structure of the verse, except that it is said: "In general, Mr Longfellow's hexameters are good.

They have, without doing any violence to the pronunciation, the mixed trissyllable and dissyllable flow, which is the character of this kind of verse." *Herman and Dorothea* is here described as "the most perfect of domestic epics, the Odyssey of the nineteenth century,—the poem more likely to be familiar with our grandchildren than any other which the past generation has produced." I cannot follow Dr Whewell in his enthusiastic estimates of German poetry; I venture to agree with Mrs Browning, who has herself made some worthy additions to the treasures which she praises so highly: "This English people— has it not a nobler, fuller, a more abounding and various literature than all the people of the earth, past or present, dead or living, all except one—the Greek people?" See the *Contemporary Review*, April 1874, page 812.

In this review of *Evangeline* Dr Whewell says of American poetry, "its beauties have been rather felicitous adaptations of the jewels of the English Muses than any new gems brought to light from the rocks of the Alleghanies, or the sands of the prairies." The last clause is, according to rhetorical rules, rather awkward, as the general term *prairies* does not well balance the particular term *Alleghanies*. Dr Whewell had the satisfaction of knowing that his approbation of *Evangeline* delighted the author.

In *Fraser's Magazine* for January 1849, pages 103...110, there is a review of Mr Clough's *Bothie of Toper-na-Fuosisch*. A copy of this is bound up with various articles which are known to be by Dr Whewell, in a volume in the Library of Trinity College; but I doubt whether it was really written by him. It is a very vigorous article, but seems to me not in Dr Whewell's style, especially the beginning and some passages on pages 108 and 109. Moreover, although in other places, as we shall see, Dr Whewell speaks highly of Mr Clough's poem, yet it is in more moderate terms than are here employed, where we read of "some of the most perfect hexameters, in our humble opinion, which we have yet seen in the English language." There is no allusion to the review in Dr Whewell's correspondence. Moreover, Mr Clough says, that "spondaic lines are almost the rule;" and the reviewer remarks, "as we humbly think, a very good rule:" now this is contrary to the principle which Dr Whewell most strenuously maintains.

In *Fraser's Magazine* for March 1849, pages 342...347, we have *Dialogues on English Hexameters*, No. II. The subject is Mr Clough's poem. *Marcus* asks, "Do you rejoice in it as a proof of the progress of hexametrical faith, or lament over it as a manifestation of hexametrical heresy?" The poem is highly praised on the whole, and many quotations are given; but *Ernest* is sorry that Mr Clough did not make all his hexameters as good as he easily could have done. At the end *Marcus* quotes some verses from *Punch;* I will venture to give three lines:

> Tall *she* moved through the bar, a sarvin' of juleps and cocktails;
> Sweetnin' the cobblers with smiles, and firin' Havannahs with glances,—
> Nathan J. Bowie's fair darter, splendiferous Miss Dollarina.

Marcus asks, "How say you of *Punch's* hexameters?" *Ernest* replies, "Ah! spare my feelings! You know I must think such pleasantries profane. But the verses are very good as to rhythm." I had always placed these verses at the head of English hexameters, and was amused to find a copy of them carefully preserved among the papers of Dr Whewell.

In *Fraser's Magazine* for April 1849, pages 454...457, there is a translation of Voss's *Luise First Idyll, The Banquet in the Wood.* It occupies about three pages, and at the end we have the words *To be continued.* I do not think however that any more was printed; but there is a rough draft of 45 additional lines among Dr Whewell's papers.

A letter by Dr Whewell signed W. W. appeared in the *Athenæum*, No. 1125, May 19, 1849. It relates to a matter which had been noticed in Nos. 1121 and 1124 of the *Athenæum*. A correspondent drew attention to the opinion which M. Philarète Chasles had expressed as to Longfellow's *Evangeline:* the eminent critic appeared to consider the poem to be alliterative. The first two sentences of Dr Whewell's letter indicate its nature. "I have been in the habit of admiring M. Philarète Chasles's acquaintance with our language and literature—and see additional reason for such sentiments in his letter to you on the subject of Mr Longfellow's 'Evangeline.' But I venture to surmise, from what is there said, that M. Chasles is not quite fully informed on the history of English hexameter poetry." The letter concludes with an example of the hexameter which M. Chasles had given, but which had apparently been wrongly divided; so Dr Whewell divided it thus:

I ăm | yoūr mŏst | hūmblĕ ănd | vĕrў ŏ | bēdīĕnt | sūrvănt.

In *Fraser's Magazine* for January 1850, pages 33...40, we have an article entitled *Göthe's Herman and Dorothea*. This is in fact a review of a new translation into English hexameters published anonymously in 1849. The new translator is blamed for having frequently done violence to the natural pronunciation and accent of English words. Dr Whewell's own translation is referred to as the *former* translation or the *earlier* translation. The new translator frequently neglects the *cæsura*, and besides his defects as to versification falls into errors as to the sense. The following passages in the review may be noted:

"The mention of the *spondee* leads us to remark that the present translator has continued the mistake by which several recent English hexameterists have marred their lines, of often placing a spondee at the end, instead of that which the English ear really demands, a trochee........."

"The former translator of the poem to whom we have referred had not learnt, at the time when his translation was made, to avoid the practice. Thus he wrote:—

Well I would never, to see such a sight, go out in the *hot day*.
But the housewife started at last, and said, as she *lookt out*.
So they went in all, and all were glad of the *cool air*.

But some of the other translations which are published in the same volume—for instance, that of Schiller's 'Walk,' and those of Homer—are by more skilful hands, and are free from this blemish."

Towards the end of the review some curious coincidences are noticed between the new translation and the earlier one. "These are so remarkable as at first to suggest the notion that the new translator must have seen the older attempt, though he nowhere, we think, mentions it :...... But we are disposed to regard these coincidences as evidence, not that the second translation owes any thing to the first, but that the problem of translating German hexameter verse into English with great closeness is very nearly a *determinate problem,* and when worked out by different persons, will give almost the same result; or at least, that this will be the case with regard to many passages." This is curious, but it may be doubted whether it is well established; at any rate those who have translated Homer into hexameters, seem to vary sufficiently in their forms of expression. Mr Lockhart, in writing to Dr Whewell, drew attention to the difference in language between his own version and that of Dr Hawtrey in dealing with the same passage, though they agreed in general effect. At the end of the review we read, "the earlier translation has already been twice reprinted in the United States; and this has been done, as is somewhat too commonly the case with our cousins on that side of the Atlantic, without any acknowledgment of its English origin."

In the third volume of Miss Winkworth's *Life and Letters of Niebuhr,* 1852, there are some hexameter translations of Latin poems of the eighth and twelfth centuries. These were written by Dr Whewell, at the request of Bunsen; the manuscripts are preserved among Dr Whewell's papers.

The late Mr James Inglis Cochrane was a very earnest cultivator of English hexameters; much correspondence passed between him and Dr Whewell on the subject although they never had any personal intercourse. The correspondence began in December, 1852; at that time Mr Cochrane had published translations of *Herman and Dorothea,* and of Voss's *Louisa;* afterwards he published a translation of *Hannah and her Chickens.* Mr Cochrane found relief in his literary pursuits during many years of illness and suffering; he translated the whole of the Iliad into hexameters, and had his version printed. He survived Dr Whewell, a few weeks, dying on the 19th of May, 1866; his translation of the Iliad has not been published, but some copies were distributed among his friends after his death.

At the request of Mr Cochrane Dr Whewell wrote an article on *English Hexameters* which appeared in the *North British*

Review of April 1853, where it occupies pages 129...150. The article is a good account of what we may call the history of English Hexameters; it notices the early efforts of Sidney and Spenser, alludes to Southey and Klopstock, and quotes several passages from Mr Clough's *Bothie of Toper-na-Fuosisch.*

A passage is given, translated by William Taylor of Norwich, from Klopstock's Messiah; and it is said, "These lines are, for the most part, not only smoothly rhythmical enough to satisfy any unsophisticated ear, but graced with a significant variety of pauses, such as we admire in Milton's blank verse." I will print the passage continuously, that the reader may exercise himself in resolving it into hexameters: "So at the midnight hour draws nigh to the slumbering city pestilence. Couch'd on his broad-spread wings lurks under the rampart death, bale-breathing, as yet unalarm'd the inhabitants wander; close to his nightly lamp the sage yet watches; and high friends over wine not unhallow'd, in shelter of odorous bowers, talk of the soul and of friendship, and weigh their immortal duration. But too soon shall frightful death in the day of affliction pouncing, over them spread; in a day of mourning and anguish; when with, wringing of hands, the bride for the bridegroom loud wails."

The book which has already been noticed, *English Hexameter Translations,* is mentioned in the article, and the names of the authors given. It is said of the pieces, "Some of them are excellent specimens of hexameters; some a little harsh; among which we may note the translation of 'Herman and Dorothea.'" From the *English Hexameter Translations* the translation of Schiller's *Columbus* is quoted, which was one of Dr Whewell's pieces. The last two lines are,

"Nature with genius holds a pact that is fixt and eternal:
All which is promised by *this, that* never fails to perform."

It is said that "The latter distich was quoted, with great effect, by an illustrious German, in speaking of the discovery of the planet Neptune by an English and a French mathematician, before it had been disclosed by observation."

In 1856 Dr Whewell printed for private circulation some

Elegiacs on the death of his wife: I have already adverted to these on page 217.

In *Macmillan's Magazine* for September 1860, pages 383...385, we have a piece entitled *Priam and Hecuba;* this consists of a translation of 92 lines from the Iliad, Book XXII, into English hexameters. Dr Whewell says that "the scene preceding the death of Hector is, perhaps, the most pathetic picture in the whole range of poetry." Whether it be owing to less care on the part of the writer, or to a stricter obedience to the rules which he had laid down, I cannot say; but these hexameters seem to me less happy than others which he wrote. The following may serve as a specimen :

> " But on the rest of Troy that grief will lightlier press if
> Thou too, my son, fall not, smit down by the spear of Achilles.
> Nay but, O son, return to the wall, that yet thou mayest save the
> Sons and daughters of Troy, nor feed the glory and pride of
> Him, Pelides, and so may'st escape the omen'd disaster.

Dr Whewell's correspondent, Mr Cochrane, complained of the use of trochees instead of spondees, and sent as a specimen the first seven lines altered according to his views. I will first give Dr Whewell's version, and then Mr Cochrane's.

> " Thus, flying wild like deer, to their city hurried the Trojans;
> There from their sweat they cool'd, and assuaged the rage of their hot thirst,
> Leaning against the crest of the wall; and *on* the Achaians
> Nearer came, with their shoulders join'd, close locking their bucklers.
> But outside to remain, his malign fate, Hector ensnarèd,
> There in front of the Ilian wall and the Skaïan portals.
> And thus then to Pelides outspake Phœbus Apollo :"

> " Flying like wild deer, back to the gate the Trojans retreated;
> Here they their hot thirst quenched, and the warm sweat washed from their
>   bodies,
> Leaning the while on the crest of the wall; but the Argives, advancing,
> Nearer approached with their shoulders joined, close-locking their bucklers.
> Destiny dire howbeit ensnar'd brave Hector to linger
> Still outside at the Scaïan portals in front of the city.
> Whereupon Phœbus Apollo addressed thus godlike Peleides :"

In *Macmillan's Magazine* for April 1862, pages 487...496, Dr Whewell reviewed the translation of the first twelve books of the Iliad into hexameters by Mr Dart. The review begins by defending English hexameters against Dean Alford, and a writer in *Fraser's Magazine*. Some praise is awarded to Mr Dart, but he is blamed for *forcing the rhythm*, especially in the case of proper names. A remark on page 488 might lead to some confusion between the names of Dr Hawtrey and Mr Lockhart. It is said that Dr Hawtrey translated a great part of the sixth Book, and that Mr Lockhart "had translated, perhaps even better than Dr Hawtrey, the Parting of Hector and Andromache...." But Mr Lockhart really translated the *largest* part of the sixth Book, namely, lines 236...516; while Dr Hawtrey translated only lines 394...504, which consist of the Parting of Hector and Andromache. Dr Whewell, in his correspondence with Mr Cochrane, speaks of Herschel's *Walk* and Lockhart's *Homer* as the best English hexameters. On page 490 Dr Whewell anticipates that a good version of Homer into English hexameters will obtain currency; but as yet this has not happened. Sir John Herschel's seems to have been but coldly received, and Lord Derby's is perhaps the only very popular translation among the many that have appeared in recent years.

I have already said that Mr Cochrane translated the whole of the Iliad into hexameters; the first Book was published during his life-time. We see from his preface to it that there was a very important difference of principle between him and Dr Whewell; for he adhered to the spondee while Dr Whewell held that writers of English hexameters erred by "hankering after spondees." The following is an extract from Mr Cochrane's preface : "The difference between a hexameter line composed of dactyls and spondees, and one composed of dactyls and trochees, cannot be better illustrated than by three well-known hexameter lines occurring in the Sacred Scriptures:

'Why do the heathen rage, and the people imagine a vain thing?'
'How art thou fallen from heav'n, O Lucifer, son of the morning!'
'Husbands, love your wives, and be not bitter against them.'

The first two lines are composed of dactyls and spondees, for as 'heathen' ends with a consonant, it may be considered a spondee; the last line of the three is composed entirely of trochees, except the fifth foot. Now, let any person, even though unacquainted with hexameter verse, read these lines aloud, and it will be at once obvious that the first two lines have a majesty and harmony of flow totally wanting in the third line: indeed the third line can hardly be considered verse at all." Still however some persons may assert that the third verse is at least as good as the first, and that there is something arbitrary in asserting that the third line is composed entirely of trochees, except the fifth foot.

In *Macmillan's Magazine* for August 1862, pages 297...304, there is an article by Dr Whewell entitled *New Hexameter Translations of the Iliad.* This is a review of two translations of the first Book of the Iliad, one by Sir John Herschel, and one by Mr Cochrane. The difference in the versification of the two is that Sir John Herschel prefers trochees, while Mr Cochrane strives for spondees; Dr Whewell, according to his established principles, prefers the former. But he is much more tolerant of the spondees than in previous articles, and he admits that some readers may prefer them as more classical. In a letter he allows that Mr Cochrane's verses are good *German* hexameters, but thinks them not good English hexameters. By a comparison of various passages of the two translations, Dr Whewell concludes that Sir John Herschel's is superior as to the rendering of the meaning.

This article is one of the most interesting of those which Dr Whewell published on the subject, and being the last may be regarded as containing his final opinion. A few sentences from the first paragraph may be quoted: "As was long ago said in the discussions on this subject [English hexameters], the fallacy of the impossibility of such verses is like the ancient fallacy of the impossibility of motion—*solvitur ambulando.* Southey was convinced of this; and intending to make English hexameter verse current, he proposed, as he writes to his correspondent Taylor, to march down upon the English public twenty thousand strong. Unluckily the result of this project was that most unhappy performance, *The Vision of Judgment.* And yet the introductory verses to that poem are allowed by all to be beautiful in rhythm as in expression." Dr Whewell took a lively interest in the progress of Sir John Herschel's translation of the Iliad, and read much of the manuscript, as appears from the correspondence between them during the years 1862...1865. The book was dedicated to him.

Some of Dr Whewell's friends seem to have amused themselves by sending indifferent hexameters to him; I will venture to print two specimens from eminent scholars and divines.

> " I, ..., to the Head of Trinity, greeting,
> Thanks, many thanks, Master, for your most acceptable present,
> Just come from Murray's shop, but sent by the Editor also,
> Brimfull of hexameters, translated all from the German,
> Except some very few from Greek original authors—
> Almost impossible, in our own inharmonious English
> Harsh words to modulate, and arrange in rhythmical order,
> Such as might please the ear of a staunch admirer of Homer.
> Quantity, which regulates all the flow of classical verses,
> With well known fixt laws, in our tongue is an element unknown,
> While accent ill supplies its place in versification.
> That which I complain of, in our English hexameters,
> Is a want of spondees, and the too great number of dactyls,
> But still more I object to the want of a natural pause, or,
> Suspension of the verse, *cæsura* we commonly call it.
> But time forbids me to extend my remarks any further,
> So no more just now from your most obedient Servant."

> " Master of Trinity College, to day the .... Feoffees
>   Met in their new School room, cast up their many accounts.
> Lo! they found to their sorrow that there was a balance against them
>   Small tho the debt it is high whenas the pocket is low.
> There stand fully against them in horrible characters marshall'd
>   Forty two pounds and six shillings and thirty four pence.
> So they humbly entreat that you would send your subscription,
>   And as much more as you may reckon it proper to give.
> Duly rememb'ring the spot where you learned hexametral music,
>   Bidding in due response nimble pentameters roll.
> Thus I write representing Feoffees and Master and Scholars
>   Who of dinner are now fuller by far than of verse.
> Thus I compose in spirit and letter stans pede in uno.
>   Thanks that I'm long with one leg—Homer was short of two eyes.
> O let our cause erect before thee stand upon two legs:
>   May for accents weak quantity come in return,
> Lame as Tyrtæus of old may I stir thy masculine mind up.
>   ' Glorious it is for a man bravely to bleed for his school.' "

These lines received a favourable reply, for in the course of a week the writer says: "Many thanks for your kind enclosure which I have forwarded to our Treasurer. I am proud of your approval of my Elegiacs. But I cannot help thinking that the Hexameter constructed according to the Latin accent would be best, if our language could provide words long enough for it—which however it cannot."

The papers of Dr Whewell include many letters from correspondents on the subject of English Hexameters; they shew that he did not make many converts to his own firm belief in the attractions of this kind of poetry. I will give extracts from three of these letters:

(1) "You and Herschel have not quite satisfied me that the ancient metres are well suited to our language. But you have satisfied me of what is much more important—that a genius for poetry and a genius for the several sciences are perfectly compatible."

(2) "But it is not a measure which I should wish to see much of in our poetry. Deal with it how we may, to my ear the cadences are too pronounced and importunate; and the fall at the close of every line is against elasticity and vigour. I think the consonantal effects hardly have fair play in it, and the rhythmic effects generally which are most favoured by it seem to me to be those which belong to visionary and picturesque poetry, rather than to that which is significant or impassioned."

(3) "I have been reading Herman and Dorothea with great delight, and found much in it of general truth and very noble ideal of manhood and womanhood—much to be thought upon... In the first place I cannot get over the hexameters. They seem to me easy to write if,—once you assume you may stop when you like in the midst of a sentence;—into an uncouth order the words of average English—forcing by pokes in the ribs—as a crowd is poked by policemen. Not without loss of sense; and the trampling down of the weakest words in the stream of the metre. But if this power were denied you (Power said I—license rather, which ends in doubt and disorder) much I muse whether verses like these could be written in English. I see there is an awkward tendency to run constantly into

dactyls—and I should not like to have to write all my letter so—but it seems to me there is little difficulty in getting on, if these licences be taken—especially of transposition ; and that the difficulty of a metre enhances our delight in the power which vanquishes it into pure English and flowing melody. But I cannot tell how much of this feeling is mere prejudice—no—not prejudice, but true liking and disliking excused by custom—not based on principle.

For the manner of the Story itself I like its Homeric tone —but I don't honour Goethe for taking an Homeric tone. I dislike in general the feeling of a resolute effort to imitate—a man may be involuntarily affected by another's style or modes of feeling ; but here there is *affectation* of parallelism. Then the story is a sufficiently absurd one—for this or the last century. Prudent young men do not bring home a strolling girl and betroth themselves to her 24 hours after seeing her, whatever may be the result of their 'enquiries as to character.' It reminds one immensely of buying a horse."

CHAPTER XVI.

SCIENTIFIC MEMOIRS.

I SHALL give in this Chapter a short account of Dr Whewell's Memoirs contributed to various scientific societies and journals. I follow the order of the Royal Society's *Catalogue of Scientific Papers*, and take the numbering from that catalogue. The papers relating to the Tides have been noticed in the Chapter on that subject, so that I need not recur to them.

1. On the Position of the Apsides of Orbits of great Excentricity. This memoir occupies pages 179...191 of Vol. I of the *Cambridge Philosophical Transactions;* it was read on April 17, 1820. The following is the proposition investigated: Let there be a central force which is constant, or which varies according to a power of the distance; let a particle be projected at right angles to the distance with a velocity which is *very small,* compared with the velocity in a circle at that distance; then it is required to determine the angle described round the centre of force before reaching a second apse. If the force is constant, or varies as any direct power of the distance, the angle is found to be very approximately a right angle. If the force varies inversely as the nth power of the distance, where n is between 1 and 3, the angle is found to be very approximately $\dfrac{1}{3-n}$ of two right angles.

2. On Double Crystals of Fluor Spar. This memoir occupies pages 331...342 of Vol. I of the *Cambridge Philosophical Transactions;* it was read on November 26, 1821. These crystals in their most common forms are cubes, but there is a peculiarity which is thus described: " ...we see sometimes a large cube, with

only one corner of a smaller one appearing above its surface; and at other times, two crystals, nearly equal, seem to penetrate and pierce through each other in a very curious manner, which is best understood by specimens." The calculation of the angles between the edges and the faces thus exhibited is effected by the aid of some simple Trigonometry, Plane and Spherical. A few remarks occur at the end respecting what are called the *integrant molecules* of crystals.

3. A General Method of calculating the Angles made by any Planes of Crystals, and the laws according to which they are formed. This memoir occupies pages 87...130 of the *Philosophical Transactions* for 1824; it was read on November 25, 1824. We have here investigations relating to the angles of planes and straight lines, conducted by the aid of Geometry of Three Dimensions; the crystals themselves which suggest the problems are not brought very prominently forward, so that the whole memoir is intelligible to a person with an elementary knowledge of mathematics. The formulæ required are principally two, which are given in the memoir forming number 5 of the present series. At the end of the memoir reference is made to a paper by Mr Levy, in which the principle at the basis of Dr Whewell's investigations is employed; but the formulæ in the memoir were mostly calculated before that paper fell under Dr Whewell's notice. Dr Whewell consulted Sir J. Herschel as to the destination of this memoir; and he decided in favour of the Royal Society. In a draft of a letter preserved among Sir J. Herschel's papers, but not sent to Dr Whewell, it is recorded that the method of the memoir had already presented itself to his mind. Sir Humphry Davy, then President of the Royal Society, spoke highly of the memoir in a letter to its author; he said, "It has appeared to our best judges on this subject to be an admirable application of mathematical to physical science, a species of investigation which the Royal Society has every disposition to encourage." A short notice of the memoir is given in the *Edinburgh Journal of Science*, Vol. II, 1825, pages 312...315.

4. On the Rotatory Motion of Bodies. This memoir occupies pages 11...20 of Vol. II of the *Cambridge Philosophical Trans-*

actions; it was read on May 6, 1822. The problem of the motion of a rigid body round a fixed point had been discussed by D'Alembert, Euler, and Lagrange; and their results were in agreement. Landen, an English mathematician of some reputation, had in 1785 and 1789 published opinions quite at variance with the conclusions of the great foreign writers. The present memoir urges an ingenious argument against Landen. A particular rigid body is taken, namely, a system of three particles in rigid con- nexion; then by using equations of motion for particles, which were universally accepted, formulæ are obtained coinciding with those of Euler, and therefore contradicting those of Landen. It seems scarcely possible that any person who might have been led astray by Landen's authority could have failed to submit to the reasoning of the memoir.

5. On the Angle made by two Planes, or two Straight Lines, referred to three oblique Co-ordinates. This memoir occupies pages 197...202 of Vol. II of the *Cambridge Philosophical Trans- actions;* it was read on November 24, 1823. Two problems in Geometry of Three Dimensions are here solved, the first by Dr Whewell, and the second by Mr Lubbock. The elements by which the oblique system is determined are not the same in the two problems; in the former they are the angles between the co-ordinate planes, and in the latter they are the angles between the co-ordinate axes. The modern elementary books when they treat the subject usually employ the elements adopted in the latter problem.

6. On the Classification of Crystalline Combinations, and the Canons by which their Laws of Derivation may be investigated. This memoir occupies pages 391...425 of Vol. II of the *Cambridge Philosophical Transactions;* it was read on November 13, 1826. A scheme of notation is given by which the forms of crystals may be denoted, and this is applied to a long list of crystals; the whole is illustrated by five plates. We read on page 395, "The demonstrations of the preceding theorems will be given in another place. To complete the subject it will also be requisite to give methods of transforming the symbols, which are employed in other systems of notation hitherto proposed, (those of Hauy,

Weiss, Mohs, &c.) into this system. This will be done by means of certain formulæ, which we may call *formulæ of transformation.*" The design here announced does not seem to have been carried into effect.

7. Reasons for the Selection of a Notation to designate the Planes of Crystals. This memoir occupies pages 427...439 of Vol. II. of the *Cambridge Philosophical Transactions*; it was read on February 11, 1827, for the date given, 1826, is obviously a misprint. This memoir takes as a basis the notation of Mohs, but alters it so as to render it less repulsive to a person of mathematical tastes, and more useful for crystallography. The subject is treated briefly in the *Novum Organon Renovatum,* pages 357...359. On the first page of the memoir there is an allusion to a certain point, and we are told that "the discussion of it must be reserved for another opportunity:" and on page 437, after an allusion to the designation of twin crystals, we are told, "but this will be considered hereafter." There seems however to be nothing corresponding to these two announcements.

8. Specimen of the use of Notation in the Analysis of Crystalline Forms. This paper occupies pages 1...6 of the *Edinburgh Journal of Science,* Vol. VI, 1827. The notation proposed in number 3 of this series is noticed towards the beginning of the paper in these words: "Such a notation is that of Weiss, which differs very little from a system which I had proposed in the *Philosophical Transactions* (Part I, 1825), without being aware that any similar method had been elsewhere brought forwards." But the present paper asserts "that it is possible to obtain, in almost all cases, the laws of crystalline derivation more simply" than by the method of the former memoir. An example is taken; but the *enunciations* only of the theorems which it is necessary to use are given.

9. On the principles of Dynamics, particularly as stated by French Writers. This paper occupies pages 27...38 of the *Edinburgh Journal of Science,* Vol. VIII, 1828. Dr Whewell maintains that we require three distinct laws of motion, in opposition to the French writers, who reduced the second and third laws to one

principle; he enforced his opinion in number 23 of this series, and in the *Philosophy of the Inductive Sciences*.

10. Mathematical Exposition of some Doctrines of Political Economy. This memoir occupies pages 191...230 of Vol. III of the *Cambridge Philosophical Transactions*; it was read on March 2 and 14, 1829. The design of the memoir is very simple. Taking certain principles adopted by some of the writers on Political Economy, it proposes to deduce, by the aid of Algebra, the necessary consequences of these principles. Thus the resources of the most generalised forms of calculation are employed instead of ordinary arithmetic; and the example thus given of the use of mathematics might be imitated with great advantage by the professional cultivators of Political Economy. Comte is said to have advised students of Natural Philosophy to become sufficiently familiar with mathematics to be able to check the abuses arising from the treatment of physics by the extreme analysts; and in like manner Political Economists might guard themselves if they anticipated any dangers from the incursions of mathematicians. The problems discussed in the memoir relate to the distribution of the burden of taxes which affect the land, between the landlord, the cultivator, and the consumer; the special circumstance which has to be regarded being, that if any burden is thrown on the land some of the poorest sorts may go out of cultivation. It is conceivable that the mode of cultivation may be changed, but whether land to any appreciable extent is ever thrown out of cultivation may be doubted; the supposition that such is the case, however, is frequently made by theoretical writers.

11. Observations on some passages of Dr Lardner's Treatise on Mechanics. This paper occupies pages 148...155 of the *Edinburgh Journal of Science*, new series, Vol. III, 1830. In his Treatise on Mechanics, Dr Lardner had spoken of Newton's three laws of motion as having little or no utility; and Dr Whewell combats this opinion. The paper is very interesting, especially in tracing the consequences of supposing that bodies, by a universal law, tend to move more and more slowly the longer they continue in motion. This imaginary state of things is alluded to in

a note to the *Philosophy of the Inductive Sciences*, second edition, Vol. I, pages 219 and 220.

12. Progress of Geology. This is mentioned in the Royal Society's Catalogue as occupying pages 242...267 of the *Edinburgh New Philosophical Journal*, Vol. XI, 1831. On examining these pages, however, they will be found to consist of passages attributed to various authors, the first and last only to Dr Whewell, which are extracts from his review of Lyell's work on Geology.

13. On the employment of Notation in Chemistry. This paper occupies pages 437...453 of the *Journal of the Royal Institution of England*, 1831. According to Dr Whewell, the greater part of English chemists had "been hitherto averse from the practice of using a technical and mathematical notation to express the chemical composition of bodies." On the continent the notation of Berzelius had been used, notwithstanding several anomalies and inconveniences. Accordingly the paper proposes "to remove the gross anomalies and defects with which the foreign notation is disfigured, and to reduce it, with as little change as possible, to mathematical symmetry and consistency." The notation recommended is very simple and expressive, and its merits are clearly set forth in the paper. The subject was briefly noticed by Dr Whewell at a much later period: see the *Novum Organon Renovatum*, pages 359...364. In the *Philosophical Magazine* for August, 1831, there is a paper by Mr J. Prideaux, the object of which is to defend the chemical notation of Berzelius against the criticisms of Dr Whewell. It is, I think, to this paper by Dr Whewell that Mr Faraday alluded in a letter dated February 21, 1831, in which he says: "Your remarks upon chemical notation with the variety of systems which have arisen with regard to notation, nomenclature, rules of proportional or atomic numbers, &c., &c., had almost stirred me up to regret publicly that such hindrances to the progress of science should exist. I cannot help thinking it a most unfortunate thing that men who as experimentalists and philosophers are the most fitted to advance the general course of science and knowledge, should by the promulgation of their own theoretical views under the form of nomenclature, notation, or scale, actually retard its progress.

It would not be of much consequence if it was only theory and hypothesis which they thus treated, but they put facts or the current vein of science into the same limited circulation when they describe them in such a way that the initiated only can read them."

On pages 224...240 of the *Journal of the Royal Institution of England*, 1821, there is a paper entitled "Description of a Mode of erecting light Vaults over Churches and similar Spaces. By M. de Lassaux. (Communicated by Professor Whewell of Cambridge.)" It may be presumed that this article was translated by Dr Whewell from the original, in French or German. It is of a technical character, suited to architects and builders. It concludes with a modest but rather timid reason, given by M. de Lassaux for his resolution of limiting himself in future to the round-arch style: "This subject is, moreover, the more agreeable, because less that is fine in it has come down to us; while, on the contrary, our buildings in the pointed style, compared with those of the ancients, always appear more or less as a miserable subterfuge."

14.   On Isomorphism. This paper occupies pages 401...412 of the *Philosophical Magazine* for December, 1831. Dr Whewell here maintains the importance and utility of the doctrine of Isomorphism against the contrary opinion which had been advanced by Mr Brooke in the *Philosophical Magazine* for September, 1831. See also on this subject the *History of the Inductive Sciences*, 3rd edition, Vol. III, page 189, and the *Philosophy of the Inductive Sciences*, 2nd edition, Vol. I, page 464. There is a note on pages 405 and 406 of this paper on Isomorphism which alludes to the remarks on chemical notation made by Mr J. Prideaux. See the account of number 13 of this series.

15.   Report on the recent Progress and Present State of Mineralogy. This paper occupies pages 322...365 of the first volume of the Reports of the British Association; it was presented to the meeting at Oxford in 1832. Much of the substance of this Report was embodied in the parts of the *History* and of the *Philosophy* of the Inductive Sciences which are devoted to Mineralogy. A remark made on page 338 relative to a mineral

examined by Mr Brooke was thought by him to be incorrect, and he wrote on the subject to Dr Whewell, who attended to the matter in a note on page xviii of the next volume of the Reports of the British Association.

16. Mathematical Exposition of some of the leading Doctrines in Mr Ricardo's "Principles of Political Economy and Taxation." This memoir occupies pages 155...198 of Vol. IV of the *Cambridge Philosophical Transactions;* it was read April 18 and May 2, 1831. Certain Postulates are taken which are the foundation of the doctrines of Mr Ricardo, and these are traced to their consequences by the aid of Algebra. Dr Whewell by no means accepts these Postulates himself, but as he says: "Perhaps, however, to trace their consequences may be one of the most obvious modes of verifying or correcting them." From page 177 of the memoir we learn the connexion between this and number 10 of the series. In the present memoir the principles of Ricardo are traced to their consequences; in the former memoir Dr Whewell followed Colonel Perronet Thompson's views in opposition to those of Ricardo, on an important point, and deduced results contrary to those of Ricardo. This memoir and also number 10 of the series are very interesting and instructive.

17. Essay towards a first approximation to a Map of Cotidal Lines. *Philosophical Transactions,* 1833.

18. Suggestions respecting Sir John Herschel's Remarks on the theory of the Absorption of Light by coloured media. This paper occupies pages 550...552 of the Report of the meeting of the British Association held at Edinburgh in 1834.

Sir John Herschel had made some remarks with the view of furnishing a possible explanation of the phenomena of the Absorption of Light on the principles of the Undulatory Theory; it is the object of the paper to add a little confirmation to the proposed explanation.

19. Remarks on a recent Statement by Berzelius respecting the Use of Chemical Formulæ. This paper occupies pages 9 and 10 of the *Philosophical Magazine* for January, 1834.

Berzelius replied to the criticisms contained in number 13

of this series; here Dr Whewell notices the reply, and maintains the justness of his original criticisms.

20. On the Empirical Laws of the Tides in the Port of London, with some Reflexions on the Theory. *Philosophical Transactions*, 1834.

21. Report on the Recent Progress and Present Condition of the Mathematical Theories of Electricity, Magnetism, and Heat. This paper occupies pages 1...34 of the Report of the meeting of the British Association held at Dublin in 1835; it is of the same valuable character as the other historical papers which enrich the volumes published by the British Association. The report is reproduced in the relative portions of the *History of the Inductive Sciences* but in a condensed form; in particular the mathematical allusions in the report are omitted in the *History*.

22. On a New Anemometer. The Report of the meeting of the British Association, held at Dublin in 1835, contains a few lines concerning the anemometer which Dr Whewell invented, and which he exhibited at the meeting: see page 29 of the *Transactions of the Sections* in the volume. The Report of the meeting of the British Association, held at Bristol in 1836, contains a little more about the anemometer: see pages 39 and 40 of the *Transactions of the Sections* in the volume. One of the anemometers was at this date erected at Cambridge, and others were about to be erected at York and at Plymouth. There is an account of the working of one of Dr Whewell's anemometers, from July 1840 to July 1841, in pages 36 and 37 of the Report of the meeting of the British Association held at Plymouth in 1841. For accounts of other anemometers, and for improvements in Dr Whewell's construction, see pages 340...346 of the Report of the meeting of the British Association held at Southampton in 1846; also pages 111 and 112 of the *Transactions of the Sections* in the volume. See also the references to various volumes of the Reports in the *Index to the Reports and Transactions of the British Association* from 1831 to 1860.

In the *Transactions of the Sections* of the Report of the meeting of the British Association held at Dublin in 1835, there are some remarks by Dr Whewell on the Tides on page 6, and on the application of physical science to geology on pages 65 and 66; there are also a few lines respecting the Tides on page 130 in the *Transactions of the Sections* of the Report of the meeting held at Bristol in 1836.

23. On the Nature of the Truth of the Laws of Motion. This memoir occupies pages 149...172 of Vol. v of the *Cambridge Philosophical Transactions;* it was read on Feb. 17, 1834. Dr Whewell maintains that each of the three laws of motion involves a necessary and an empirical element. Thus, for instance, with respect to the first law he finds the necessary element to be, velocity does not change without a cause; and the empirical element to be, the time during which a body has already been in motion is not a cause of change of velocity. The doctrine is substantially the same as was afterwards enforced in the *Philosophy of the Inductive Sciences;* and the memoir is reprinted in the Appendix to the second edition of that work.

24. On the Results of Tide Observations made in June 1834, at the Coast Guard Stations in Great Britain and Ireland. *Philosophical Transactions*, 1835.

25. On the probable importance of Tide Observations at the Cape of Good Hope. South African Quarterly Journal, Vol. ii, 1835, pages 367...372. I have not been able to see this.

26. Remarks on a Note on a Pamphlet entitled "Newton and Flamsteed," in No. cx of the Quarterly Review. This paper occupies pages 211...218 of the *Philosophical Magazine* for March, 1836; it appeared also in the *Cambridge Chronicle* of the period, and is reprinted in the second edition of Dr Whewell's pamphlet entitled "Newton and Flamsteed," which we have already noticed. The next two articles in the *Philosophical Magazine* are on the same subject; the first is by Professor Rigaud, and the second is signed C. S.: both agree with Dr Whewell against the Quarterly Review.

27. Researches on the Tides. Fourth Series. On the Empirical Laws of the Tides in the Port of Liverpool. *Philosophical Transactions*, 1836.

28. Researches on the Tides. Fifth Series. On the Solar Inequality, and on the Diurnal Inequality of the Tides at Liverpool. *Philosophical Transactions*, 1836.

29. Researches on the Tides. Sixth Series. On the Results of an extensive system of Tide Observations made on the Coasts of Europe and America in June, 1835. *Philosophical Transactions*, 1836.

30. On Tides. A few remarks on Tides occur on pages 4 and 5 of the *Transactions of the Sections*, in the Report of the meeting of the British Association held at Liverpool in 1837. Also on pages 32 and 33, there are a few remarks on the principle of Dr Whewell's anemometer.

31. Researches on the Tides. Seventh Series. On the Diurnal Inequality of the Height of the Tide, especially at Plymouth and at Singapore, and on the Mean Level of the Sea. *Philosophical Transactions*, 1837.

32. Researches on the Tides. Eighth Series. On the Progress of the Diurnal Inequality Wave along the Coasts of Europe. *Philosophical Transactions*, 1837.

33. Account of a Level Line measured from the Bristol Channel to the English Channel during the years 1837...38 by Mr Bunt. This paper is printed in the Report of the meeting of the British Association, held at Newcastle-on-Tyne in 1838. Pages 1...11 are by Dr Whewell; they are followed on pages 11...18 by Mr Bunt's explanation of the details of the process.

34. Report on the discussion of Tides. This occupies pages 19 and 20 of the volume just mentioned.

35. On the Results of Observations made with a new Anemometer. This memoir occupies pages 301...315 of Vol. VI of the *Cambridge Philosophical Transactions;* it was read on May 1, 1837. The memoir contains a brief description of the machine with directions for observing with it. In a plate we have an exhibition of the integral effect of the wind, as determined by one of the machines at Edinburgh and two of them at Cambridge, during the month of March, 1837. I may observe that *Directions for observing with Whewell's Anemometer* were printed separately

on an octavo page; they agree substantially with what we have in this memoir.

36. Researches on the Tides. Ninth Series. On the Determination of the Laws of the Tides from short Series of Observations. *Philosophical Transactions,* 1838.

37. On Tide-calculation. A few remarks on the Tides occur in the Report of the meeting of the British Association held at Birmingham in 1839: see pages 17 and 18 of the part of the volume devoted to reports, and pages 11 and 12 of the *Transactions of the Sections.* Also in the Report of the meeting held at Glasgow in 1840, we have on pages 436...439 a report on the application of part of a sum of money, voted at the Birmingham meeting, for work on the Tides; this is followed by a letter from Mr Bunt, giving an account of his calculations under the direction of Dr Whewell.

38. Remarks on Dr Wollaston's argument respecting the infinite Divisibility of Matter, drawn from the finite Extent of the Atmosphere. These occur on page 26 of the *Transactions of the Sections* in the Report of the meeting of the British Association, held at Birmingham in 1839. Dr Whewell maintains that Dr Wollaston's argument is quite baseless: see also the *Philosophy of the Inductive Sciences,* 2nd edition, Vol. I, page 436.

39. Rapport sur les progrès de la géologie, principalement en Angleterre, pendant l'année académique, 1837...38, L'Institut, Vol. VII, 1839, pages 175, 176, 183, 184, 211, 212. This is a translation of that part of Dr Whewell's Address to the Geological Society, delivered in February 1838, which relates to the progress of geology during the preceding year. Some details are given on points to which Dr Whewell only alluded.

40. Researches on the Trdes. Tenth Series. On the Laws of Low Water at the Port of Plymouth, and on the Permanency of Mean Water. *Philosophical Transactions,* 1839.

41. Researches on the Tides. Eleventh Series. On certain Tide Observations made in the Indian Seas. *Philosophical Transactions,* 1839 and 1840.

42. On the relations of Tradition to Palætiology. This paper occupies pages 258...274 of the *Edinburgh New Philosophical*

Journal, Vol. XXIX, 1840. It is taken from the *Philosophy of the Inductive Sciences;* see the second edition, Vol. I, pages 680...699.

43. Rapport sur les progrès de la géologie, principalement en Angleterre, pendant l'année académique, 1838...39, L'Institut, Vol. VIII, 1840, pages 4...6, 15, 16, 34, 35. This is a translation of that part of Dr Whewell's address to the Geological Society, delivered in February 1839, which relates to the progress of Geology during the preceding year.

44. On the mean Level of the Sea. This paper occurs on pages 321...325 of the *Philosophical Magazine* for November, 1840. The first paragraph indicates its nature and its connexion with the operation noticed in number 33 of this series. "In the Philosophical Magazine for August last, page 134, and the following, are some remarks by Mr Richard Thomas, on the Account of a Level Line, published in the Transactions of the British Association for 1838. The level line was carried from the Bristol Channel to the English Channel by Mr Bunt, under my direction: the conclusion which I drew from the operation was, that the *mean* height of the sea was at the same level at the two extremities of the line ; and this conclusion Mr Thomas disputes. I beg to offer a few observations on the subject, suggested by his remarks."

45. Researches on the Tides. Twelfth Series. On the Laws of the Rise and Fall of the Sea's Surface during each Tide. *Philosophical Transactions*, 1840.

46. Report on Discussions of Bristol Tides, performed by Mr Bunt. This paper occurs on pages 30...33 of the Report of the meeting of the British Association held at Plymouth in 1841. It contains the results of calculations to determine if any effect on the heights of the tides at Bristol is produced by atmospheric pressure. The average effect appeared to be about 15 inches depression of high water to 1 inch rise of mercury in the barometer.

47. Report on the Discussion of Leith Tide Observations, executed by Mr D. Ross. This paper occurs on pages 34...36 of the volume which contains number 46.

48. Demonstration that all Matter is heavy. This memoir occupies pages 197...207 of Vol. VII of the *Cambridge Philoso-*

phical Transactions; it was read on February 22, 1841. The memoir does not seem to me very convincing; and I have made a remark on it already: see page 138. The memoir was reprinted in the Appendix to the second edition of the *Philosophy of the Inductive Sciences,* and subsequently in the Appendix to the *Philosophy of Discovery.* Dr Whewell's opinion was controverted by Sir W. Hamilton, and Dr Whewell's reply is given in his *History of Scientific Ideas,* Vol. II, pages 37 and 38, and also in his *Philosophy of Discovery,* pages 330 and 331.

49. Discussion of the Question:—Are Cause and Effect successive or simultaneous? This memoir occupies pages 319...331 of Vol. VII of the *Cambridge Philosophical Transactions;* it was read on March 14, 1842. The discussion was suggested by some remarks made by Sir J. Herschel in his review of the *History and Philosophy of the Inductive Sciences.* The memoir decides with justice that strictly speaking cause and effect are simultaneous, that is the instantaneous effect or change is simultaneous with the instantaneous force or cause by which it is produced. But at the same time it explains how if "we consider a series of such instantaneous forces as a single aggregate cause, and the final condition as a permanent effect of this cause, the effect is subsequent to the cause." The memoir was reprinted in the Appendix to the second edition of the *Philosophy of the Inductive Sciences,* and subsequently in the Appendix to the *Philosophy of Discovery.*

50. On Glacier Theories. This and the next may be taken together.

51. Remarks on Mr Hopkins's reply to "Remarks on Glacier Theories." Three short papers relating to glacier theories were published by Dr Whewell in the *Philosophical Magazine,* Vol. XXVI, 1845: A controversy had arisen between Mr Hopkins and Professor J. D. Forbes; and Dr Whewell intervened at the request of the latter. The first paper is entitled *On Glacier Theories;* it occupies pages 171...173 of the volume. The second paper is entitled *Additional Remarks on Glacier Theories;* it occupies pages 217...220 of the volume. Mr Hopkins is not named; but he understood the papers to refer to himself; and he replied in

pages 334...342 of the volume. Dr Whewell's third paper is entitled *Remarks on Mr Hopkins's Reply to the previous Remarks on Glacier Theories;* it occupies pages 431...433 of the volume. Some copies of Dr Whewell's three papers were printed together so as to form an octavo pamphlet of eight pages.

52. Method of Measuring the Height of Clouds. This paper occurs on pages 15 and 16 of the *Transactions of the Sections* in the Report of the meeting of the British Association, held at Southampton in 1846. A person observes the angle of elevation of a cloud, and also the angle of depression of the image of the cloud seen in a lake; the latter angle is rather greater than the former: the height of the eye above the level of the lake is also supposed known. Then the height of the cloud can be determined by a simple Trigonometrical formula.

53. On the Wave of Translation in connexion with the Northern Drift. This paper occupies pages 227...232 of the *Quarterly Journal of the Geological Society* for August, 1847; it begins thus: "The great geological problem of the 'Northern Drift' has been attacked in various ways; and the diffusion of Scandinavian rocks and northern detritus over a vast area in the northern part of Europe has been ascribed to various kinds of natural machinery. Of late a large part of this operation has been attributed to 'Waves of Translation' produced by the sudden upheaval of the bottom or shore of the sea. This view is advocated in the 'Geology of Russia' by Sir Roderick Murchison. There are some very simple numerical calculations which belong to this subject, and which may throw some light on the probability of such a theory. These calculations must necessarily be hypothetical as to their quantities, but as to their quantities only: and even these will be capable of correction by a more careful survey of the facts." The calculation involves only simple mechanical principles; the conclusion is, that "450 cubic miles of water raised a mile high would produce an effect equivalent to the dispersion of the whole body of northern drift." And moreover "no action except such as is of a paroxysmal character could produce the effect."

54. Researches on the Tides. Thirteenth Series. On the

Tides of the Pacific, and on the Diurnal Inequality. *Philosophical Transactions*, 1848.

55. On the Fundamental Antithesis of Philosophy. This memoir occupies pages 170...181 of Vol. VIII of the *Cambridge Philosophical Transactions;* it was read on February 5, 1844. The doctrine of this memoir plays an important part in the *Philosophy of the Inductive Sciences*, and may be described as the basis of that work: see the Preface to the second edition. The memoir was reprinted in the Appendix to the second edition, and subsequently in the Appendix to the *Philosophy of Discovery*. In a letter to Mr Jones Dr Whewell refers to paragraph 29 of the memoir as an answer to the main argument of Mr J. S. Mill against his doctrine. On pages 614...620 of the same volume of the *Cambridge Philosophical Transactions* we have a Second Memoir on the Fundamental Antithesis of Philosophy; this was read on November 13, 1848. The substance of the memoir is reproduced in Chapter XXIV of the *Philosophy of Discovery;* see the note on page 300 of that work. There are two mistakes in that note; instead of 1840 read 1849; and omit the words " and the next." I have already said on page 188 that this memoir contains, in the form of a suggestion, an anticipation of the main position of the *Essay* on the Plurality of Worlds, namely, that man is the principal object of creation.

56. Of the Intrinsic Equation of a Curve and its Application. Second Memoir on the Intrinsic Equation of a Curve and its application. The first of these memoirs occupies pages 659...671 of Vol. VIII of the *Cambridge Philosophical Transactions;* it was read on February 12, 1849. The second occupies pages 150...156 of Vol. IX of the *Cambridge Philosophical Transactions;* it was read on April 15, 1850. The intrinsic equation to a curve is an equation connecting the length of the arc between a fixed and a variable point with the angle between the tangents to the curve at the two points. Dr Whewell introduced the subject to the notice of mathematicians in these two memoirs which are illustrated by numerous figures. The subject has since passed into the elementary text-books used at Cambridge.

57. On Hegel's Criticism of Newton's Principia. This memoir

occupies pages 697...706 of Vol. VIII of the *Cambridge Philosophical Transactions*; it was read on May 21, 1849. Hegel depreciated Newton in comparison with Kepler; but, as Dr Whewell shews, Hegel did not understand the subject sufficiently to render his opinion of any value. The memoir is a model of temperate and decisive criticism. As I have already said on page 135 Coleridge is as much to blame as Hegel in the matter. The memoir is reproduced in the Appendix to the *Philosophy of Discovery*.

58. Researches on the Tides. Fourteenth Series. On the Results of continued Tide Observations at several places on the British Coasts. *Philosophical Transactions*, 1850.

59. On our Ignorance of the Tides. This paper occurs on pages 27 and 28 of the *Transactions of the Sections* in the Report of the Meeting of the British Association held at Ipswich in 1851. The object of the paper is to point out those coasts on the surface of the globe along which observations of tides had not yet been made: these coasts consisted generally of all except those of Europe, of the eastern side of North America, and of the eastern side of Australia. I may mention also that on pages 134 and 135 of the Report of the meeting held at Oxford in 1847, there is a report drawn up by Dr Whewell and Sir J. C. Ross, recommending an expedition for the purpose of completing our knowledge of the Tides. Also on page 28 of the *Transactions of the Sections* in the Report of the Meeting held at Liverpool in 1854 there are some remarks entitled *On Mr Superintendent Bache's Tide Observations:* the remarks point out that much is still unknown as to the Tides.

60. On a new kind of coloured Fringes. This paper occurs on pages 336 and 337 of the *Philosophical Magazine* for April, 1851; it is also in the *Proceedings* of the Cambridge Philosophical Society for November 11, 1850. Dr Whewell had observed these fringes many years previously, and communicated an account of them to M. Quetelet. Recently attention had been drawn to them by M. Mousson of Zurich. The fringes have since been discussed on the principles of the Undulatory Theory by Professor Stokes in a memoir published in Volume IX of the *Cambridge Philosophical Transactions*.

61. Criticism of Aristotle's Account of Induction. This memoir occupies pages 63...72 of Vol. IX of the *Cambridge Philosophical Transactions;* it was read on February 11, 1850. A passage in Aristotle is translated and explained, some emendations being made for the purpose of rendering it intelligible and consistent. Dr Whewell in sending a copy of it to Dr Meyer, said, "The paper had its origin in a discussion between Mr Hallam and me, and has excited a good deal of interest among our scholars here." Mr Hallam's letter which led to the memoir is preserved, and also that in which he acknowledged the receipt of it. A distinguished Member of the University printed a paper on the subject, in which, I think, he controverted some of Dr Whewell's views; but I have not seen the paper. Dr Whewell's memoir is reproduced in the Appendix to the *Philosophy of Discovery.*

62. Mathematical Exposition of some Doctrines of Political Economy. This memoir occupies pages 128...149 of Vol. IX of the *Cambridge Philosophical Transactions;* it was read on April 15, 1850. The language of Algebra is here applied to questions which relate to the connection of demand, supply, and price, whether in the same country, or in different countries. The principles assumed are those of Mr J. S. Mill. Dr Whewell says towards the end: "In the preceding investigations, I have done little more than put into a general and algebraical form the reasonings which Mr Mill has presented to his readers in numerical examples." I have ventured to speak highly of Dr Whewell's other memoirs in connection with Political Economy, and the present is still more interesting, and well deserves the attention of those who are devoted to the subject. A person in Dr Whewell's position, however, might have found very easily an accurate and intelligent friend who would have read the memoir in proof before it was finally struck off; some very troublesome misprints might then have been corrected, and the reasoning in some places rendered more clear and convincing. Dr Whewell himself attached much importance to this memoir. In a letter to Mr Jones he says, "But the paper really does contain a refutation of certain vaunted theorems of John S. Mill on international trade; and shows them to be true, even on their mathematical

assumptions, within very narrow limits." In a later part of the same volume of the *Cambridge Philosophical Transactions* we have a memoir entitled *Mathematical Exposition of Certain Doctrines of Political Economy. Third memoir.* It occupies seven pages; it was read on November 11, 1850. The numbering of the articles is continued from the former memoir in the same volume. The special point considered is the effect of the import of the precious metals.

63. On the Transformation of Hypotheses in the History of Science. This memoir occupies pages [139]...[147] of Vol. IX of the *Cambridge Philosophical Transactions;* it was read on May 19, 1851. The memoir is very interesting; it maintains the following proposition: " When a prevalent theory is found to be untenable, and consequently is succeeded by a different, or even by an opposite one, the change is not made suddenly, or completed at once, at least in the minds of the most tenacious adherents of the earlier doctrine; but is effected by a transformation, or series of transformations, of the earlier hypothesis, by means of which it is gradually brought nearer and nearer to the second." This is illustrated in the case of the Cartesian hypothesis of vortices and the Newtonian doctrine of universal gravitation, and more briefly with respect to Magnetism, and to Theories of Light. The memoir is reproduced in the Appendix to the *Philosophy of Discovery.*

64. On Plato's Survey of the Sciences. This memoir occupies pages 582...589 of Vol. IX of the *Cambridge Philosophical Transactions;* it was read on April 23, 1855. It is reproduced in the Appendix to the *Philosophy of Discovery;* and also nearly entirely in pages 303...318 of Vol. III of the *Platonic Dialogues for English Readers.* In the same volume of the *Cambridge Philosophical Transactions* there are two other memoirs, also reproduced in the *Philosophy of Discovery,* namely, one on Plato's Notion of Dialectic, and one on the Intellectual Powers according to Plato. The former occupies pages 590...597, and was read on May 7, 1855; the latter occupies pages 598...604, and was read on November 12, 1855. In the former Dr Whewell considers the *Sophista* to be not by Plato, but by some Eleatic opponent:

Professor Thompson maintained the genuineness of the *Sophista* in a memoir which occupies pages 146...165 of Vol. x of the *Cambridge Philosophical Transactions*. A memoir by Dr Whewell on the Platonic Theory of Ideas occupies pages 94...104 of Vol. x of the *Cambridge Philosophical Transactions*; it was read on November 10, 1856, and is reproduced in the Appendix to the *Philosophy of Discovery*. The memoir relates mainly to the *Parmenides*, which Dr Whewell holds is not by Plato. Professor Jowett maintains the genuineness of the dialogue; he mentions Ueberweg as being of the contrary opinion: see Jowett's *Plato*, Vol. iii, page 227.

After the list of Dr Whewell's own memoirs in the *Royal Society Catalogue of Scientific Papers*, a Report is mentioned drawn up by him and J. W. Lubbock. The Report occupies pages 108...113 of the third volume of the *Proceedings* of the Royal Society; it was read to the Society on March 29, 1832. The Report relates to a memoir by Professor Airy, *On an Inequality of long period in the Motions of the Earth and Venus*; it furnishes an account of the memoir, which it compares with the investigations on the great inequality of Jupiter and Saturn, as originally given by Laplace. The Report, as might be expected, is adapted only for such readers as are well acquainted with Physical Astronomy. From the papers preserved by Dr Whewell it appears that much correspondence took place between himself and his colleague with respect to the important memoir which was referred to their consideration.

On pages 121 and 122 of the same volume of *Proceedings* there is a Report on Mr Lubbock's *Researches in Physical Astronomy*, by Messrs Whewell, Peacock, and Coddington. There is evidence that Dr Whewell on various other occasions took part in reporting on papers submitted to the Royal Society; but for a long period the custom of *printing* such reports has been abandoned.

In addition to the scientific memoirs which have been noticed, it should be mentioned that in the Index to the Reports of the British Association from 1831 to 1860 the name of Dr Whewell sometimes occurs in connexion with Reports where he was one of a committee and signed in conjunction with the rest. These formal productions are not recorded in the Royal Society's Catalogue of Scientific Papers, and it does not appear necessary to draw attention to them, especially as I think from examining them that they were not written by Dr Whewell.

CHAPTER XVII.

SERMONS.

I SHALL give in this Chapter notices of Dr Whewell's sermons, mainly of those which he published, though at the beginning and at the end of the Chapter I shall speak of some which exist only in manuscript.

In 1827 he preached a course of four sermons in the University Church at Cambridge, on February 4th, 11th, 18th and 25th; these sermons, with various rough drafts connected with them, are preserved among his papers. The texts are Isaiah xxxiii. 6, Romans i. 20, Proverbs iii. 5 and 6, and Job xxxvi. 3. The subject may be said to be the Relation of Human Knowledge to Divine. The sermons seem to have attracted great attention when they were delivered, and a few extracts from them may find an appropriate place here.

Towards the end of the first sermon he notices an opinion "that that which has a right to be believed, has also the power of instantly and by its mere application compelling belief;" and then he proceeds thus: "Nor, though this opinion be to the greater part, both of divine things and of human, most certainly inapplicable, is it perhaps much to be wondered that it should possess the minds of many who have been busy with the knowledge of modern times. For the entire tendency of the most admired labours of the reasoners of these later days has been so to exhibit our knowledge, that it should consist in part of a chain of deductions, compelling men's belief by the inevitable force of demonstration, and, for the remainder, of certain facts also compelling men's assent by the irresistible evidence of sense :—and in giving to what they teach this form on which more truly and peculiarly the name of *science* is bestowed, they have wrought with surpassing success, and have won high, and so far, deserved, honour. And in so many of the chambers of human learning has this bright and shining light been set up, that it is little to be marvelled if men refuse to enter those apartments which are not so illuminated, and hardly believe us when we tell them of the treasures which are there contained—if they draw back, even with averseness and mistrust, from a dimness which to their eyes, used to that new and blinding light, seems utter and hopeless darkness.

"The time would fail me if I should endeavour to show here, how unwise and inconsistent it is, to expect that all substantial and enduring truth may be, and should be, conformed to that type which has been best approved for the physical sciences. But if there be any who incline, on that account, to slight or turn from divine truths, them I would for an instant remind, that this is not a disposition which grows, naturally or invariably, with the growth of the powers and views of a philosophical mind; but that the most capacious intellects of the Christian times have found room for the love of knowledge without expelling the love of God. Especially I would point to the two names, by the consent of all, the greatest in the records of this modern philosophy. These were men who have often walked in the courts of this our Zion, and sat within these very walls; and these have left us no uncertain or feeble testimonial of the devout humility with which they bowed themselves to the Lord whom we serve in this his temple. These, the great Conqueror and the great Legislator in the realms of science, rebelled not against their Maker. To the former it was given to carry the excursion of man's intellect through the vast spaces of the material universe—and of his reverent and childlike piety, of his patient searching of the Scriptures, all men know. And if we turn to the pages of the other, who traced ordinances for our philosophy almost before its birth, it will seem as if he wished to protest with his mighty and eloquent voice to all ages against any who should make of the doctrines which he taught, a pretext for perverting or rejecting the teaching of a higher wisdom. Such expressions of piety as the following seemed to him fitting for the contemplative reason; and this is the language in which among many other devotional exercises he clothes a prayer, which he terms the *Student's Prayer*." Then follows the first of the two prayers which are reproduced in the *Philosophy of the Inductive Sciences*, 2nd edition, Vol. II, page 251.

The following is the conclusion of the second sermon : "But we may look still further—our prospects end not here. There will come a time when our views and perceptions will no longer be limited to that which can be received through the inlets of this perishable frame, or attained by a soul laden with such a weight. We may hope—through Him who is our hope—to be admitted, when this fleeting world shall have passed away, to the presence of God : to behold Him as He is, and to see and know, I will not say, even as he knows, but far beyond all that is given to the wisest of men on this side of heaven. It is true, that as repentance and faith and holiness here are the conditions of this blessedness, so adoration, and the joy of gratitude will form its essence there. It is not Knowledge but Love. Yet how should it be that all our powers and faculties should not there be purified and elevated : and if even the earthlier and lower ministry of the senses of this mortal body is not judged unworthy to describe and prefigure that happiness, how should it be that the gratifications of the intellect may not mingle in it. If the songs and harpings, the sapphire and gold of the eternal city are presented to our imagination, may we not well believe that our purged understanding will drink in new knowledge of God's ways from the waters which flow round that throne ? What is begun here will be perfected there, of Wisdom as well as of Holiness. There abides the fulness of our widening views : and we may cheer ourselves in this our pilgrimage with the thought that all which seems now dark and perplexed and discordant, will then be resolved into joy and harmony —that all obstacles which impede our purer and brighter prospects of God's creation and dispensation will melt and disappear, when there shall be granted, not only that 'unlocking of the gates of sense, and kindling of a greater natural light' of which that truly wise man spake, but when the eternal portals of Time and Space shall be unbarred, and there shall flow in, in glory unutterable, the Power and the Majesty, the Wisdom and the Goodness, the creating and redeeming Love, of the everlasting and infinite Godhead."

From the third sermon, a passage may be taken which can be compared with the preacher's later views in his *Essay* on the Plurality of Worlds : "That the earth which man inhabits is one among a multitude of worlds which roll on their appointed paths in the populous universe, at intervals which it exhausts the imagination to span—with resemblances and subordinations among them suggesting a possibility at least that other wandering spheres besides our own may bear with them their crowds of sentient, and for aught we know, responsible beings—these are not the reveries of idle dreamers or busy contrivers;—but, for the main part, truths collected by wise and patient men on evidence indisputable, from unwearied observation and thought; and for the rest, founded upon analogies which most will allow to reach at least to some degree of probability. Now it may be natural that a believer who had been accustomed to fix his thoughts on the relation between God and man, would, upon thus learning as it were the population of the whole universe let in upon him as joint claimants of God's notice and objects of his care, feel, so to speak, alarmed and dispirited ; as if it were too much to expect that his preservation and redemption and sanctification should occupy the mind which is charged with the care of ten thousand worlds and systems. Going further in knowledge than David went, he might go further in despondency. 'Lord what is man?' 'Art thou careful of him?' 'Dost thou regard him?' And yet if this thought be entertained in trust and hope till our worthier notions of Almighty God rise to the level of our knowledge, all cause will disappear for doubting or for less depending on, His parental care and kindness. What if we had in our thoughts for a time brought down the perfections of the Godhead to the measure of our own conceptions? What if we had in our imaginations limited the moral dealings of the Almighty to our own species?—perhaps even forgetting all that is revealed or hinted in scripture of other races of spirits whose God and Master He also is—or what if we had supposed His providence and

goodness confined to those living things of which we know ? Shall
we, when this our self-devised imagination is broken in upon
and disfigured, refuse all returning comfort from wider and wiser
thoughts of Him ? Does not even science herself compel us to
expand our notions of God's vivifying and cherishing care, when
she unlocks to us the world of things, from their smallness before
unseen, and shews us the myriads of animals that live in the blade
of grass or in the drop of water ? As if we were not everywhere
and by all things taught, that however far we follow in our minds
His greatness, His wisdom, His preserving goodness, we come but
in sight of the need we have to think of its extending immeasur-
ably further still. As if we would forget too that with Him, all
worlds and all the inhabitants of worlds, of whatever order and
nature, are not objects of perception only, but instruments of
action—that He is the principle of their being, the breath of their
lives, and that in His will are the laws by which their existence
proceeds.

"Hastily and imperfectly we pass over this portion of our
subject—a matter already touched by other hands than ours.
But there is another of the departments of modern knowledge to
which, with reference to its bearing on the scriptures, men's
minds seem, mainly perhaps because it possesses more of novelty,
more actively and anxiously to turn. Let us suppose it possible
that as we have in the heavens proofs of other and distant worlds
different from our own, so we should have on earth proofs of the
former existence of tribes of brute animals different from those
which now people its surface. Let us suppose such races to have
been destroyed, and other generations, more than once, to have
succeeded them, before man was placed here as the subject of God.
Let us suppose the evidence of provinces of God's creation thus
separated from us in *time*, to be no less convincing than the
demonstration which all allow us to have of other provinces of
His empire separated from us in *space*. If to any there seem
aught rash and unhallowed in the boldness of such suppositions,
let them reflect, that to others, sober and pious men, these seem
inferences undeniable from what they everywhere see. And let
us consider whether here also, as in that other case, aught of

repugnance and alarm which some good men may feel, arise not from our clinging rather to assumptions and interpretations of our own than to the real content of the revealed word."

The following passage occurs near the end of the third sermon: " Let us not deceive ourselves. Indefinite duration and gradual decay are not the destiny of this universe. It will not find its termination only in the imperceptible crumbling of its materials or clogging of its wheels. It steals not calmly and slowly to its end. No ages of long and deepening twilight shall gradually bring the last setting of the sun—no mountains sinking under the decrepitude of years, or weary rivers ceasing to rejoice in their courses, shall prepare men for the abolition of this earth. No placid *euthanasia* shall silently lead on the dissolution of the natural world. But the trumpet shall sound—the struggle shall come—this goodly frame of things shall be rent and crushed by the mighty arm of its Omnipotent Maker. It shall expire in the throes and agonies of some sudden and fierce convulsion ; and the same hand which plucked the elements from the dark and troubled slumbers of their chaos, shall cast them into their tomb, pushing them aside that they may no longer stand between His face and the creatures whom He shall come to judge."

Dr Whewell preached for the last time on Feb. 11, 1866, in his College Chapel. On this occasion he introduced the above passage substantially though not literally, beginning with the words *no mountains.* Probably it was supplied by memory and not copied from what had been written nearly forty years before.

The following is the commencement of the fourth sermon : " The book of Job, a work most remarkable on many accounts, invites our notice particularly here by its reference to the train of thought which in the present series of discourses we have attempted, however imperfectly, to excite. This reference indeed, it may be, is discerned dimly and unsteadily through the long and deepening shade of so many intervening ages, removed from us by almost the whole extent of human traditions, and by habits of life and manners far different from any thing with which we are familiar ; yet still may it be perceived and traced by an attentive eye. And this book comes down to us, freighted apparently with no small portion of the knowledge of that early age ; speaking to us not merely of flocks and herds, of wine and oil, of writings and judgments ;—but telling us also of ores and metals drawn from the recesses of the mountains ;—of gems and jewels of many names and from various countries ;—of constellations and their risings and seasons and influences. And above all it comes tinged with a deep and contemplative spirit of observation of the wonders of the animate and inanimate creation. The rain and the dew, the ice and the hoar frost, the lightning and the tempest, are noted as containing mysteries past man's finding out. Our awe and admiration are demanded for the care that provides for the lion and the ostrich after their natures ; for the spirit that informs with fire and vigour the war-horse and the eagle ; for the power that guides the huge behemoth and leviathan."

The following passage from the same sermon may be compared with section 30 of the chapter on *The Argument from Design* in the *Essay* on the Plurality of Worlds: " But again. Not only these connexions and transitions, but the copiousness with which properties, as to us it seems, merely ornamental, are diffused through the creation, may well excite our wonder. Almost all have felt, as it were, a perplexity, chastened by the sense of beauty, when they have thought of the myriads of fair and gorgeous objects, that exist and perish without any eye to witness their glories—the flowers that are born to blush unseen in the wilderness—the gems, so wondrously fashioned, that stud the untrodden caverns—the living things with adornments of yet richer workmanship that, solitary and unknown, glitter and die. Nor is science without food for such feelings. At every step she discloses things and laws pregnant with unobtrusive splendour. She has unravelled the web of light in which all things are involved, and has found its texture even more wonderful and exquisite than she could have thought. This she has done in our own days—and these admirable properties the sunbeams had borne about with them since light was created, contented as it were, with their unseen glories. What then shall we say ? These forms, these appearances of pervading beauty, though we know not their end and meaning, still touch all thoughtful minds with a sense of

hidden delight, a still and grateful admiration. They come over
our meditations like strains and snatches of a sweet and distant
symphony—sweet indeed but to us distant, and broken, and
overpowered by the din of more earthly perceptions—caught but
at intervals—eluding our attempts to learn it as a whole, but ever
and anon returning on our ears, and elevating our thoughts of the
fabric of this world.   We might indeed well believe that this harmony breathes not for us alone—that it has nearer listeners—more
delighted auditors. But even in us it raises no unworthy thoughts
—even in us it impresses a conviction, indestructible by harsher
voices, that far beyond all that we can now know and conceive
the universe is full of symmetry and order and beauty and life."
One sentence of the above passage may recall to some hearers a
favourite expression of one of the most eminent of living scientific
lecturers—" Newton unravelled the texture of the solar light."

Besides the four sermons delivered before the University in
February, 1827, _a fifth_ was prepared, but I think not delivered.
The text is from James iii. 13.   The design was bold, namely,
to reprehend what the writer maintained to be errors of some of
the political economists of the period.   Dr Whewell sent an outline or a draft of the proposed sermon to his friend Mr Jones,
whom he always acknowledged as his guide in political economy;
Mr Jones seems to have feared that the sermon would amount to
a premature disclosure of some of the views which he was himself
preparing for publication, and Dr Whewell then probably gave up
his idea of preaching it.   The following passage will shew one of
the principal topics which the sermon proposed to discuss : "It
has, for instance, been maintained, and this doctrine has produced
and does still produce a powerful impression and manifest tendency in the speculations of those who even now reason concerning the laws which regulate the prosperity and riches of
human societies, that the fiat of His will by which the Creator
ordained the increase and multiplication of men, impelled them
in a career leading by a course not to be stopped or deflected,
to want and degradation, to vice and misery.   It has been passed
from pen to pen and from lip to lip as a great discovery, that the
tendency of mankind to replenish the earth ever pushes them on

till the sharp discipline of pain, the iron hand of want and its deadly concomitant crime, drive them back or at least forbid their further progress. That whenever the large bounty of nature opens some new supply, pours out some new store of nutriment, this fierce and indomitable property of human societies springs forth instantly upon the offered food and devours it with wolfish rapidity, leaving the spot that seemed thus enriched as bare and hungry as it had been. That thus the more depressed of the orders of mankind, those portions of society that win their bread by the labour of their hands and eat it in the sweat of their brows, are destined to eternal and irredeemable degradation—fated to increase in numbers as the fruits of the earth allow of their increase, and as it were condemned to become more numerous *lest* they should become more happy. That this, or something like this, is the representation often given of the necessary course of states and nations by those who most loudly call our attention to their success in speculation most will recognize and know. That the proclamation of such a doctrine, represented as a demonstrated truth and the fruitful source of many truths besides, shook and startled the minds of pious and benevolent men, and seemed like an oppressive and disquieting thought forced in among their belief and trust in God's goodness, like a funereal and menacing light thrown upon the fair face of nature, many who bear in mind the youth and first appearance of these doctrines and their operation on the minds of men, will still recollect."

The mode in which the preacher would have handled the subject may be seen by consulting the Preface to Mr Jones's work on Rent. It may be well supposed that allusion is made to Mr Jones in the following words which occur early in the sermon: "To point, with a brief and general indication at some of the forms of error, its causes and the grounds of its refutation is here all that is permissible—all that is required. To other times, to other hands must be left the labour and the honour, the precious and blessed labour, the high and enduring honour, of combating in particulars those errors from which foolish and unhappy men have drawn or may draw inferences which darken the face of moral nature and divine truth."

I now proceed to Sermons which are in print.

A Sermon preached on Trinity Monday, June 15, 1835, before
the Corporation of the Trinity House, in the Parish Church of
Saint Nicholas, Deptford, and printed at their request. By the
Rev. William Whewell, M.A., Trinity College, Cambridge. London:
printed by Charles Whittingham, 1835. This is in quarto,
and consists of 21 pages. The sermon is dedicated to the Elder
Brethren of the Trinity House, three of whom are named, the
king William the Fourth, the Marquess Camden, and John
Henry Pelly, Esquire. The preacher had been appointed by the
Marquess Camden, at that time Chancellor of the University.
The text is John vi. 21. The sermon has been much and justly
admired. Professor Selwyn, preaching before the University on
the occasion of the death of Dr Whewell, alluded to it in these
words: "...read the Master of Trinity's noble sermon before the
corporation of the Trinity House in 1835, and you will rise up
with a better feeling of what Englishmen ought to be." The
Professor, whose loss the University has to lament just as this
sheet is sent to press, told me that being unable to procure a
copy of Dr Whewell's sermon in any other way, he transcribed
the whole of it himself. A passage may be quoted, the last sentence of which gives the prevalent idea of the sermon: "Life in
the Christian's eyes is the seed of immortality: shall we allow
this precious gem to be crushed or squandered while it is perhaps
immature for the change? The heathen indeed might feel, that
it was sweet to stand in security on the shore of the stormy
ocean, and to witness the ruin of the sinking mariner; but the
Christian's first impulse would be, 'save him, he is my brother.'
No depreciating estimate of this man's value to others, no alienating considerations of difference of race or nation, would obliterate this spontaneous and overpowering sentiment. To provide
all that can be provided, for the safety and comfort of those who
go down to the sea in ships, and occupy in the great waters, may
be desirable on account of private or public gain; may be dictated
by prudence or by policy: but, it is a duty before it is a profit; it
is commanded by benevolence before it is commended by reasoning; and before policy and prudence have made their calculations,

and brought out their result, Christian love is already stretching forth her torch to light the beacon, and lifting her voice to proclaim the dangers of the shore."

Four sermons on the *Foundations of Morals* were preached before the University in November 1847, and published shortly afterwards: they have been already noticed in connexion with the other works relating to Moral Philosophy.

The Christian's duty towards Transgressors. A sermon preached in the chapel of the Philanthropic Society, St George's Fields, on Sunday, May 16, 1847, being the annual commemoration of the Society's Establishment. By the Rev. William Whewell, D.D., Master of Trinity College, and Professor of Moral Philosophy, in the University of Cambridge. Printed by permission for distribution among the friends of the Institution. This is in 27 octavo pages. The text is Ephesians iv. 28. The title indicates the object of the sermon. A passage may be extracted which is characteristic of the author's style: "But yet, within the very circuit of the community so Christianized prevail vices and crimes, and incentives and preparations to crime, which were never surpassed in foulness and virulence by the vilest times and scenes of heathen depravity. Law utters her voice, Punishment waves her scourge in vain. The scourge of Punishment becomes as it were a schoolmaster's rod, under which unhappy creatures are taught to learn more and more completely the lesson of crime."

The Bulwarks and Palaces of Zion: a sermon (preached at Wormenhall, near Oxford, Dec. 14, 1847, on the occasion of the re-opening of the church after its restoration) by the Rev. William Whewell, D.D., Master of Trinity College, Cambridge. Re-printed (by request) from the Church of England Magazine for May, 1848. London: printed by Joseph Rogerson, 24, Norfolk Street, Strand, 1848. This is in 19 small pages. The text is Psalm xlviii. 12, 13. The following extract recalls the author's researches in architecture: "The course of centuries is in this country marked in the various forms which were given to our sacred edifices, by the builder's art constantly struggling to produce something more beautiful, more adorned, more manifold, more gorgeous than what had preceded. The course of time in our history is not more distinctly marked by the reigns of our sovereigns, than it is by the successive modes of building and adorning the houses of prayer, which we inherit from our ancestors. The life-time of Christ's church in this nation has extended through many, many revolutions of the national taste in building; and the different types of the Christian sacred edifice are, as it were, measuring points in the duration of our national Christianity."

Sermons preached in the Chapel of Trinity College, Cambridge. By William Whewell, D.D., Master of the College, 1847. This is an octavo volume containing xii and 392 pages. It is dedicated to the Reverend William Carus, M.A., Senior Dean of Trinity College. There are in all 22 sermons; 19 of these had been preached by Dr Whewell in the College Chapel, in virtue of his office of Master. The other three had been preached by him as a Fellow; these are the College Commemoration sermons for 1828 and 1838, which had already been printed, though not published; and a sermon called the *College Student,* published a few years previously as one of the "Sermons by xxxix Living Divines of the Church of England." We read in the Preface, "The Second and Third of the Sermons were composed when the preacher entertained the intention of introducing into the Series of Discourses which he had to deliver, an exposition of the place and Office of Religion in a State. This intention was afterwards abandoned." We see here additional evidence respecting a point which appeared in examining Dr Whewell's writings on Moral Philosophy, namely, that he was much perplexed with the difficult problem of the relations between the Church and the State.

Without professing to draw attention to the contents of the volume in detail, we may notice one or two particulars. The ninth sermon, entitled *Admiration of Men,* is very striking; on pages 270...274 we have, as in the author's writings on Moral Philosophy, a vigorous opposition to the exclusive claims of the Romish Church; and on pages 321 and 392 we have the two prayers of Bacon which Dr Whewell so much admired. The Commemoration Sermon for 1828 is in a more ornate style than the other sermons, which belong to a later period. Wordsworth, in a letter to the preacher, said of the sermon: "It is worthy of the occasion, though I will take the liberty of saying that it opens rather inauspiciously, the first sentence being unusually cumbrous for so few words.

> But where the subject rises proud to view,
> With equal strength the Preacher. rises too."

In this sermon and the notes to it the author refers with just praise to the honourable manner in which the College was represented among the translators of the Bible into English; had he lived to the present day he would have enjoyed the satisfaction of seeing that, among the revisers of the authorized version, the College, so dear to him, had more than maintained its ancient fame. It would appear from the Commemoration Sermon of 1838, page 360, that at that time sermons were rarely delivered in Trinity College Chapel, and the preacher suggests the advantage that would follow if these were made a regular accompaniment of the services.

A Sermon preached before the University of Cambridge on the day of the general thanksgiving, Nov. 15, 1849. By William Whewell, D.D., Master of Trinity College. Cambridge: printed at the University Press. J. Deighton, 1849. This is in 24 octavo pages. The text is Psalm civ. 29. The general thanksgiving was for the cessation of the cholera; the sermon also contains a reference to Dr French, Master of Jesus College, who died suddenly three days previously : it is dedicated to Dr Cartmell, the Vice-Chancellor. The following passage is from the first paragraph of the sermon: "We are as children, weak, ignorant, helpless ; but even by the light of nature we can see that we are *His* children ; 'so one of your own wise men has declared'— said St Paul to the Athenians; and doubtless the hearts of all the better natures among his hearers responded with reverent satisfaction, when he added, to this text of natural reason, his noble comment, ' In Him we live and move and have our being.' They would recollect to have heard expressions of the like import— utterances of similar feelings—floating among the groves of Academus and echoing along the columns of the Portico, though, in that former time, not given forth with the power and authority which belonged to him who then stood on the Areopagus."

In the course of the sermon the preacher urges on his hearers the duty of humility; he says: "This moral is on the surface of the occasion, and surely it was not given to us till it was called for. For have not men of late been prone to a temper of self-gratulation and self-admiration, for which a lesson of humility was much needed?......If our ancestors could see us, they would be filled with admiration; they would look with comparative contempt, or with humbled vanity, on the scanty inheritance of human power and human knowledge which was allowed to their generation. Such sayings have been frequent among us. Perhaps, even in their first aspect, not wise sayings; for what can be wise which involves so much of pride and self-applause? Perhaps not wise sayings: for is not this thought,—How much others would admire us if they could see us,—rather the thought of a vain child, arrayed in some new and gaudy vestment, than that of a wise man who knows that spectators do not so easily transfer their admiration from themselves to others; and to whom it may occur, that our ancestors may have thought as much of what they did in their time, as we think of what we have done in ours; and with equal reason :—to whom it may occur, also, that there have been ages in which the amount of manifest progress of man was far greater than it is in ours : for instance, the age of Columbus and Gutenburg, or the age of Luther and Galileo."

Strength in Trouble. A sermon preached in the Chapel of Trinity College, Cambridge, February 23, 1851. By William Whewell, D.D., Master of the College. Printed by Request. London: John W. Parker, West Strand. Cambridge: J. Deighton, 1851. This is in 24 octavo pages. The text is Isaiah xxx, part of the 15th verse : *In quietness and in confidence shall be your strength.* The trouble to which the sermon relates is the theological dissension of the time, connected with the Gorham controversy. In the course of the sermon the desire of the Israelites in some of their difficulties to go back again to Egypt suggests the parallel of the wish of some perplexed members of the Church of England to seek refuge in Rome : " We have thus come forth out of the land of Egypt, by God's blessing, once and for ever. Who would return thither? Who would tread back the steps of our Exodus? Even if at times we are vexed with contentions and struggles, with doubters and gainsayers, with obstinate questionings, it may be, beyond our power to answer; with the weakness of authority which naturally grows up among those who are, above all things, jealous for freedom;—even if such evils at times assail us, who would seek to escape them by returning back to slavery? by again submitting ourselves to a human authority which silences all questionings and decides all controversy by its own absolute sentence? Yet such a disposition may sometimes be discerned among us, amid the whirl of opinions on matters of religious concern. Men, impatient to know more than God has taught in His Scripture, would rush back to the teaching of men who assume the place of God. Men, offended at the conflict of opinions which they see going on around them, would fain have among them an infallible judge of controversies. Men, dissatisfied with the simplicity of the means by which in our common worship we endeavour to raise our hearts to God, desire to behold again Egyptian pomp and splendour. Men, disturbed at some of the circumstances of the Exodus by which we were brought to our present condition, are ready to speak evil of that great and blessed event in our history." With all his reverence for Law and Order Dr Whewell had an abiding detestation of slavery, bodily, mental and spiritual; and he would have been as unwilling to recognize papal supremacy as his great predecessors Barrow and Bentley. His unpublished sermons of this date contain numerous exhortations to the maintenance of Christian freedom; he cautions his hearers against "all weariness of the use of our own reason which makes some yearn for the guidance of an authority which claims to be infallible; all impatience of the seemingly imperfect organization of our church, which makes men ready to bow before the gigantic golden image erected on the ruins of ancient Rome."

And every man that hath this hope in him purifieth himself, even as He is pure. A Sermon preached in the Chapel of Trinity College, Cambridge, February 3, 1856. By the Rev. W. Whewell, D.D., Master of the College. Cambridge: printed at the University

Press, 1856. This is in 16 small pages. The text consists of the words given in the title, taken from 1 John iii. 3. The sermon was preached very soon after the death of Mrs Whewell, and is especially noticed in Mr W. G. Clark's memoir of Dr Whewell. The short sentences which form the first two pages tell plainly of the emotion with which the writer engaged in his task; though as he proceeds it is obvious that the discharge of his duty becomes a relief to his deep feelings. After speaking of the purification to which the text refers, the preacher continues thus: "Those who seek heaven, cherish such tempers as these. And if we see persons cherishing such tempers as these, and growing in them more and more, we cannot doubt that they are on the road to heaven. Their heaven seems almost begun, even here on earth: and when their earthly tabernacle is dissolved, and they are translated to the region of Departed Souls:—when the bodily frame which was interposed between the Soul and the presence of the Saviour, is interposed no longer;—the change must be, as if we had been living in a region of light and of glorious sunshine, but immured in a close hut or prison-house which admitted only a few rays through its chinks;—and as if, by some mighty blast from above, awful and agonizing for the moment, but yet gracious in its purpose and effect, the earthly hovel were blown down and carried away; and the happy Tenant of it left rejoicing in the new-found glow of the celestial light, by which, though dimly seen, she had been so long in truth entirely surrounded."

Then follows an application to establish, if proof were needed, the fact of "an hereafter for the Souls of men," namely, that we cannot suppose the soul to have passed through a long course of development only to sink into nothingness. The consideration is urged with great impressiveness; and as the author said in a sub-sequent work it forms an important and weighty line of argument: see the *Platonic Dialogues for English Readers*, Vol. I, page 439. On page 443 of the same volume an argument drawn from the affections is regarded as weak, namely, that "when those whom we love die, we cannot believe the separation final;" and consistently with the opinion there is no appeal to such an argument in the present sermon.

On January 29, 1860, Dr Whewell preached a sermon in St Paul's Cathedral by the invitation of the Bishop of London, now Archbishop of Canterbury. The text was Luke xviii, part of the 8th verse; *Nevertheless when the Son of man cometh, shall he find faith on the earth?* From Dr Whewell's correspondence I learn that this sermon was printed in the *Clerical Journal*, but I have not seen this and do not know whether it was taken from shorthand notes : Dr Whewell's own manuscript does not seem to have been in the hands of a compositor. It is a noble sermon ; the preacher takes the opportunity of rebuking the national hardness of character and indifference to the pain of others, especially as manifested by our conduct in India. He indicates in unambiguous language his fears that blame for some of this evil might attach to our great public schools : " But among ourselves in those schools where the sons of our families of the higher and middle classes spend their boyhood, and which may therefore be regarded as forming and exhibiting the best part of the national character, what is the demeanour of the stronger to the weaker ? I speak not as of any special knowledge : but I ask what are the pictures of the life of our large schools which have been in recent years presented to us in books by those who were familiar with it, and who most assuredly did not wish to vilify our national character ? Are there not such books, and not one merely, but several, which represent the life of Englishmen in their boyhood at such institutions as a progress from the sufferings of the weak to the tyranny of the strong?" The bishop's experience would render him a good judge on the subject thus handled ; the preacher's correspondence shews that it was one on which he himself felt very deeply.

A few words may be given to the manuscript sermons left by Dr Whewell. It appears from his correspondence that he was never very willing to preach, except in the course of his duty as Master of the College, and this function of his office he discharged with great satisfaction. Some of his manuscripts are undated, but in the majority of cases the dates and places of delivery are carefully recorded. There are few manuscript sermons before 1834, none between 1834 and 1840, nine between 1841 and 1844, and none between 1845 and 1847 : many of the sermons delivered in these later years were inserted in the volume published in 1847. From 1848 onwards there are three or four sermons for each year; many of these are very interesting, superior, I think, to those of the printed volume, and it may be hoped that a selection from them will be eventually published. The allusions to moral philosophy are numerous and impressive. It is the opinion of those who were familiar with his preaching that his delivery failed to do justice to his matter, and this agrees with my recollection of the only sermon which I myself heard, namely, that preached before the University in 1849. The manuscripts of his later sermons were obviously executed in a most rapid manner; an appropriate number of leaves was taken; they were folded, and then filled on both sides with such characters, that although intelligible to the careful scrutiny of one familiar with his style and handwriting, it is a marvel that the preacher could have deciphered them with the readiness necessary for delivery.

I will extract a few sentences from some of these sermons:

In a sermon preached on May 6, 1855, there is an allusion to the recent death of Mr J. C. Hare. "Such an affection, the devout Christian of whom I speak, had for the seventeenth Psalm, and when this Psalm was read to him a few hours before his spirit departed, he thanked those who had thus chosen the words of Scripture which he so especially delighted in....... With these sounds of glory ringing in his ears, our dear friend fell into that sleep from which he was to wake in the likeness of Christ. His deepest thoughts on the things which employed his mind, his acute searching into man's nature and powers and actions, he had been accustomed, in a spirit of humility, to call *Guesses at Truth*. He is gone where he will no longer have guesses, but certainties: —no longer dim glimpses as through a glass darkly, but full vision, face to face:—here he knew in part, and he prophesied in part, and therefore he guessed: there he will know even as he is known, and will guess no more."

In a sermon preached on October 24, 1858, not long after his second marriage, there is an allusion to his own circumstances: "The love of mother and father, of sister and brother, and the ties of friendship more tender than any to which accident or caprice could give rise, are, I am sure you all feel, more precious than any other earthly possessions can ever be. If there be any who have by some calamity been prevented from enjoying in their fulness the sweetness of these relations, or who, having enjoyed them fully, have had them taken away, such persons will still feel, not less than others—perhaps will feel more keenly than others—how great these blessings are. If there be any of us who having been for a time deprived of these blessings of the united life of families have afterwards had the precious possession given back to them ;—who, after years of solitude and lonely sorrow, have been restored to domestic companionship and to the enjoyment of habitual affection, they may well say, What shall I render to the Lord for all his benefits towards me ? And if, as I trust is the case, most of you are in the full possession of the affection and confidence of the family circle to which you belong, be well assured that there is no blessing of heaven for which greater gratitude is due ; and that you may well exclaim What shall I render to the Lord for all his benefits ?"

A sermon preached on Oct. 16, 1859, was obviously prompted by Mr Mansel's well known Bampton Lectures, which it strongly opposes : " As I have said, my anxiety to warn you against this doctrine, false as I hold it to be in philosophy, does not arise mainly from its bearing upon philosophy, but upon religion—not because it makes Natural Theology impossible, but because it makes Revealed Theology equally impossible. If we *cannot* know anything about God, revelation is in vain. We cannot have anything revealed to us if we have no power of seeing what is revealed. It is of no use to take away the veil when we are blind. If in consequence of our defect of sight we cannot see God at all by the sun of nature, we cannot see him by the lightning of Sinai, nor by the fire of mount Carmel, nor by the star in the East, nor by the rising sun of the resurrection. If we cannot know God, to what purpose is it that the Scriptures, Old and New, constantly exhort us to know Him, and represent to us the knowledge of Him as the great purpose of man's life and the sole ground of his eternal hopes ?"

A sermon preached on Feb. 16, 1862, speaks of the abolition of

West Indian slavery by the English nation in these terms : " For though to put an end to this buying and selling of men, and this holding of them in slavery, was a step which involved a vast sacrifice of material wealth, the Nation resolved that nevertheless it should be done. And it was done, and a loss of treasure represented by many millions of golden coin, was cheerfully, eagerly undergone. I think that those who write the histories of Nations, speaking as they ought to speak, in the most calm and unprejudiced manner, cannot hesitate to say, and they do not hesitate to say, that this was the noblest national act of which history contains a record." It may seem ungracious, but still it is just to remark that the nation did not *pay* the twenty millions required to effect the emancipation of the negroes ; they raised the money by loan, and thus threw the burden on posterity.

The following passage in a sermon preached on Oct. 23, 1864, alludes to the death of Mr Romilly, a much esteemed member of Trinity College : " Most of you probably will stay in this place but a few short years, and will then be removed by your various lots to other spheres of action. But some of you may continue to be resident members of this community for many years, or it may be for the whole of your lives. Such a lot may, at the present moment, appear to you dreary and desolate, but it need not be so. Such a lot is not at all inconsistent with the constant culture of all kindly affections, with an enduring enjoyment of literary pleasures, and with a truly Christian spirit, enduring all things and waiting calmly for the end in faith and hope. Such has been the character and career of many who have spent their lives here and died while still active members of our body. Such a one has been removed from among us during the time that we have been separated. To some of us known during the whole time of our abode here, and always known as a kind and faithful friend, always looking on the better and more cheerful side in every character and in every event ; to others known by his share in the councils and action of this our body, and of the larger academical body of which he was so long a valuable and diligent officer : to many not belonging to our body known by his works of love and charity—by aid rendered to the sick and needy, by persevering and widely spread benevolence:—but wherever known, loved and respected, and seen by all to be of a truly Christian spirit."

In a sermon preached on May 14, 1865, soon after the death of his second wife, there is obvious allusion to her: "When, for instance, we have for a long time enjoyed the blessing of intimate intercourse with a true servant of God and disciple of Jesus Christ:—one who habitually and constantly felt this service and discipleship:—one whose thoughts were in heaven as the only region where real things were to be found, and to whom prayer and praise were the natural and habitual expression of thought.— We may have admired and rejoiced in the spectacle of such habits—of such a nature, or rather of a nature so transformed, elevated, and sanctified by the Divine Blessing. We may have seen day by day, the purity, the sweetness, the wisdom which dwelt in such a character and were displayed in a thousand incidents in the course of a daily life and conversation. We may have dwelt day by day upon the lessons which such a character offers to those who have the privilege of living near it."

Dr Whewell's last sermon was preached in his College Chapel on Feb. 11, 1866, from the text Revelation i. 8. The following extract will shew what was the nature of the subject: "And, undoubtedly, there have been men who have busied themselves with these portions of human knowledge, who having discovered, as manifested by undoubted evidence, vast cycles of change succeeding each other, and occupying such vast intervals of time that our ordinary powers of conception and imagination are overwhelmed with their magnitude, have been led to express an opinion that the spectacle of the universe suggested nothing but this beginningless and endless series of changes. Astronomy, the study of the stars in their courses, and especially the study of those stars which revolve round the sun and which we call planets, a study which dates from the earliest times of man's intellectual history, shews us them revolving to-day as they did six thousand years ago—as we may believe them to have revolved sixty thousand years ago. Hence, it has been said, rashly

I think we shall soon be led to judge, here we see no trace of a beginning,—no promise of an end. And of the changes which can be discovered to have taken place in the history of the earth in the course of past ages, as manifested in the materials of its hills and mountains, and in the bones which men there discover of the former inhabitants of the land and of the water, it has been said, with a boldness borrowed from the elder study of the heavens, it has been said, we see no trace of a beginning,—no promise of an end. A perpetual series of changes past is all that we can discern. A perpetual series of like changes in future is all that we can anticipate. Such reasoners would abolish the terms Beginning and End. They would erase the Alpha and the Omega from the Alphabet of the Universe; and so would deprive of meaning such passages as that which I have made my text."

Against these opinions Dr Whewell maintains that the fact of a creation is really suggested by the teaching of history and science; his arguments are in substance the same as he had published in the Book on Palætiology in the *Philosophy of the Inductive Sciences*. Then he proceeds to suggest that we must judge in like manner of the *end* of the world: "In what way this end may come, we can by no means by the light of nature discern: whether by a gradual decay of the energy of natural forces, or by some great catastrophe, such as many geologists have thought they could trace the evidence of in the arrangement of the earth's materials. But the event, the end of the world, must be out of the common order of the history of the world. It is constantly spoken of by Christ and his Apostles as an event which is to be sudden, violent, and overwhelming. According to that teaching it will not be a process of slow decay." Then follows the passage which I have mentioned on page 328, and which is quoted by Dr Lightfoot in his funeral sermon upon Dr Whewell.

CHAPTER XVIII.

[I HAVE already alluded to three sets of manuscripts preserved among the papers of Dr Whewell, namely the collections made for the *Bridgewater Treatise* and for the *History* and *Philosophy* of the Inductive Sciences, the lectures and discussions on Moral Science, and the Sermons. Besides these there are others which may be arranged in three sets, namely, Notes on Books read including some very early essays on various subjects, Poetical pieces, Miscellaneous Prose pieces. I shall publish extracts from these three sets, taking in the present Chapter the first set. This collection has been thus described by Mr W. G. Clark : " This consists of a vast body of notes on the books which he read from the year 1817 to 1830—books in almost all the languages in Europe, histories of all countries, ancient and modern, treatises on all sciences, moral and physical. Among the rest is an epitome of Kant's 'Kritik der reinen Vernunft,' a work which exercised a marked influence on all his speculations in mental philosophy." I place the extracts in chronological order, beginning with some which are undated, but almost certainly belonging to the years 1816 and 1817. In this and the following two Chapters my own remarks will be enclosed in square brackets, and all the rest belongs to Dr Whewell.]

[*Personal Identity*]. Nothing is more absurd than to talk of a person's identity, according to any imaginable criterion of identity, but most of all the criterion of consciousness. Every person must be conscious that he knows no persons so different from each other as he knows himself to be from himself in another situation. Alone and in company—making himself master of the ideas of others or giving himself up to his own—in a happy and in an unhappy state of mind. Not only is the whole system of his present perceptions and feelings completely different, but the whole face of the world changes before him. The relations of its objects among themselves and above all the relation of all of them to him is perpetually altering, altering so as not to leave a single vestige of their former condition. Sometimes he sees himself one of the innumerable individuals which compose the mass of animated nature and feels as if it were impossible that his relations with any other of them could ever swell into such importance as they do, so as to seem to hide from him the whole universe. Sometimes he sees around him beings animated with the same moral and social feelings with himself capable of giving enjoyment to each other by the participation and interchange of sentiments or even of looks, and feels as if he could accost each of them by the name of brother. And sometimes he retires into himself, sees the world around him as a series of phenomena which affect not him—sees beings in a shape similar to his own moving before him—sees faces of business or of feeling that pass before him and yet feels that he has nothing to do with them, that he has nothing in common with them—they are unrealities— phantoms—shadows which move along the wall of his chamber. But an enumeration is useless unless it could be given with such fidelity and vivacity as to excite the moods which it attempts to describe. But what is still more remarkable is that not only are we ourselves and the objects present to us changed but our past identity is altered. We recollect ourselves as different beings. We recognize at different times different parts and frag- ments of our former consciousness. We seem to have several identities and each as we enter into it brings with it the recollec- tion of so much of our former existence as has been spent in it. As when for instance by some combination of situation or inci- dent the reminiscences of our boyhood come back to us—when all the careless and enthusiastic temper, all the visions of incipient thinking, all the elasticity of the spirit before it was pressed by the load of observation and knowledge—or strained by the effort of acquiring them—rush back upon us. "I feel the gales that

from ye blow"—then we seem to awake from a sleep in which
we had been plunged during the intermission of these feelings—
to resume an existence which we had laid down when we suffered
our youthful visions to slide away from us. Of all the visits of
old friends the most agreeable and the most affecting is the
visit from a man's former self. Agreeable because there is
nobody of whom we know so much and with whom we have
so many subjects of common interest—and affecting because
in truth it is rather more like a visit from the ghost of an old
acquaintance than from himself—we know that he is gone for
ever, we see something supernatural about his return—we know
it is only momentary (like angels' visits)—we know that all
our efforts cannot detain him long, that as soon as the broad glare
of daylight breaks in upon us he will vanish and be lost. We
strive in vain to remain what we once were, and though we may
feel ourselves for a small space the same person whom we recollect
running about after butterflies our feelings roll away from us
like smoke—the busy realities of life reappear round us, and we
are left like Mirza when the vision vanished and the bare hills
of Bagdad spread drearily around him. Thus we have as many
identities as we have suits of clothes and verify Locke's idea of
a man with a sleeping and a waking set of consciousnesses, and if
Kehama had sent all the different selfs of which we are conscious
up the streets of [there is a blank in the manuscript, but the word
must be *Padalon*, see *Curse of Kehama*, XXIV] not only would he
have had as many as the service required but no one would have
recognized them for the same person.

As a further illustration of this we may observe the complete
change of identity which seems to have taken place if we throw
our reflections back at random upon the extent of our past years—
the chance is great that among the bundle of personages which
we have brought forward to the present time we do not at first
pitch upon the one which at present we are, and therefore that
we stumble upon a person who is to us at the present moment as
great a stranger as if we had never been in habits of intimacy
with him. There is something rather melancholy in this unceasing flux of mental as well as corporeal individuality—this

perpetual leaving out of old and taking in of new substance.
Perhaps it reminds us too strongly that transitory as are the
forms and materials of the external world, our sensations and
reflections which are the materials of happiness are at least as
transitory. I would willingly be much more than I am indebted
to my former intellectual habits for my present enjoyments—
but alas! it may not be; the blossom must wither as the fruit
ripens.

Obs. Some of these identities are much more agreeable and
much more valuable than others. It might be worth while to
examine which and what are the means of getting in to them.

Edinburgh Review of Swift, No. 53. The credit of the writers
of Queen Anne's age is attacked—that is while credit is allowed
them for their polish and elegance they are denied the praise of
all the higher qualities of poetry—its enthusiasm—its feeling—its
sublimity. This is a review which must produce a strong effect
upon public taste—precisely because it comes to tell people what
they already begin to know or rather to feel—to embody in words
and to arrange in a consistent theory the perception of unsatisfac-
toriness and even of weariness which every body must have begun
to feel in reading and considering the writings of that age—the
craving after more powerful passion—more sweeping interest than
those productions can exhibit. The review will produce its effect
because people are ready for it. It is not indeed the first display
of such opinions we have had—but it is the first from a judicious
and disinterested quarter. All the other opponents have tried to
talk down that taste because they had something to set up in its
place. Their extravagancies indicate that a change is taking
place in the national taste. They are the straws whose whirling
shews that the stream is turning.

[Compare the remarks on Locke, *Philosophy of Discovery*,
page 202.]

Rimini. There is no standing out against this poem—in spite
of affectation and vulgarity the truth and nature and vivid-
ness of the poetry compel admiration—love. There are hundreds
of words and phrases which insulated are ridiculous and disgust-
ing but in reading the poem they are what one would scarcely

wish away—they shew that the man has got something in his mind and is trying to express it. He is not seeking a meaning but a word. The scenes and personages seem to be present to him *with the vivacity of an ocular spectrum*. His peculiarities that grate upon the feelings for a moment as they pass—that give you a slight convulsive twitch as you read—are like the snappings of an electrical machine as it sends out its flashes. Such writing has obviously a tendency to impregnate words with meaning. In fact besides the meaning of a word which can be locked up in a definition there is a certain feeling arising from sound—derivation—resemblance to other words—poetical uses and associations of all kinds, which floats about it like a fragrance. This part of the impression conveyed by words he seems to have much felt, and in his use of words he has often sacrificed the other to it—which accounts both for the defects and merits of his diction. His scenery reminds every body of something he has seen and felt. His moral characters are feelingly drawn. Paulo *is* a 'glorious being.' His feelings after the deed are those of the best of human beings who could commit it.

Coleridge's Lay Sermon. The metaphysical part of it is a professed attempt to set enthusiasm above reason, which he calls setting reason above understanding—a most imprudent attempt to make reason commit suicide, to induce her by names to support a cause which involves the disclaim of her authority, and if she were not immortal the termination of her existence. The Reason of Coleridge seems so far as can be made out from the unintelligible mysticism of his language to be the principle of enthusiasm—the principle by which things which cannot be proved are asserted, things which cannot be asserted are believed, and things float across the mind which can neither be proved, asserted, nor believed.

Quarterly Review, April, 1817. *Rise and Progress of Popular Disaffection.* This article I suppose is Southey's. When a man of real talents falls into the ranks of party writers he gives a very different tone to their cry. He gives the word and they all repeat it after him. He invents for them arguments and nick-names and abuse for their enemies and they most diligently utter them. He takes care of their consistency—he does not go blundering on from argument to argument and from sentence to sentence—not knowing what he is about or where he is, and caring for nothing but the attack or the retort of the moment. He sees the past the present and the future—what his principles have proceeded from and what they tend to. Southey found many passages of our history considered by the world in general in a light not precisely consistent with the opinions he wanted to support, and he has accordingly been mending them—and what is more working them into a whole so that from this time any tory will know what characters and transactions in the history of his country he ought to admire and why—to the great benefit of the consistency of his majesty's ministers. Southey has the good talent of turning to account, and good account, every thing that he reads—and of writing with so much spirit and freshness of feeling that he seems but just to have read the books on which he forms his opinions— perhaps he has. But still his science is bad and absurd. His metaphysics is bad, his political economy is bad. He does not understand either—and yet he is resolved to abuse modern philosophy. He is a terrible bigot—and a great believer in conspiracies, literary and political, which is nonsensical. His abuse of Voltaire is quite irrational. He is however a man of sufficient possession of general views and particular facts to make him with his talents a formidable political writer.

July 25, 1817. *Coleridge's Biographia Literaria.* We do not get so clear an idea of C's character from this book as we should from an exposition equally frank by any other person—Apparently because C's views of human character are so different from those of any body else—we must understand something of it to begin with and by this means get a nearer approximation and so on. The foundation of much of his peculiarity of character seems to have been an opinion that he was peculiar—a resolution not to find sympathy in the mass of mankind and not to recognize as his the ideas and feelings which fashion has established. Our present state of society is so artificial that he who does this will find himself very different from the beings about him. He will however very probably, as C. does, fancy that the difference is much greater than it really is. He will think there is much less feeling, much less sympathy, much less of the original structure of human nature under all the improvements and modernizations, than he would find if he were to go rightly to work. But he does not. Most people take for granted that the rest of mankind are something of the same nature with themselves and proceed to examine *how*. C. did not immediately see *how* it was so, and so he came to the conclusion that it was not so. And hence he seems to himself to be much further from the general tone of society than he really is. He will not use its poetical diction—its prose diction—its metaphysics—its humour. And therefore he makes words for himself—uses old words in new senses—makes distinctions where there are none—puts old thoughts in new language— says whimsical things and then proves them to be humorous—and all this with the air of a man quite sure of his originality and only doubting whether he can make people understand things so much out of their common track of ideas. His metaphysics for instance when divested of the patch-work of old and new language in which he has disguised it would be found simple, often superficial, not always true. His humour is heavy and uncouth— taking for granted that his perceptions of humour are quite different from those of the rest of the world—he brings out every thing that strikes them however silly, and explains every thing however obvious. As for his wit—that kind of wit which is or was more peculiarly so called—the wit of similes with a point— he possesses it more perhaps than any other of our present writers. But this too is heavily managed—he seems too much in earnest— he seems to love his similes as real existences and to believe in them as real arguments. He has no lightness and versatility of mind or manner. He carries his egoism even to the heart and believes most fully in the malignity and bad feeling even without temptation of the reviewers and apparently of all but his friends. This appears too in his abuse of the reading public and of the literary character of the age.

July 27, 1817. Hare's brother was in Holland (I think in Brussels) and seeing a portrait of Barnevelt hanging in the room of the inn observed to the landlord that he was a great man and that it was wrong to put him to death. "Oui Monsieur mais il n'était pas orthodoxe sur la grâce efficace." Two hundred years are not long enough to cool popular fanaticism.

Aug. 2, 1817. *Montaigne's Essays* i. 46. "Mais encore à la vérité est-il commode d'avoir un nom qui aisément se puisse prononcer et mettre en mémoire." Alas then for Whewell!

Sept. 18, 1817. Morality should be a real animated and active person and not merely a wig-block to dress sentences upon.

Some people are obstinate from fickleness—and keep their opinions only when reason is against them, like the weathercock which points firmly in one direction so long as the wind blows steadily in the opposite.

Xenophon's style is too prolix—the ends of his sentences hang upon you long after the effect is produced—like the smoke that floats about long after the bullet has been discharged.

Criticism can only spoil poetry. Bray a peacock's feather in a mortar to analyse the colours and you will get only dirt and dust for your pains.

Critics seem to think poets made for them—poetry is to them the grist to keep their windmill turning.

There are people who are never so obstinate as when they know themselves to be wrong. Opinions stick in their heads like nails in a block—the firmer the more crooked they are.

Nov. 8, 1817. *Johnson's Preface to Shakespeare.* How could such a style ever be admired? Some of his illustrations and antitheses are happy; but the parade of truisms—the stateliness of his march along passages that lead to nothing is absolutely laughable. He seems perpetually employed in giving point to a wooden dagger.

Nov. 26, 1817. Supped with M. Biot at Peacock's—who is come to get information about the *system* of Cambridge—he expressed unbounded veneration for Newton. Professor Monk was roused by some observation of his that at Oxford the buildings were more magnificent which was agreeable to the spirit of the place as all the poetry of the nation was there—and began to enumerate *our* poets. Biot seemed to care very little for the poets. "Ah! but you have Newton; you want nothing more."

Nov. 28, 1817. Biot talked of Shakespeare and Racine—the beauty of Racine was his apparent simplicity and real boldness. Take, said he, the dialogue of Achilles and Agamemnon—read Achilles' speech—lay down the book—what shall Agamemnon say? He does not seem to have a word—take up the book— read—Agamemnon says just what you feel is right, "he has reason." Shakespeare is inimitable in his great characters. When Lady Macbeth says "Who would have thought the old man had so much blood in him?" it is terrible. Prince Henry preserves his dignity among all his low society—he never gives them a direct answer—he talks with a perpetual irony. Biot admired the Scotch novels exceedingly—he was asked if he could understand them. "O I had Callum Beg for my servant two months. He never went straight forwards, always obliquely and by shifts. I pulled him back into a straight line but he soon was in his oblique direction again." He said he understood Scotch.

There was something unpleasant in seeing such a man as Biot talking bad English—with a good physiognomy—dignified and expressive self-possession, and with a reputation for distinguished talent. When he was silent you looked at him with veneration— the moment he began to speak the ridiculous mingled itself with your impressions. His bad English and false accent gave you a superiority over him. He seemed transformed and disguised— under the influence of a spell. You could fancy some enchanter had condemned him to appear in a degrading shape and you were obliged to repeat to yourself "This is Biot." You could not do justice to what he said till you had repeated it to yourself divested of peculiarities of pronunciation.

Nov. 29, 1817. Miss Baillie's Plays. Basil (Love).

The Dialogue catches the attention—the characters are consistent, and the poetry very often exceedingly impressive. Basil's address to his mutinous soldiers is very eloquent. Victoria is a more worthless character than the author intended her to be The language is spoiled by an abominable fancy of inverting the words to prevent its being too like prose:

> But they beyond their proper sphere *would rise;*
> Let them their lot *fulfil* as we do ours—
> Society of various parts is *formed;*

This is detestable and runs through all. The thing that strikes most as a merit is the simplicity and unambitious clearness of the thoughts which is singularly contrasted with this restless violation of simplicity of language.

Sep. 23, 1818. It must have been very melancholy to one who had lived with and taken an interest in the writers of the last century to see them all dropping away in the last years of it. It would be like putting out the candles at the theatre when the extinguished lights remind the spectator that he cannot stay there much longer. Johnson died in 1784—his friend Garrick had died in 1779 and Goldsmith in 1774. Sir J. Reynolds survived till 1792 and Burke till 1797. Mrs Carter though a friend of Dr Johnson comes a little later, she died very old in 1806. Mrs Chapone died in 1801, so did Mrs Montagu—so did Dr Warton. Warburton died in 1789, Lowth in 1797, H. Walpole died 1797 and Gibbon 1794. Cowper in 1800. Cumberland survived till 1811, one of the last remains of the authors of the eighteenth century. These great and busy minds are now all still—their plans, their quarrels, their circles where they gave brilliance to the conversation and almost fixed the character of the age are past. In general, reflections on the mortality of our nature seem most excited by the death of heroes and kings. To me it would seem that the death of the poets and reasoners of the age is far more affecting—so much more mind seems to die with them. The reputation of a general is but like the blast of last night's wind: it roared and is now silent. It is transient by its nature and when it ceases it seems but to fulfil its destiny. But he whose mind turns on it the attention of men—who excites a commotion among them by his thoughts, who twines his ideas with theirs till he seems to have an existence in the breast of each—he seems as if death had no hold upon him, and we feel when he goes that there is no part of us twined around this system of mortality too strongly to be separated. Even the surviving of his works increases the solemnity of the prospect.

The fruit lies before us but the tree has vanished. However much we may know of his history and habits, yet that which perishes is so much more than that which is remembered, that the productions of his hand stand before us like the pyramids of Egypt—the works of skill and genius, but their author is buried in oblivion. We cannot call him back to ask him what was the meaning or the occasion of any passage or work—it remains what it is—and we must find its key in itself. And however great may be his command over language, we know well that it conveys to us but a very small part of the ideas which passed through the mind of its author at the time when he was writing it.

March 7, 1819. First proof of Mechanics. Agreed with Deighton for copyright for three guineas a sheet for first edition and two guineas a sheet for succeeding editions, with three guineas a sheet for additional matter.

June 25, 1820. *Milman's Fall of Jerusalem a Dramatic Poem.* It cannot be considered as a tragedy for the situations and characters are not or are very slightly subservient to a connecting plot. It must rather be looked upon as a series of poetical pictures of character, interspersed with lyric poetry. The characters are introduced in succession as is convenient for this purpose and several of them make their appearance only once or twice. The poetry in which they are clothed is some of it beautiful and imaginative but the work of one who has felt and read more than observed. Fancy and analysis predominate over knowledge of human nature and what may be called moral imagination or the power of entering into the feelings of the different classes of human character. A curious consequence of this penury of moral materials is this (and it is by no means peculiar to Mr M. but on the contrary very extensively and strongly characterises the narrative fiction of the present day)—that in order to make feelings appear strong the author wrests them into something beyond human feeling—some exaggeration amounting to a contradiction. They take the crookedness of the staff for an indication of its strength—they put in too large a charge to give the ball its greatest velocity. Thus the old man who was carried by the natural contagiousness of popular feeling to join the crucifixion of Christ is made to exult even during the darkness and the earthquake. Simon is made to rejoice in taking some bread from a famished Jewess who is preparing it for her children contrary to the orders of the besieged city and to say

> We were wiser than to bless with death
> A wretch like her.

There are, however, some passages of great tenderness, which is Milman's happiest strain. The second meeting of Javan and Miriam at Siloe is very beautiful—her parting address to him is touchingly solemn and affectionate. The fanatical ferocity of Simon and the sneering malignity of John are written for contrast but their quarrels are not made very interesting or intelligible. Salone too who is more than half mad with the combined effects of enthusiasm, hunger, and love is very abrupt in her transitions. Miriam is the most naturally written character and indeed appears to have flowed warm from Mr M's pen, though she is sometimes too reflective. The lyric poetry is harsh and often feeble and flutters very low from the rest.

Dec. 10, 1820. Wordsworth's Excursion Books 6, 7, 8, 9 finished. Either the latter part is worse than the first or I am not in a proper frame of mind to enjoy it (I do not suppose that having seen the author has any thing to do with it). This part of the poem appears to me very prolix, prosaic, and in some places nearly unintelligible (beginning of Book 9). The general objections to the whole plan of the poem—its want of interest and incident—its childishness—its enthusiastic but I think quite false philosophy are burdens which press it more as it draws towards the close. Still there are some beautiful touches of description and feeling—especially in Book 6 where the vicar gives the histories of the tenants of the different graves. The discussion of the effects of the manufacturing system does not shew any great power of general reasoning, which is a fair criticism upon it because that is what it pretends to.

Sunday, Feb. 10, 1822. It was said of Socrates that he withdrew the philosophy of his age from subjects entirely speculative and unprofitable, the mere matters of curiosity and exercises of acuteness which had employed the reasoners of Greece up to his time, and directed it towards practical principles. His study was the rules of judging of human conduct and of analysing the motives of human action, and he taught that the most important use of knowledge is to regulate our daily conduct and habitual feelings. If we consider those men as more particularly the instruments of the superintending Deity, who, gifted by Him with extraordinary talents and virtues, have tended to elevate the tone and character of those among whom they have lived, to purify their manners and ideas, and by their powerful influence to purge away as it were some of that mass of selfishness and grossness which by the daily intercourse of vulgar minds and passions is perpetually accumulating in the habits of men;—then we may look upon this teacher as the unconscious messenger of a wakeful Providence, and as the preacher of a doctrine which a benevolent God would not suffer to be extinguished. And accordingly his labours were not useless, and his philosophy was long felt in its kindly operation upon the views of those who followed him in moral speculation.

But we stand under another dispensation than this. We have not been left to the lessons which human reason and human virtues, however exalted and purified, could teach us. We have not to trust for our illumination to those casual and intermitting stars of genius which in the usual order of human affairs blaze forth from time to time like the brilliant meteors of a polar winter. For us the long darkness is passed, and the superior and unvarying light of the risen sun fills our hemisphere. Doctrines fitted to refine and elevate the soul, motives to the most spotless and energetic virtue, means of purifying and renewing the heart, far beyond all that man could before imagine, have been imparted to us; and of these we do not conjecture but are certain that He who taught them had them from God, that He knew what was the ministry in which He was engaged, and we may conceive that the manner in which the laws of the material world were suspended at His bidding was emblematical of the miraculous place which His religion was to hold in the world of human thoughts and feelings. Other systems may have checked in a few persons, for a short time, the vicious propensities of human nature; but Christianity was to change the views and influence the conduct of the world. It was not to improve only, it was to transform the heart; it was not to restrain the actions only, it was to reach the motives; and it was not to seek admission at the reason of a few speculative men only, but it was to claim, as of right and by the authority of the Creator, dominion over the minds of all.

If Heathen Philosophy could boast to be not a merely curious and useless pursuit, but a course of moral discipline and practical self-government, surely Christianity should still rather and more justly bear this character. Its great and ultimate purpose is with the hearts of men; and it uses their reason only as a passage to that object. It promises no discoveries except those which concern the relation of man to God, it does not require as essential any exercise of the ingenuity or acuteness—it makes no display of food for the curious or the profound; but it says " my son give me thy heart"—it requires a submission of the will and the hope to God, and it promises a regeneration of the moral man—it is in short, simply and solely, a practical religion.

Would it not then be strange if possessing and professing such a Religion, being the disciples of such a school, we should still require teachers to do for our philosophy, if we may call it so without offence, that which an unassisted reasoner did for the ancient world—bring it home to our bosoms and to our daily business and remind us that it is neither a mere matter of speculation nor of form; that it is neither a science nor a controversy but a Religion?

May, 1822. *Fortunes of Nigel*—sadly stupid—scarcely a good passage or character to redeem it.

Feb. 18, 1823. *Peveril of the Peak.* Extravagant in story, though the author has moulded historical facts without any ceremony to answer his purpose. Characters hard and dry, circumstances forced and barren compared with the earlier works. Supernatural parts beyond poetical belief—still, better than Nigel.

Elia (Essays by Ch. Lamb). Ingenious, and shewing a kind and good-humoured disposition in many places; but conceited and strained. The singular narrowness of his materials and mode of speculation forces him to follow out all the phraseological, moral, and metaphysical, relations of his subject in the narrow field which his habits of combination and observation give him.

Feb. 22, 1823. *Burke. Discourse on Taste, Introductory to Essay on Sublime and Beautiful.* Taste the faculty which is affected by works of imagination and the elegant arts. In simple sensations the sensation is the *same* to all (as sweet, bitter). Also the same sensation is originally *pleasant* to all—though it may be otherwise by *habit* (as tobacco). The *imagination* possesses a power of recalling and combining images of things—and here besides the pleasure of *sensation* there is the pleasure of the perception of *imitation.* In the first men agree as appears from what has been said—in the second they may differ by their differences of *sensibility* and of *attention* (or experience) but in this case it is the judgment which determines. *Sensible qualities* are subjects on which men judge alike—so are *passions*—but *manners* and *characters* are subjects on which we judge differently— and we learn to judge of these when we learn "morality and the science of life—skill in manners, the observances of time and place and of decency in general." Hence Burke infers that taste depends upon *sensibility* and upon the exercise of the common faculty of judgment.

The objection to this is that the enumeration of the pleasures of the fine arts above does not appear to be complete. For it appears that there are many objects which please where we cannot reduce the pleasure either to that of sensation or to that of imitation. Can we resolve into these elements the qualities which we call elegant, graceful, picturesque ; the pleasure which we receive from a beautiful urn ; from a rose as compared with other flowers ; from a thatched cottage with a broken outline ? It would seem not. And on these qualities men do not agree.

Again, with respect to character and manners, it appears that as objects of taste they are judged of by other principles than matters of morality, and apparently much less certain ones. Of morality we have measures and standards. Of taste we have none except those which arise from taking principles generally assented to and enunciating them as rules; and these vary much with time and place.

Finally, can taste be placed in the exercise of the judgment till it can be shewn what those principles are with which we judge objects to agree or to differ?

April 2, 1824.   Wilberforce's Practical View of Christianity. Evangelical views enforced with great clearness and strength, much acuteness and skill in detecting the shades of worldly feeling and the modifications of self-deceit, and remarkable elegance of style and often eloquence.   See particularly Chap. 4 § 4, on the error of substituting amiable tempers and useful lives in the place of religion—the paragraph on the short and precarious duration of such qualities.

There is an error which appears to belong to all persons of this class of opinions of degrading human nature too much.   To say that we are utterly corrupted, and that our best actions have no merit, is merely to substitute words for ideas, for our feeling of moral approbation cannot be got over.   By instances and inferences they prove that man has many bad feelings and propensities; in the same way you may prove that he has many good ones; both are true—and without religion the bad will in the bulk of mankind predominate.   They ought to be content with this.

Sep. 26, 1824.   Southey's Roderick—as touching and beautiful as ever.

Feb. 26, 1825.   Goethe's Wilhelm Meister.   A collection of speculative generalities, which are put in the mouths of all the characters without exception.   No incidents, no intelligible reasons why the persons should feel and act as they do, no character in the conversations, little wit, little description which is at all striking, and in short nothing to interest.   The description of a German dramatic party is curious in its way if it be exact, and the sequel of Mariana's story is pathetically told.   Mignon is unintelligible and unpleasant.

Sunday, Nov. 13, 1825.   Reflexions on God.   The view entertained by those who do not see the necessity for a perpetually sustaining and preserving Providence in the world, seems to be this, that the material portions and elements of the universe may have certain *laws* impressed upon them—or certain properties in their *nature*—according to which they operate and combine and produce all the phenomena which we see. And these persons think that they can easily conceive that these laws and properties may be so balanced and related to each other that they may give rise to a perpetually proceeding series or cycle of physical events and appearances, always varying but never tending to absolute derangement, and therefore not requiring any interference to keep them in continued play. Such laws, for instance, are the laws of motion, of the action of light, of heat, the chemical properties of substances, and others. And to such persons it seems that when we have reduced natural phenomena to these laws we have done something in the way of simplifying and illustrating the explanation of nature as an assemblage of matter merely, and have removed the necessity for considering an immaterial and intelligent agent one step farther from us, while we have brought the idea of the world as a machine much more distinctly into our view.

The discovery of such laws is undoubtedly an employment which gives a strong and peculiar pleasure to a speculative mind. Their beauty and satisfactoriness consists in their being generalizations and simplifications; we are able to comprehend a number of particular facts so as to consider them as instances of the same rule—or we are able to explain a complex appearance as the operation of one or two properties which at first seems to depend on a combination of many. And when we can carry on this process from one stage to another till we arrive at laws of very great generality and simplicity, we see innumerable facts of the most remote and dissimilar kind arranging themselves under one class, and becoming as it were only different exhibitions of a principle which we represent to ourselves as the most simple and single imaginable. Thus the few laws of motion and force being once obtained, the steady and perpetual march of the heavenly bodies in their orbits and the motion of the slightest atom which is carried about by the wind are alike exemplifications of them. And it is felt by every contemplative person that these combinations when they can be thus considered, seem in a remarkable manner to give us that *knowledge* which it is the object of our enquiries into nature to obtain. We are satisfied with our acquired views as far as we go, and what we wish is to go still farther in the same direction—to bring our general principles under some one still more general; to reduce our simple laws to one still more simple

Why we receive this great pleasure from the success of such speculations is perhaps more than we can expect to answer. But the fact will probably be allowed by those with whom we are concerned. And we may then ask why we can after such discoveries and views more easily believe the course of nature to be self-subsisting and self-permanent than before. That we can conceive and can explain it more easily is granted, but why should this affect our belief as to the agency by which it is produced? The particular events are the same as before, only we have learnt to consider them in a new point of view. We have discovered no new power, no new agency in nature; we have only found the rule according to which the results are produced. Such a rule satisfies us as a way in which we can consider objects, but such a rule so discovered by us cannot be an active and positive agent. Except in the mind of an intellectual spectator such a rule is nowhere and nothing. It exists in nature just as the knowledge of a language exists, by means of which we are enabled to understand the characters of a book presented to us. The letters are there and we know what thoughts and feelings are expressed by them, but these thoughts and feelings have not produced the letters. So in nature, her facts are the same whether or not we know her language, that is the laws which she expresses by them, but these laws cannot have produced and cannot reproduce the facts. They classify them, they give significance to them, they make me read and understand and rejoice in the contemplation of them, but they cannot alter my views of the agency which produces each particular event, except they make me believe, like the deluded astronomer in Abyssinia, that I myself am the ruler and mover of all the system.

But we cannot stop here. If these general laws which we

thus discover in the universe exist only in our minds, and operate not upon matter but upon our way of considering and viewing its properties; and if nevertheless we feel this irresistible conviction that they are truly and rightly deduced from nature, and are really the rules to which she is subjected; and if this conviction is accompanied with a glow and an exultation as if we had obtained something which indeed deserves the name of knowledge, we cannot but suppose that they have some meaning and origin beyond ourselves. This indeed is manifest. For these laws are assuredly not arbitrary and conventional; not taken up according to caprice or accident; neither are they mere classifications and indexes of phenomena; but along with the possibility of their deduction comes always the evidence that they have a meaning and a reality of their own proper kind. Now it appears that their only meaning and reality consists in their existing in the mind of an intelligent being, and thus becoming the rules according to which he contemplates or acts. And here therefore the conclusion forces itself upon us that these general laws must have prevailed in the mind of an intellectual being by whom the material world is governed. There is no other manner in which they could exist as they undoubtedly do. They are there, and we can trace them to no other origin. With this view they become intelligible and conceivable. They are the laws according to which the Author of the world works. They are the characters which He has stamped upon His productions and which we are enabled to read. With us, as was said before, they are the knowledge of the language in which the book of nature is written. And this language is the expression of His thoughts, and we can read it so far as our minds are of the same nature with His. As soon as we suppose this book to have had an intelligent author our knowledge of the language does enable us to attribute it to Him, to understand Him, to a certain extent to go along with Him. The laws of nature, then, are no longer laws which inert matter *observes,* as the phrase is used, and *obeys* in a manner apparently unintelligible but in fact contradictory; they are laws according to which the Creator causes matter to act and work—objects of thought to Him as to us—rules of operation with Him; trains of delighted.

contemplation to us. He is the agent, and every single fact—the leaf that falls to the ground, the bubble that bursts in the stream, the comet that sails through the planetary orbs, are under the immediate guidance of His finger. It is He who directly and immediately paints the western sky, and streaks the tulip, and scents the rose, and gives its flavour to the anana. And it is most unwise and inconsistent when we forget this, because we can go a step or two farther and trace the rules of action of this Divine mind, and see the permanency of His ways, and feel how agreeable and suitable they are to the highest intellectual nature. Still more strange and insane is it, when fixing our attention solely on these rules we forget the very Being in whom alone they exist, and deify as it were the maxims of His government while we deny that He governs at all.

But having got so far we can now perhaps give some reason at least for the strong and peculiar pleasure which we receive from such discoveries as have been spoken of. It is a feeling which we may well exult and rejoice in, that we can obtain to some extent a view of the rules by which the Highest Intelligence forms and guides the world; that we can see their beauty depending upon this very circumstance, that they are so few, so wide, so simple, so sufficient. It is not wonderful that every step which we make in this direction, every new success, every additional glimpse should come upon our sense with a sweetness all its own, and seem the perfection of our speculative nature. It is natural that as we cultivate this speculative spirit we should be more and more desirous of going on in this career, that whenever numerous and various facts present themselves we should struggle to take hold of their general laws, and thus to understand what it is that God is expressing by them, that we should seize with eagerness such fragments and discovered glimpses as we can catch, like phrases of the sublimest eloquence in a language which we imperfectly understand, and that we should remain unsatisfied and restless till we can connect these fragments into a whole, these broken phrases into an intelligible declaration. Every new law that we discover does thus, in proportion to its generality, bring us to a nearer and nearer knowledge of the thoughts of God as they

regard the material world, and the pursuits of science acquire a
dignity and an importance which almost invests them with a religious awe. The pride of discovery, the elevation of spirit which
we feel when a new principle dawns upon us, the triumph with
which it fills our mind and the energy with which it impels our
enquiries, all these feelings are as it were hallowed and sanctified
because they arise from that tendency which draws us towards the
Divine mind, and makes us seek the perfection of our nature in
Him. And it must be thought to add no small dignity to our
intellectual nature that it thus appears that it so far resembles His
—so far that something of what He does we can understand when
done; that the laws by which He delights to act we delight to contemplate; that we can consider nature till the material shell and
frame-work falls away and we see the Divine ideas, almost we
might say before they are embodied in matter.

This is not so far applicable to the moral world, because the
laws which we discover are of much more limited generality and
more numerous, and in no instance reach an *elementary* simplicity
and extent. Indeed there seems to be no example in which we
can go even *far* in this process, much less near the end of it.
The speculations concerning *final causes* in the moral world are
the best specimens, and these are not of the certainty and extent
of the others. Hence it would seem that we were not intended to
speculate upon the phenomena of the moral world by means of
our reason alone.

The examples are however enough for the argument, and these
therefore lead us demonstrably to the moral qualities of God.

March 19, 1826. Shakespeare's description [of Cleopatra's
passage up the Cydnus] contains an allusion to the *fuga vacui*:

> Antony...did sit alone,
> Whistling to the *air;* which, *but for vacancy,*
> Had gone to gaze on Cleopatra too,
> And made a *gap in nature.*

Sept. 30, 1828. *Edinburgh Review. On History* (I suppose by
Macaulay). Extremely brilliant and elegant Essay writing like
his preceding reviews, but not very profound. Perhaps this is
an advantage and tends to the effect. His thoughts are a little

beyond commonplace and vulgar opinions, and by bestowing all his pains to illustrate them he obtains at the same time the praise of originality and of clearness. His opinions in themselves are not stamped with great thought, but the detail and illustration of them is so varied and pointed that they cease to be superficial. He takes a common thesis—that poetry declines as criticism advances—that authors reason better when they are to be *read* than to be *heard :* and these doctrines become so clear that they seem to be new by their ceasing to have the indistinctness of commonplace. He is in prose what Byron is in poetry, *one step* before his readers ; and thus he fills the eye and strikes the senses more than if he were farther in advance.

CHAPTER XIX.

POETICAL PIECES.

[DR WHEWELL'S manuscripts include some poetical pieces; from these I select a few, all belonging to his early years.]

I.

O GENTLE shepherd tell me why—
When evening opes her dewy eye,
Why still with folded arms you love
And head reclined to seek the grove,
And wander pensive where the breeze
Sighs sadly through the waving trees;
If thou art wrung with woes like me
Then, shepherd, I will go with thee.

If Avarice grieve for riches lost,
Or if thou mourn Ambition crost,
Or if it be Remorse that speaks
In the wan hollow of thy cheeks,
If on thy lonely thoughts attend
An injured parent or a friend,
Then thou, 'thank heaven, art not like me;
No, shepherd—I go not with thee.

If some fair nymph who once was kind
Now slights thee, changeful as the wind,
Or if, thy absent charmer gone,
Thy days and hours drag weary on,
Then envy might pervade my breast
For thou, my friend, hast *once* been blest;
Yes, heaven's more kind to thee than me
Nor, shepherd, will I go with thee.

But if thou go'st to tell the grove
Thy mournful tale of HOPELESS LOVE,
Of love that still would silent lie
Save in the frequent-bursting sigh,
For one as fair as eve's fair star
And ah! beyond thy reach as far,
Then thou art yoked in grief with me
And, shepherd, I will go with thee.

—To haunt with ling'ring step the place
Where first thou saw'st her angel face;
To watch where thy bright goddess lies
If one chance glimpse may bless thy eyes;
But never, never yet to dare
To tell thy passion to the fair—
Hast thou done this? ah! how like me—
Yes, shepherd, I will go with thee.

Together let us seek the shade
By black'ning pines and poplars made,
Beneath the beech together lie
And answer mutual sigh for sigh,
And on the willow carve their names
('Tis all we can) to soothe our flames;
Thus wilt thou share thy griefs with me?
Then, shepherd, lo! I go with thee.

June, 1813.

II.

O THOU who bid'st th' adventurous spirit roam
Beyond the narrow boundaries of home;
Who view'st the distant shores with piercing eye
That hid by half the world's big convex lie;
Who mak'st the East and West join hands to pour
Their mingled produce on each other's shore;

Thou that can'st tread the sea and ride the gale,
O heaven-born GENIUS OF DISCOVERY, hail!
Inspired by thee, since first the world began
Man seeks acquaintance with his brother man;
Phœnicia's early sons, inspired by thee,
Ranged every shore that bounds their inland sea;
And while fond self-complacent Athens own'd
Alcides' pillars earth's remotest bound,
Thy favoured Carthage sought on Albion's shore
Her precious mineral and her sparkling ore:
And though when clouds of Gothic gloom opprest
The setting glories of the darkened *West*,
Though then amid the night of science chain'd
Barbarian fetters long thy flight restrain'd,
Yet the first breath of Europe's balmy morn
Shook off thy bonds; beneath thy influence born
Pleased Commerce bask'd on thy Venetian isles,
And wealthy Genoa felt thy cheering smiles;
Beneath thy sheltering wing did Vasco brave
The storm-clad spirit of the Indian wave;
But chief thy soul-expanding power possest
Its spacious temple in COLUMBUS' breast;—
Ill-starr'd Columbus!—with a soul to embrace
The rounded earth hung in th' abyss of space,
Yet doom'd the petty arts of courts to try
And freeze beneath neglect's ungenial eye,
To stoop to sue the gifts of kings to gain;
And,—O caprice of fate!—to sue in vain:
Yes—the great veil that o'er the Atlantic spread
Concealed the secrets of its western bed,
That veil, which first Columbus dared to raise,
Had fallen again with him:—In future days,
Europe, denied her niggard princes' aid,
In ignorance of a sister world had laid.
What parent Genoa fitter had supplied,
What France and Spain and England had denied,

The power to execute his great design,
Thine, Isabel, the glorious gift was thine;
His plans derided and his hopes forlorn,
'Twas thine to bid him rise above their scorn,
To bid his prow explore the wish'd-for sea
And half we owe the western world to thee.

And now the busy shore behind them cast
Three little barks obeyed the eastern blast;—
Night rolls on night—and now their recent keels
The wild waste of the mid Atlantic feels.—
Sweet on the wide expanse the moonbeams sleep
And faintly throbs th' interminable deep;
Proudly the ships above the waters glide,
The sparkling ocean murmurs at their side;
The whitened canvas swelling in the gale,
The sportive moon-beams silver o'er the sail,
Where ne'er the moon had seen a sail before,
Nor heard the surge about a vessel roar;
Five thousand years the moon that sea had seen
And many a lovely night had smiled serene;
But till that hour ne'er did her radiance rest
On aught but ocean's vast unvaried breast;
Ne'er had she heard, borne on the midnight breeze,
Aught but the drowsy hum of endless seas.

July, 1813.

III.

In early days when social life began,
When nature's wants were all the wants of man,
A day of hunting or of war gone by
The naked tribe enjoy'd the evening sky.
Then rose the early bard and pour'd his song,
Attentive silence held the list'ning throng;

Rude, nervous, bold, th' untutor'd numbers flow'd,
Thick in the strain the daring image glow'd ;
The crowd enraptured saw at his command
Each absent scene before them vivid stand ;
And, (if the soul swept onwards in the strain
Left reason pow'r to exercise her reign,)
Perhaps they marvell'd how the poet's ken
Could catch those traits that scape the gaze of men ;
Which each one owns to mark the object strong
And wonders how they shunn'd his eye so long.

But *then*, where'er the poet turn'd his view,
Burst on his sight some scene, some object new ;
And when th' enthusiasm of poetry
Had pour'd its influence on his raptured eye,
Whatever prospect met his heav'n-purged sight
Was clothed in colours fair and glitt'ring light.

The Muse as yet was young, and like a child
Where'er she turn'd her eyes with wonder smiled ;
Amid romantic glens she careless stray'd,
And cull'd her flow'rets in the deep'ning shade ;
With all the world before her trackless spread
In others' steps she need not fear to tread.

But when bard after bard arose to sing,
And still the same trite themes awaked the string ;
To praise the hunter or the battling host,
To soothe the warrior's or the maiden's ghost ;
The muse indignant spurn'd the well-worn sod,
And sought more verdant fields and plains untrod ;
And hence with love of *Novelty* imbued
In every path the flying form pursued ;
Stalk'd through the lengthen'd *Epic's* sounding song,
Or pensive walked with *Elegy* along,
Or sported in the *Pastoral*, or rode
In wild career upon th' unbridled *Ode*,
Or vice to reach with *Satire's* lashes strove,
Or rambled in the myrtle bow'rs of *Love :*

Yet flying still, the ever-changing shape,
Though often grasp'd, was grasp'd but to escape.
    But when, no more a heav'n-selected few,
Succeeding poets still more num'rous grew;
When those who felt no muse's holy fire,
Dared with unhallow'd hands to touch the lyre;
When wider still the itch of verse did steal
On all who could, or could not think and feel;
When flatt'ry too, audacious, dared to claim
The sacred sanction of the muse's name;
And when ev'n ruffian vice unblushing prest
With hand defiled to touch her heav'n-wrought vest;
Despairing then she stood with tearful eye
And saw her plains in desolation lie,
While servile bands rapacious ranged around,
Claim'd the waste fields and seized the ruin'd ground.
Here *Imitation* shakes her leaden chain,
And there stands *Plagiarism* disguised in vain,
And *Nonsense* rages by no rule confined,
And mad *Bombast* and drowsy *Dulness* join'd:
Hence with disgust the goddess heav'n-ward flies,
And spurns the worthless crowd and seeks her native skies.
    Yet though 'mid fields of light she dwells on high
She looks on scenes below with gracious eye,
And, when among the sons of men she sees
One whom for her the will of heav'n decrees,
Stoops, as the bird that guards th' Olympian king,
And bears him upward on her eagle wing;
And bids him view the scenes that lie below
With glories new and rainbow colours glow;
And, as he sees the world beneath him roll,
Feel loftier raptures swell within his soul.

Sept. 17, 1813.

IV.

FRIEND of my youth! my dearest, earliest, best!
My Muse! companion of my boyish years!
On whose fond bosom once I loved to rest
Each opening feeling life's fair spring endears!

O well thou know'st how raptured oft I sought
With thee to rove—to seek the shade with thee—
Warm on thy ear to pour the rising thought,
And gaze on all around in extacy.

Beloved Enchantress! while with thee I strayed,
Responsive to thy smile all nature smiled,
And every opening bud of life seemed made
To fling its fragrance round thy favour'd child.

O loved departed days! but thou—so soon—
Art thou too gone with boyhood's rosy morn?
Must youth's joys vanish at th' approaching noon,
Like early dew drops on the glittering thorn?

With unsuspicious hope—with feeling warm—
With fancy's vivid powers—art thou too gone?
With thee is vanished Nature's every charm—
A wilderness around—and I alone.

In vain I seek thy voice—ah! well known sound—
In all the haunts thy soul was wont to love,
When spring in fragrant whispers breathes around,
When autumn sighs along the flutt'ring grove—

E'en when I roam great ocean's shore along—
And hear his nightly roar—and see him roll
The bright moonbeams his restless waves among,
And all his boundless voice comes o'er my soul—

E'en at that hour—to thee of old so dear—
Scarce can I catch thy sweet familiar tone;
Amid the roar it faintly strikes my ear,
Like a departed friend, with distant moan.

O yet, my friend, return—O speak again—
  With all thy dear delusions hither haste;
Bring all my early joys—thy smiling train—
  Change the drear present for the glowing past.—

For all but thou is full of fraud and guile—
  The world is fair, but joyless is its band;
The heartless pleasure and the unfelt smile,
  The withered flower that flies the grasping hand.

To thee—my earliest love, to thee I come,
  O ope thy arms, and let me on thy breast
Forget the lures that drew me from that home,
  And lull my disappointed hopes to rest.

BURLINGTON, Sept. 1816.

V.

June, 1817.   Schiller's Works, Vol. I. p. 49.

[*An Minna.   Schiller's Sämmtliche Werke*, 1844, Vol. I. p. 71.]

Am I dreaming? am I doating?
  O'er my eye-sight sure's a blot:
Can my Minna by me floating
  Pass along and know me not?
On the arms of coxcombs leaning—
  Playing antics with her fan—
Vain, and of herself o'erweening,
  Minna this?   It never can.

Plumes, that on her hat waved lightly
  When I gave them, wave there yet:
Knots that deck her bosom brightly
  Cry—ah! canst thou then forget?
On her breast and hair are braided
  Flowers that grew beneath my care,
But that heart to me is faded
  Though the flowrets bloom so fair.

Go—around thee call thy smilers;
　Go—and think no more on me;
Prey to sordid mean beguilers
　How my heart despises thee.
Go—for thee a heart once fluttered,
　Throbbed in earnest, but in vain,
For a silly woman uttered
　Vows—but it can bear its pain.

Beauty was thy heart's undoing—
　Beauty—shame to have it said;
Lo! to-morrow sees its ruin
　Like the flower whose leaves are shed.
Swallows come when spring days brighten,
　Go again when north winds roar,
Lovers will thy autumn frighten
　But thy Friend returns no more.

In the ruins of thy sweetness
　All forlorn I see thee stay
Weeping at thy spring time's fleetness,
　Mourning for thy hours of May.
All that with desire once burning
　Snatched thy kisses when they glow,
All thy faded beauties scorning
　Scoff thy days of winter snow.

Beauty was thy heart's undoing;
　Beauty, shame to have it said,
Yet to morrow sees its ruin,
　Like the rose whose leaves are shed.
Then may I despise and spurn thee—
　Spurn thee! God above forefend;
No—ah deeply shall I mourn thee,
　Bitter tears shall weep thy end.

CHAPTER XX.

[A few pieces were put in a packet and marked *unpublished*, either by Dr Whewell himself, or by those who looked over his papers; these I will now mention.

On the nature of the operation of thought by which we form Metaphors. This is a very early essay, consisting of 16 pages; it is unfinished.

Draft of a letter without superscription or signature, enquiring about the optical researches of M. Fresnel. This was probably meant for M. Biot or M. Arago; it consists of 2 pages.

Remarks on Style and on Keble's Christian Year. This consists of 6 pages; it is unfinished.

Analysis or Review of the History of the Inductive Sciences. This is written in a playful style; it consists of 18 pages.

A long paper on the subject of communication between the Earth and the Moon. This is the most curious of all the unpublished pieces, and I will give an analysis of it. The paper takes the form of a fictitious story in which the narrator, who had long felt a desire to communicate with some other part of the Universe, as the Sun, the Moon, and the Planets, describes the amount of success which had attended his wishes. The manuscript is in four parts; not only however is the story left unfinished, but each part also seems to terminate abruptly. The whole would occupy about 70 octavo pages. The handwriting and the treatment of the subject shew that it was composed after the *Essay* on the Plurality of Worlds.

Contents of the First Part. Desire to communicate with some other part of the Universe. The Fixed Stars, the Sun, Mercury, Jupiter, and Saturn cannot be supposed inhabited; but Mars, Venus, and the Moon may be. It is hopeless to get to Mars or Venus, but perhaps not to get to the Moon. Engineers can accomplish anything now if they have adequate funds; they might perhaps project a man, or build a structure, from the Earth to the Moon. The expense of such a structure would be enormous, though perhaps not greater than is now wasted in wars. But the want of air above a certain point would be an insuperable difficulty; and this sends us back to the notion of a projectile. Then however we find that the velocity required at starting would be enormous; but that to come from the Moon to the Earth would not be so extravagant. This leads to the consideration of meteoric stones, which may have come from the Moon. Astronomers have generally held that the Moon is uninhabited, but the account of the shape of the Moon recently given by Professor Hansen suggests that there may be inhabitants of the Moon on that side of it which is turned away from the Earth. This is a fresh incentive to the examination of meteoric stones.

Contents of the Second Part. This begins with a digression relating to E. D. Clarke, the traveller and Professor of Mineralogy at Cambridge. The story is resumed with the narrator at Pekin, where he is introduced to an extraordinary person named Mono, who like himself is deeply interested in meteoric stones. The two resolve to proceed together to the spot in Siberia where the famous meteoric stone was discovered of which Pallas speaks in his travels. On the journey they have a long conversation respecting the appearances which the Earth and the Heavens would present to a spectator on the Moon.

Contents of the Third Part. Mono reveals who he is and whence he came.

Contents of the Fourth Part. The narrator obtains some explanations from Mono as to portions of the extraordinary revelation; and then they agree to visit Peru in hopes of finding some traces of a brother whom Mono has lost and deplores.

I print the digression relating to E. D. Clarke, and also the whole of the *Third* Part. The two preceding Chapters consisted mainly of compositions of Dr Whewell's earlier years; the remainder of the present Chapter belongs to his later years.]

As the result of the train of reflection which I have been describing I resolved, as I have said, to devote myself to the study of meteoric stones—stones which had fallen or were supposed to have fallen from the skies upon the earth. This subject acquired a new and fascinating interest for me in the recent years of which I have mainly to speak and was the occasion of the strange adventures which I have to describe. But of such stones I had known something for many years and of the oscillations of scientific opinion concerning them. When I was an undergraduate at Cambridge about 1813, I attended the mineralogical lectures of the celebrated Edward Daniel Clarke, then just returned from his travels which had extended from the Baltic to the Crimea and the Mediterranean. Certainly Clarke was one of the most striking characters belonging to the Cambridge life of that my early time. He was very eloquent:—I should say the most naturally eloquent man I have heard : that is, he gave to what he said all the charm that fluency, earnestness and fine delivery could give, independent of its meaning and purport, which often could not bear a close examination. He was not a profound or exact man of science, but he had a good knowledge of what was doing in the world of science, and undaunted courage in endeavouring to take his share in it. He very nearly blew himself to pieces once or twice in his experiments with his oxyhydrogen blowpipe. He on returning to the University after his travels began to deliver a course of lectures on Mineralogy, which were very attractive, for in them he introduced stories and discussions about all that he had seen and heard of in the course of his travels. Among other things he spoke of meteoric stones. The celebrated mass of meteoric iron which Pallas had seen in Siberia and had described in his *Travels*, had then recently drawn general attention to the subject. Clarke had of course a theory on the subject of these stones. I do not know if any one now maintains that theory. He held that as all substances can exist in a gaseous state, the components of these stones might occur, in a gaseous state, in the higher regions of the earth's atmosphere; might there, owing to some natural event or other, combine; of course with explosive violence, noise and fire, and might then fall to the earth. I do not know if his theory made many converts; some of us certainly laughed at it; and one of my friends said that it seemed to him just as likely that the air should drop biscuits from time to time in the neighbourhood of a flour mill. But the accounts of the peculiarities of the known masses of meteoric iron were very curious. There was a specimen of Pallas's Siberian specimen in the Geological Museum which had been bequeathed to the University of Cambridge by Woodward; a man ridiculed by Pope who was his contemporary, but who was far in advance of his age in perceiving the importance of collections of organic and other fossils. Among other curiosities he had managed to procure a specimen of this Siberian mass of meteoric iron. And there were indeed two specimens of this mass in the collection; for one of the Woodwardian Professors, Hailstone, had procured a specimen after the mass became known through Pallas's account of it; and had placed it in the Woodwardian Museum to shew how vigilant and judicious Woodward had been in making his collection of curiosities.

[We now pass to the *Third Part* of the piece.]

We now approached the place where the celebrated mass of meteoric iron had been found and where it was supposed to have fallen on its arrival at the earth. This spot is,. as we are told by Pallas, an elevated point of a slate-mountain, between the rivulets Ubei and Sisim, both which spring from the side of the wild mountains between Abakansk and Balskoi, or Karaulnoi Ortrog, and fall into the Yenisei: the spot is four wersts from the former, and six wersts from the latter place; and about twenty wersts from the Yenisei. The mass when found was entirely detached and exposed; and the inspector who had his attention drawn to it by the Cossack who discovered it (in 1749), took care to ascertain by careful examination that there was no trace of mining or smelting operations in the neighbourhood.

As the mass had been removed to Petersburg and other places (indeed I believe there is hardly any considerable mineral collection in Europe which does not possess a specimen of it,) I could not understand why my companion Mono should feel so strong an interest in seeing the spot where it fell. That his interest in the result however was intense was manifest as we drew nearer and nearer to the place. His whole frame was agitated in the most violent degree; and his utterances to himself were deep and passionate, though they no longer seemed to be in the language in which he and I had conversed. We found the hill designated by Pallas, and though the designation of the locality was too vague to guide us to the exact spot, yet I thought it not unlikely that we might determine the place with precision, by finding in the neighbourhood some fragments of the transparent mineral—olivine or peridot, the mineralogists have called it—which was found in the cavities of the cellular iron mass, and of which some globules remain in the existing specimens, while others have fallen out loose. On my suggesting this to Mono he responded with vehemence—"Yes, yes find me some of the minerals which have come in company with the iron mass, and you will make me supremely happy." Though somewhat surprized at this vehemence, I sought about with great industry ; at last, in a small plain near the top of the mountain, I saw a deep cavity such as might have been made by digging. The sides of this cavity had recovered from all trace of the violence which had produced it, and were already grass-grown or moss-grown; but it might still be the indentation in the surface which had been made by a large mass falling with a great velocity a hundred years ago. I sought with the most careful scrutiny the sides of this cavity ; and sure enough, I did find some grains of a transparent glassy substance exactly resembling the grains of olivine which I had seen in the cabinets of mineralogists. I exclaimed to my companion: "Here without doubt, is the place where the stone fell: here are grains of olivine :" he cried out with extreme vehemence : " Ha! olivine!— that is well. But asbestos? do you find no asbestos? Look well in the neighbourhood of the place. If you can find asbestos, I may be happy. I may no longer be a desolate solitary exile. I may once more see my beloved brother. I may once more have some one with whom I can speak of my beloved home: which alas I shall never see more." I was altogether at a loss to conceive how these consequences could be connected with our finding asbestos in this place; but I sought with as much diligence as if I had seen a reason for this examination. I was obliged to announce to Mono, however, that though I found many more grains of olivine, I could discover nothing like asbestos. Mono seemed to acquiesce in this as if the opposite result had been a greater good fortune than he could reasonably expect. "No," he said. "It was not to be hoped! the chances were too much against my finding the place where he fell. And if this had been the place, so great a quantity of asbestos could not have disappeared. There must have remained fragments of the cords and of the sheets which were used. No—my last hope is gone. I shall never see my brother again!"

These exclamations, and this apprehension of consequences, utterly perplexed me. What connexion there could be between our finding masses, or, as he said, ropes and sheets of asbestos, there, and his seeing a brother again, I could not conceive. But whatever might be doubtful, Mono's grief was real and profound. He lay on the ground in a state of the most abject despondency and despair; sighs and groans bursting at intervals from his breast, and his noble head buried in his delicate hands. During our intercourse I had really come to be exceedingly fond of my companion, and to think very highly both of his intellect and of his heart. I was desirous of consoling him, so far as it might be possible to do so; but unless I knew something more of the circumstances to which he referred, no attempt at consolation was possible. I ventured to enquire therefore how it was that he saw in the circumstances of our new discovery anything to call forth such grief as he shewed. I said: "I have seen enough of you, my dear Mono, to be convinced that there is reason and right feeling in your present emotions. But they are to me so incomprehensible, with my present knowledge of your condition and history, that I cannot attempt either to comfort or to help you. If you would confide to me the story of the misfortunes which you lament, you would at least have a most sympathizing listener. Whether you would have a helpful friend, I cannot tell till I hear what you need: but at least no endeavour of mine would be wanting: this at least I trust you will believe."

This address seemed to soothe him a little, and he said in the solemn tone which he sometimes assumed: "Child of Earth, I thank you for your sympathy. From you—from any one now — there can be to me no comfort and no help. But to creatures whose happiness depends in any degree on sympathy —and mine does so no less than yours—it is something—it is a softening of irremediable ills — to have drawn forth the sympathy of a being so far removed in origin and nature as you are from me. I speak not to disparage you in comparison with myself. But there *is* a vast and strange distance between your origin and your nature and mine: a distance so strange that you have never known or suspected the like, and even now I doubt whether I shall, by revealing it to you, run the risk of awakening your incredulity, and if I do not do that, of turning your sympathy into repugnance; for what sympathy could you feel with a being of another species; a monkey for example? And as for your disbelieving my story—I feel that however confidingly you may now feel towards me, what I could tell you, and must tell you to enable you to understand my misery, would go beyond the limits of the belief which you are willing to accord to me."

I replied: "Do not doubt me so much. As for your differ-ence of origin and nature, it cannot be of such a kind as to disturb the admiration and affection which I feel for you. After what I have seen, if your modesty will allow me to speak so, of your clearness of intellect and goodness of disposition, if I were to think that you could be of a different species from myself I should incline rather to suppose you of a higher than of a lower nature. And as for the incredible nature of your story— what can you tell me so strange that I shall not believe it? What regions of the earth have you traversed ? With what adven-tures ? With what sufferings ? You speak of a brother whom you have loved, and whom you have now had reasons for despairing of finding. When did you leave him and how? Tell me something of him. When did you see him last? How did you part from him?"

Mono appeared deeply agitated and almost convulsed at these questions of mine. He said: "I shall task your confidence in vain if I tell you. It is better that I should die here upon this earth, as I must die, without alienating the heart of the creature who has been more a friend to me than any other child of earth. Ask me when I last saw my brother. Shall I venture to utter it to you? Why should I not? I can at this moment point out to you — so that you can see it — the place where that parting took place, though it is a place where you will never come, and where I, alas! shall never be again. Will you believe me if I really show to you that fatal spot where my beloved brother left me. He ardent in the pursuit of know-ledge, fearless of danger, entered upon a voyage such as you have never conceived. I, devoted to him, resolved to try to rejoin him or to perish in the attempt. I can see the region —the very spot—where this tragedy took place; tragedy as it now seems to me—a gay and hopeful adventure as I then thought it. Then O then, brother beloved, was the greatness of your character seen! Then it was that I last saw your dear form! Never alas, now to be seen again by me."

As he said this he gazed steadfastly and intensely at the moon which was shining above us. I confess that I was in no small degree bewildered. That he had parted with his brother in the moon, literally and physically speaking, did not occur to me as a rational story. I supposed, in spite of the direct and positive expressions which he used, that this was some figurative mode of speaking:—that he had made some sentimental parting by the light of the moon, as lovers and as friends sometimes use to do; and that this was what, in his phrase, he meant to describe. I said to him,

"It is true, Mono, that the moon shines alike on all parts of the surface of the earth; and your brother, if he be alive, may at this moment be gazing upon it as we are, and thinking as you are of the moment when you parted from him in its light. But I meant to ask you to determine the place of your parting on the surface of the earth. In what region did it take place? In Europe, Asia, Africa or America? What countryman are you? If you have no objection to give me some insight into your past history, it would be a great pleasure to me to know something of the country and travels of one for whom I have conceived so great a regard." He replied:

"My friend, you must take my words literally, strange as they may seem to you. It is not without a struggle that I have resolved to impart to you my history wonderful as it is. Do not let me have the additional pain of finding you disbelieve or misunderstand me when I tell it. You ask me from what country I am—I am from the moon. I am a native of the orb which you see yonder in the sky. There was I born. From thence did I come. There did I part with that dear brother whom alas! I shall never more see."

I still could not believe that a sane man, as Mono seemed, was telling me seriously and gravely that he came from the moon. I tried to find some figurative interpretation of a statement which seemed so absurd. But I thought it best to learn what this figurative meaning might be, by urging upon my companion the absurdities which were involved in the literal signification of his words. I was not without a suspicion that Mono, intelligent and even wise as he appeared to be on other subjects, might be a monomaniac labouring under a delusion on this point. So I said:

"And so you really come from the moon, Mono! But you will not wonder that I find a difficulty in believing this. You know the vast difficulties which a voyage from the moon hither must involve. How did you overcome them? How were you conveyed from thence to this earth? How did you acquire a velocity sufficient to carry you hither? How did you survive the shock which must have taken place when this great velocity was given to you. It must have been you know a more violent shock than that of a person shot from the mouth of a cannon. How did you survive the still greater shock of falling upon this earth at the end of such a flight? How were you supplied with air to breathe during your flight? If you consider that these and other points of difficulty naturally offer themselves to my mind, you will not wonder that I have some scruples about the truth of your account."

Mono smiled sadly and said: "Of course it is fit that I should satisfy you on these points, and when I resolved to impart to you my origin, I made up my mind to give you information on all these points. But my narrative will necessarily be of some length, and will involve many matters which you do not know, and which you must accept on my credit. I must beg of you, therefore, to give me a patient and candid hearing; and when you have taken into your mind the general outline of what I have to tell you I shall be glad to answer any questions of detail which my general narrative may suggest. Listen, then, to the tale of a Native of the Moon."

You know from your astronomical calculations, that the hemisphere of the moon which is turned from the earth, and is always invisible to you terrestrials, is a cavity twenty miles deep. This cavity is filled with air, so that it can be occupied and in fact is occupied, by creatures not much different from the men who inhabit this earth, of whom I am one. You will easily understand that the surface of this hemisphere thus depressed below the remaining surface of the moon appears to us, the inhabitants, to be surrounded by a ring of mountains twenty miles higher. These mountains like your highest mountains on the earth's surface, rise into regions of great cold, and then, like your terrestrial mountains, are clothed with perpetual frost and snow. Each hemisphere of the earth, the northern and the southern, may be considered as a circular region of warmer temperature, with a pole of cold, and a region of ice and snow in its centre; the outside hemisphere of our moon on the contrary is a circular region of warmer temperature surrounded by a border of mountains of ice and snow, to which your Alps are hillocks, and by a degree of cold to which your polar cold is moderate. For not only does our ring of mountains rise into the region of perpetual ice and snow, but far beyond it, into a region where our atmosphere ends, and of course the ice and

W.   25

the snow end with it. The atmosphere does not rise so high as the tops of our mountain border. And accordingly your astronomers have ascertained that beyond that border, on the side of the moon visible from the earth, there is no air, no water, no ice, no snow. The surface is as your telescopes show it, a surface in the condition in which its igneous origin has left it—a vast volcanic region; one of your writers has called it, not without some reason, a huge cinder.

Wonderful is the contrast between the blank, burnt up, cinderlike, desolate, airless, side of the moon which alone is visible to the inhabitants of the earth, and the other side of the moon, with its varied surface covered with vegetable life and tenanted by animals and by *moon-men*: for by that name I think I may best describe to you the race to which I belong, approaching so near as we do in our nature to you human beings on this earth. Wonderful I say is the contrast between the death on that side and the life on the other side. We may call them for brevity the dead and the living sides of the moon. On the living side we have as you have, hills and dales, rivers and valleys, clouds and showers, and indeed all the variety which you have on the earth; except that inasmuch as the moon's axis is very slightly oblique to the ecliptic instead of being greatly oblique as the earth's axis is, we have no difference of climates corresponding to your tropical and polar: the whole surface with us is in the temperate region. But while we have all these beautiful things on the living side of the moon, there is on the other side an object more striking than all these; namely, your earth, hanging constantly in the sky, so much larger than our moon is in your sky, and constantly keeping the same place—vertically over the centre of the dead side of the moon,—resting upon the horizon or slowly heaving there, to those that live among the border mountains, and changing its phases month by month, as our moon does to your eyes.

You have yourself studied this matter, and have described the appearance which your earth presents to the inhabitants of various parts of the moon; and your account is so exact that I need not add anything to it nor dwell upon the subject.

But there remains this to be added. You have described the

appearances which would be seen by the inhabitants of the hither side of the moon, if there were any inhabitants. But this is the dead side of the moon. There are *there* no inhabitants. In that airless region they cannot live.

How then, you may ask, do I presume to confirm, as a witness to the facts, the inferences which you have drawn from astronomical theory?

To answer this, I must give you a portion of the history of us moon-men. We, like you, have cultivated astronomy with great success. Ages ago our astronomers (inhabiting the living side of our moon) had made out the motion of the body on which they lived. At first as with you, the ancient astronomers had held that the moon which they inhabited was fixed and immovable, and that the sun revolved round it, bringing to us our day and night of 30 of your days. It was held that the sun moved in a circle round the moon, in the course of a year: and then it was discovered that the path was not a circle but an ellipse (for we, like you and before you, had our science of conic sections,) and the excentricity of the ellipse and the position of the axis were made out by means of instrumental observations of the sun, and the laws of the elliptic motion were determined. But when instrumental observation became more exact, an irregularity or rather a regular inequality of the sun's motion was determined. It was found that when allowance had been made for the elliptical motion of the sun, there remained still to be accounted for an oscillatory motion of the luminary of which each oscillation occupied one of our days. The sun, besides his regular motion in his orbit, was found to go back and forwards during each day by about one-third of a degree before and behind his mean place. Of course the obvious way to account for this, or rather the exposition of the fact was that the sun revolved in a cycle round his mean place, while the mean place was carried round the moon, the center, by the elliptical motion.

But in the course of time we had our Philolaus, our Copernicus, as you had yours. A grave philosopher arose among us moon-men, and taking into account the motions of the planets, which we had for centuries carefully observed, taught us how much more simple it would be to suppose that all the planets, and our moon among them, revolved about the sun, each in its elliptical orbit, than to ascribe the motion to the planets about the sun, and the sun about the moon, as we had been accustomed to do.

I must say to the credit of us moon-men, that this heliocentric system was introduced among us with far less of struggle, anger, and persecution than it had to make its way through among you terrestrials. Indeed, I think I may say in general that we are of a more philosophical character than you are. But of this perhaps I may speak hereafter.

But the sun being thus at rest, and our moon and the planets revolving round him, what did our astronomers say of that daily cycle (daily according to our moon days) which they had discovered in his motion?

Plainly on the heliocentric system this must now be ascribed to the moon herself. Instead of the sun going backwards and forwards every 30 (terrestrial) days the moon must now go backwards and forwards every 30 days. Instead of the sun being carried by a cycle, in addition to his elliptical motion, the moon must be carried by a cycle in addition to her elliptical motion. The observations would be accounted for if the moon, besides her elliptical motion round the sun, revolved in a circle of which the radius was about one two-hundredth of her distance from the sun.

This was a very curious astronomical discovery. Strange that the moon should thus go pirouetting round the sun, instead of following her elliptical path tranquilly and steadily. For many years there were persons who refused to believe in this pirouette motion. They said it was inconsistent with the grave and orderly character of our mother moon. They said that an irregularity of a third of a degree might be an error of our instruments, and the like. But the doctrine that our moon moved in a cycle about our mean place gained ground more and more.

And as our instrumental observations became more exact, it was found that along with the change of place of the sun by this backward and forward motion, there was also a very slight change of his apparent diameter, such as would arise from his being nearer and further off alternately by about one two-hundredth of his distance, or by the moon being nearer to and further from the sun by the same fraction of the distance.

This was of course confirmation, or rather demonstration, that the moon not only had an oscillatory motion, but that she had a *circular* motion round the point which described an ellipse about the sun. This was a capital discovery, and deservedly immortalized among us the moon-man who made it.

Of course this discovery may seem to you an obvious one, for you see at once that this circular motion of the moon is her motion round the earth. But you are to recollect that our astronomers lived on the other side of the moon and never saw the earth: and I think it was very much to their credit to have thus divined its existence from astronomical phenomena. It was a step very like what happened among you terrestrial astronomers, when your Adams and Leverrier divined the existence of Neptune by feeling his operation upon the interior planet.

But in truth our moon-men astronomers did not, at least for a long time, exactly divine the existence of the Earth. What they did infer was, that in the direction of the diameter of the moon passing through the middle of the inhabited surface, there was a *point* round which the moon revolved once in a lunar day, this point being carried round the sun in an elliptical orbit.

But this point—what was it? A mere geometrical point for the moon to revolve about? This was a strange hypothesis, yet no other suggested itself.

Of course our astronomers knew that if they could travel beyond the border mountains of the living side of the moon, they would come within sight of this point round which the moon revolved in her cycle. But the inaccessibility of that region was too well established to be doubted for a moment. Even to climb the mountains covered with ice and snow, which are the first steps of that vast stair, was an undertaking which tasked the utmost powers of our moon-men. We had our lovers of mountains, like your Alpine club-men; and they had scaled many of the summits in every quarter of the border. But when they came to the limit of our atmosphere, and of course far below that, their powers of proceeding higher were exhausted.

They were obliged to give up as desperate all hope of crossing the
horizon to the other side. Yet if they could achieve this what
discoveries awaited them! This they well knew, though they
did not guess the full glories of the spectacle which awaited
them there.

This then was the problem which our moon-philosophers long
had before them, as interesting in the highest degree and on
many accounts:—to find some means of travelling on the dead
side of our moon. If this problem could be solved, some great
astronomical discovery must follow in its train, and probably
discoveries in several other of our sciences.

For we had other sciences as well as astronomy, and in some
of these we had gone further than you have yet gone. At least
there was one discovery which our chemists made, which enabled
us to solve the problem which I have mentioned, and thus enabled
me to come hither, and to have the pleasure of making your
acquaintance. This I must explain to you, for it is really the
turning point of my narrative.

You heard me inquiring whether you saw any asbestos in
the neighbourhood where this meteoric stone fell. In fact the
substance which I would fain have seen here is not the same
substance as your terrestrial asbestos, but it resembles it so much
in appearance that you would have supposed it to be that mineral.
But our asbestos has very peculiar properties, which make it
invaluable for the purpose of which I am speaking. One kind
of it is of a spongy texture, and in fact very like a fine sponge,
and has this property, as one of our great chemists in a happy
moment discovered; the spongy mass absorbs and retains the
vital part of atmospheric air with such energy, that a small bunch
of it applied to the mouth and nose enables us to breathe in a
vacuum for some hours. When this great discovery was made,
our mountain travellers were not slow in turning it to account
for their purposes. With their heads wrapped in large masses of
this spongy substance, holes being left for their eyes only, they
mounted many thousands of feet higher than they had ever
done before. They mounted in fact far above the regions of perpetual snow and ice; they ascended even into the airless space

above the surface of the atmosphere. And they soon proceeded to do this in a progressive and systematic manner. They established depôts of this vital-asbestos, at stations successively a few thousand feet above each other, and by this means in a few years they ascended to the horizon of the moon's hither surface and came in view of your earth.

It was an ancestor of my own, named Tisiri, who had the privilege of making this discovery, the greatest event in our lunar history, and I think, the greatest and most striking scientific discovery ever made by any inhabitant of the solar system. This I think you will allow if you consider for a moment. For no discovery of a new star, nor even of a comet, whose tail stretches across the sky, even if we were to come suddenly upon it, can at all compare with the discovery of a luminary thirteen times as large as the moon is to you : and then this discovery was made, not by increasing the power of vision by optical means, but by climbing into a region previously inaccessible, where the heavenly body so discovered was as manifest as the sun is in the sky. You may try to conceive my ancestor's feelings when, climbing the last ridge of the boundary mountains, he saw resting, as it seemed, upon the mountains of the opposite horizon this vast luminous orb : for it so happened that at that time the earth was *at full*, as you say of the moon : that is, her whole disk was enlightened. Tisiri, in his travels which he wrote and which we possess in our family, gives vent to his feelings in the most rapturous language. "I saw," he says, "that mysterious center which we had long felt, as an invisible power, shaping and controlling the movements of our planet, expanded before my eyes into a vast luminous orb, vying in splendour with the sun and far larger than he. Considering how vast an influence this orb exerts upon our world I was disposed to fall down and worship it, but better thoughts prevailed, and I knelt down and thanked him who made both it and our planet, for having permitted me to see it, first of moon-men, and to disclose its existence to my fellow-creatures."

After Tisiri's great discovery, the knowledge which my fellow moon-men acquired of the dead side of our moon, gradually but rapidly expanded. Depôts of vital-asbestos were established at various points, and roads constructed from place to place over the volcanic soil, till at last the whole was explored. You know the surface of our moon as it has been mapped and named by your astronomers ; and therefore I may say to you that the point where my ancestor first made his way to the hither hemisphere was on the right border as it appears to you, near the ring-formed mountain which has been called Cleomedes. From this point our explorers travelled by the two expanses which you have named Mare Crisium and Mare Tranquillitatis (for the flat spaces which you have called seas in the moon are much the easiest for our travelling), and then still further south by the Mare Nectaris till they came into the region where vast diverging ridges, proceeding from the crater Eratosthenes, occupy nearly a third of our moon's disk to your eyes. This series of discoveries which our travellers made on this dead side of the moon—the struggles by which they made their way from point to point and solved problem after problem, resembled in many ways the history of your Arctic discoveries. And you must recollect that not only had our discoverers to winter, if I may call it so, that is to spend many months in regions of cold much more intense than that of your poles, but also to live there without air, except what they carried with them as they did other food. However in the end the task was finished, as seen in its general features ; and we have our maps of that side of our moon, which of course closely resemble your best lunar maps, those by Mädler and Beer, though they contain many particulars which your astronomers had no means of discovering.

There is, however, one discovery which your astronomers made, and then, as seems to me, let it slip away from them again. I must dwell upon this because it is a matter intimately connected with my history, and indeed is the means of my being now with you here upon this earth. Upon the dead side of the moon, which is, as I have said, and as you have guessed, a region of extinct volcanic action, there are still a few volcanoes still active with long intermissions of their activity. I say that your astronomers obtained a view of this fact and then allowed it to slip away from them again : for at present, I think, they none of them hold the existence of active volcanoes in the moon. Yet

Ulloa on his journey to measure a degree of latitude in Peru saw a volcano on the moon's surface during a total eclipse of the sun (Chladni, p. 415). [I do not know to what book reference is here made; Ulloa's observation belongs, I think, to a much later date than his labours in Peru.] Herschel saw one, as he tells us in the *Philosophical Transactions* for 1787. Piazzi saw one in 1803. Schroter several times saw appearances in the moon which proved the fact. On the 28th of September, 1788, he saw on the border of the *Mare Imbrium* a luminous appearance, and soon afterwards, on the 12th of October, he saw a new crater. And again between Jan. 7 and April 5, 1789, according to his observations, two new craters had appeared within a circuit of eight miles. These observations demand far more notice, I may say more credit, than they have obtained. Our discoverers found active volcanoes, though, as I have said, very intermittent in their action, and these they carefully studied.

They found that some of these intermittent volcanoes have a considerable regularity in their action, and make their eruptions at fixed intervals, so that the eruption can be foreseen and reckoned upon. In particular, near the center of the moon's volcanic disk there is a smooth plain which your selenographers have called Sinus Æstuum, and rising from this plain is a ring-formed crater. This crater was found to make its eruptions at intervals, for some years at least, of once in five of our days (your months). These eruptions were observed with great care by our philosophers; the matters ejected were observed with instruments constructed for the purpose, and followed by the eye as far as they could be seen. The velocity with which they were projected was so great that they went off into space and returned no more. You must recollect that there is no air to impede their motions, as there is in your terrestrial volcanoes: so that even small masses, cinders and ashes, thus projected, went away and returned no more to the surface of our planet. But moreover our observers discovered that along with smaller masses, there were ejected larger ones; also it seemed, in a certain order. Every seventh eruption a large mass was observed to proceed from the crater and to mount up into the zenith till it disappeared from their vision.

That these masses were metallic and consisted in a great measure of iron, they judged from an examination of fragments which flew out obliquely with a smaller velocity and fell to the ground at a small distance from the volcano.

This working of the central crater was observed, as I have said, for several years, till there was no doubt of the fact in our minds. And then a new train of thought suggested itself. As this crater was near the middle of the side of our moon which turned towards the earth, the earth-luminary was always in or near the zenith of the place where these eruptions took place. Seeing, as we saw (for I have taken part in these observations), masses shoot vertically upwards from the crater, they seemed to be shot at the earth, and to have a good chance of hitting it. Our mathematicians, who are at least as skilful as yours, found means of calculating the velocity of these masses; and as they ascertained by this means that they would ascend to a height of a thousand diameters of our moon before losing all their velocity, they inferred as probable that they would reach the earth. They had not at that time determined the relative attractive power of the moon and the earth. Your astronomers, who have done this, know that the earth being above 60 times as great as the moon in attractive power, the neutral point at which the attractions balance each other is about one eighth from the moon of the whole distance, that is about 30,000 of your miles: and if a body projected from the moon were to rise so high as this above the surface it would from that point fall to the earth by its gravity. We were not, I think, quite so much in a state of clear knowledge as you are: but we thought it in the highest degree probable that many of the masses thus emitted by our central volcano would reach the earth.

And now I must tell you of the wonderful use which we made of this portion of our knowledge; and you will then allow I think that though you deservedly admire the boldness of the designs of your own engineers, and their skill in accomplishing them, the performances of our engineers far surpass them. It was suggested by one of our clever moon-men that if a vast sheet were spread across the aperture in the volcanic crater from which the masses were ejected, it would be carried away by the eruption, and if a rope were attached to it, might be made the means of catching the ejected mass; and if materials and powers sufficiently strong could be devised, of drawing it back. As to the materials the difficulty was overcome. Our invaluable aid the asbestos, capable as it is of resisting fire, was found to be applicable also to this purpose. Another sort than the spongy kind which was the vehicle of our breathing air, and which I have called vital asbestos—a fibrous asbestos—was found capable of being spun and then woven into sheets of vast extent and strength. It was a memorable day when one of these sheets was laid across the orifice of the volcano a little before the time when, as we knew, the ejection of a large solid mass might be confidently expected. A rope of many miles in length was laid on the ground, attached to this sheet, and coiled so as readily to uncoil when the projectile flew upwards, after the manner of the ropes which are thrown over a wreck by a gun, in the apparatus often used on your coasts. I was present at this memorable experiment; and well do I remember the event. When the time came the well known thunder of the volcano was heard rumbling deep in the heart of the mountain. The explosion took place, the sheet rushed into the air [but by supposition there was no *air* on this side of the moon] so rapidly that it was soon scarcely visible, and the rope was uncoiled with a mad impetuosity. The end of the rope had been loaded with heavy weights; and it was thought possible that these might be sufficient to retard and finally to draw back the projected mass. Our philosophers and chemists wished greatly to have such an opportunity of examining it. But the experiment, so far as this purpose went, was utterly defeated. As soon as the rope was uncoiled so far by the upward flight of the projectile that there came to be a strain upon the weights with which it was loaded, it snapt in an instant, the weights which it had scarcely raised falling back to the ground, and the projectile enveloped in its sheet soaring instantly out of sight. This experiment was several times repeated but always with the same want of success for the principal purpose.

But though we could not pull the volcanic projectile back

when it had once set out upon its journey, we found that we could, by making the rope stronger, compel the projectile to carry with it the weights with which we loaded the cord, like the tags on the tail of a kite. This we completely succeeded in doing, and several projectiles thus loaded I have seen soaring up to the zenith and disappearing there, as you may have seen a balloon; although the immense velocity with which our volcanic projectile went upwards, makes the resemblance, in the outset at least, not very close.

As I have said, it was a question with us whether the projectile, which we thus sent upward, would go to your earth or not. As your earth-luminary was in the zenith of our volcano, and as you know, subtended an angle of two degrees, all these projectiles *seemed* to go to the earth. They entered within its visible disk before they finally disappeared. Hence the prevalent opinion among us certainly was that they did go to the earth. And various were the proposals made among our philosophers, that having this means of sending to you what we had, we should impart to you some of our knowledge: send you geometrical diagrams, or some objects which might awake in you some intellectual sympathy. I believe some of our wise men did expedite to you some such objects; but I have never, since I have been on the earth, been able to hear that any of them have come to hand —have been found by any inhabitant of the earth and noticed as something different from a rude metallic mass. Of course we were aware that it would be far more difficult for you to send anything to us than for us to send objects to you, even if the material means were the same; since the force with which objects must be projected must be much greater in your case, in consequence of the greater attractive power of the earth, which makes it so much more difficult for projectiles to leave it.

But the interest belonging to such projects of communication between our Moon and your Earth was quite extinguished by the promulgation of a grand and daring project of my noble and lamented brother. I have said that we had repeatedly seen our volcanic projectiles, with various masses attached to their tails, soaring into the zenith like balloons, and disappearing only when within the visible disk of the earth. My brother dared to conceive the possibility of using these projectiles as your Mr Glaisher uses your balloons; of ascending in, or rather attached to, one of them, and in this way actually and bodily visiting your earth.

Of course the difficulties belonging to such a scheme are manifest and most formidable. You mentioned some of them when you provoked me to give you this account of myself. How was the traveller through the void space to be supplied with air? But this problem had already been solved by the discovery of our vital-asbestos; and this material only required, for a longer journey, to be supplied on a larger scale. The whole body of the traveller must be wrapped in it, instead of his head only. How was the traveller in leaving the surface of our moon to be secured from injury by the intense shock which he must receive on being thus lifted from the surface? This difficulty could, it was conceived, be overcome by an improved method of coiling the rope so that the velocity should be communicated to the traveller, not by a single shock, but by degrees. This you, I know, are engineer enough easily to see to be possible. How was the traveller to bear the shock of falling down with so great a velocity upon the earth? This too, it was conceived, could be provided for by our unfailing aid asbestos. The great mass of this spongy material in which the traveller was to be enveloped, amounting in size to one of your ordinary haystacks, would, it was hoped, break the shock of his fall and make it harmless. There seemed to be no insurmountable physical obstacle to the execution of this plan, bold as it seemed.

My brother conceived this plan in his mind and proceeded by calculations and experiments to prove that it was, as I have said, physically possible. He himself made a contrivance by which he was enveloped in a large mass of the spongy asbestos, was precipitated from a great height with a very great velocity upon the ground, and in this experiment he received no injury. The contrivance by which the velocity was gradually communicated to the masses which the volcanic projectiles carried with them was, as I have said, a matter of engineering, and was entirely successful. The vital asbestos was, by various chemical applications, made capable of supplying the traveller with nutriment, as well as with respirable air. It was argued by him, and with great truth, as I afterwards found by personal experience, that when the great velocity was once acquired, the traveller moving through a space without resistance, with a velocity changing by slow degrees, would feel no more shock or shake, than a traveller in one of your rapid railway trains; indeed far less: for there would be no jar of wheels or oscillation of springs, or winds or showers or dust. The voyager through blank space would move smoothly on with no more consciousness of motion than the inhabitant of our planet or of yours has of his motion round the sun. My brother was enthusiastic and impressive in explaining to our moon-men how smooth and untroubled the motion would be; and, as I have said, he had proved by experiments that the arrival at your earth and the falling upon it might be accomplished without any violent destructive shock. He invited in the most passionate manner some scientific adventurer to undertake, by this means, a voyage to your earth, promising him an easy passage and a safe arrival.

But it was not easy to find any one whose enthusiasm would go to the length of undertaking such a voyage. Indeed this you cannot consider wonderful, if you consider what the consequences would be. The explorer of the region to which he was thus invited must be regarded as leaving for ever his native home. There was no chance of returning: for though some persons held that the inhabitants of your earth, inasmuch as it was greatly larger than our moon, could probably have surpassed us in their engineering skill and would be able to send him back, the greater part of our moon-philosophers held that this was not at all likely, and that once arrived there, even if uninjured, he would never be likely to return. It was not the adventure of an explorer who is to return and reap his harvest of glory among his countrymen, but the desperate venture of a person who must depart at once and for ever, with no prospect of consolation except his joy in the knowledge which he would himself acquire. It was not wonderful that my brother found no one to venture upon this forlorn hope.

Long and long did he urge the claims of this avenue to new knowledge upon the lovers of knowledge among us. He made many speculative converts to the belief in the success of his plans, such success as he promised, but no practical believer, who would undertake the adventure. The retort was very obvious. If you think it so important that some one should go, why do you not go yourself? And in truth this retort had no small force, for besides that my brother was of all moon-men the most deeply impressed with the conviction of the possibility and importance of this scheme, he had fewer ties than most of them to prevent his undertaking it himself. We—he and I—were a good deal alone in our world. Our parents were dead, and we were neither of us married. It is true he had numerous admiring friends who looked upon him with a sort of enthusiasm on account of his knowledge, accomplishments, and the vigour of his character; and with some of them a final parting would be very sorrowful: but when he had once propounded his scheme of a voyage to the earth-luminary, and had so confidently shown that it was possible, some of them began to look at his carrying it into effect with complacency, as his proper destiny; and were prepared to regard his removal from among them as a sort of apotheosis, which would leave a recollection of him as something above moon-manhood. And as he was really full of courage and an ardent lover of knowledge, these aspirations at length took possession of him also, and he resolved to carry into effect his own plan of an earth voyage.

The preparations were made in the most complete manner and on the most ample scale. We reckoned that the voyage to the earth would occupy about one-third of one of our days, ten of your days. The mass of vital-asbestos which was prepared, which, as I have said, supplied the traveller with nutriment as well as with air to breathe, was such as would suffice for a hundred of your days, though it seemed scarcely possible that more than ten days supply would be needed: for if the voyage was successful, he must reach the earth in that time. A few hours more or less, in consequence of the possible error of our calculations, was all the difference which could occur. If, indeed, the projectile discharged from our volcano should miss the earth altogether, the prospect was very awful. The traveller must go moving on through the cosmical spaces between the planets, carried by the meteoric mass to which he was attached, and without any chance of coming to a resting place. He must in that case needs perish by famine; and his body must go revolving round the earth as long as its substance should endure. But we did not much dwell in our thoughts on this possibility; we were confident in our calculations, and thought we could not fail to hit the earth. On what part of the earth, indeed, the traveller from our moon might fall, we could not venture to predict. We were aware of the possibility of his falling into water; for in our planet too there is water; and we could see the earth, as I have told you, with sufficient clearness to judge that its surface was partly solid and partly fluid. But the asbestos-mass had in it contrivances by which the voyager, if he fell into an expanse of water, would float and would be enabled to reach the shore.

Deeply engraved in my memory is the day when my brother having made all preparations for his voyage, entered his asbestos chariot and waited for the predicted hour of the volcanic explosion. The well-known thunder was heard beneath the ground; the explosion came; the sheet which lay across the orifice was lifted up and carried towards the zenith with immense velocity, and rapidly diminished in visible size as it rose. The connecting rope which drew the asbestos car gradually communicated to it its velocity. It moved at first slowly, but rapidly rushed into greater speed, and was soon clear of the ground and mounting toward the zenith, amid the breathless gaze of our moon people. It was soon lost to our gaze, and certainly seemed to our eyes to have plainly entered into the earth-disk. We could see it no longer. We saw it no more. But just as it vanished in that disk, we saw something—a small object—falling, as it seemed from that point vertically downwards. This object fell not far from the place where we were who had witnessed the departure of the traveller. It was picked up and was found to be a plate of lead, such as is commonly used for writing on with a steel pen: upon this sheet of lead was written, in my brother's well-known hand—

'To Mono. The travelling is pleasant. Success certain. Shall I ever see you again? Makomo.'

You may imagine how the feeling of awe which my brother's departure had inspired in all of us was deepened by this communication from him, after he had, as it were, entered upon another world. To me it sounded like a summons to follow him, nor could I get rid of that impression till I obeyed it as you see; for you see me here.

Of course we heard and saw no more of Makomo. The philosophers, who had come from the living to the dead side of the moon to effect or to witness this great undertaking, returned over the border to their usual abode on the living side of the moon, and there of course we did not even see that part of the sky into which we had seen Makomo enter. The grand orb of the earth was then no longer visible.

My loss of my brother hung heavily upon my heart. We had lived and worked and speculated together, so that my life was utterly marred by losing him. I could no longer work. I could no longer speculate about anything except his fate. I could no longer think of anything except him. And the sense of desolation which his removal left upon me was augmented by my not being able in my usual habitation on the living side of the moon to see the orb which was, I knew, his present abode. I pined for the sight of the earth-luminary. I made provision for mounting into the dead hemisphere. I travelled thither, and wandered with a melancholy gratification over that volcanic surface; having the earth-luminary shining above me, going through its various phases month by month. I travelled over a large portion of the volcanic hemisphere, so that I saw the earth-luminary in all those positions which you have well described: slowly heaving in the horizon among the mountains, steadfast in the zenith, or fixed no less in some intermediate position. But ever as I gazed on the luminary my thoughts of my brother became deeper and deeper, more and more impatient, and at last I was quite overmastered by the desire of following my brother, and, like him, seeing this earth. Such an act was like the suttee of widows which I found prevailing in India:—a resolution to follow where my beloved had gone before. Like that, it was an expression of indomitable love—

in me, a motive most real, however often it may be otherwise
in the widow's case, but in my case, as in the widow's,
supported by a considerable force of public opinion; for many
persons thought it was fit and graceful that I should follow my
brother in that path which had been marked out by our
common speculations and our common decisions. It was, no
doubt, a sort of suicide; but it was a suicide which offered a
considerable probability of rejoining the beloved person whom
I had lost. And I was haunted by the thought that his
volcanic car might have missed the earth, and that he might
be wandering on through the blank spaces of the universe, never
finding a resting-place. And though, if this were so, my following
him could not be of any use to him, it seemed to be a sort of
consolation to follow his fate whatever that might be; and at
any rate, such a course offered the only chance of my ever seeing
him again.

These feelings impelled me and determined me to follow my
brother to the earth. I adopted the same means which he had
used—the foreseen volcanic explosion; the asbestos sheet with
its appended cord; the vital asbestos provision for air and food;
the contrivances for departing and arriving at the end without
violence. I had all this apparatus worked by our best philosophers, and I set out upon my voyage upwards amid the
acclamations of thousands of moon-men from the middle of that
disk which you now see shining above us.

I need not describe to you my sensations in rising from the
moon. You will easily conceive them. They were, as I have
said, very nearly the same sensations as those which are experienced by a traveller on a very rapid railroad: the gradually
increasing velocity felt at first—the great velocity afterwards
acquired, which is even in railways almost insensible, became
more nearly insensible in my case, because there was no jolt
or resistance of any kind. I spun or rather glided through
space, feeling as if I were at rest. My breathing went on
tranquilly; my nutrition regularly; everything was provided
for in my apparatus. The principal inconvenience which
I experienced was from the sun which shone upon me in unbroken

and unintermitted splendour, with no shade from any solid mass
or any spreading cloud. It is true the pack of spongy asbestos
in which I was enveloped protected me in a great measure from
the heat of the sun's rays as it did from the cold of the sur-
rounding vacant space: but, as I have said, the asbestos-pack had
two perforations for my eyes, and when the sun shone in at
them the light was blinding. You will easily conceive that
my asbestos did not move quite steadily without some angular
motion. By inevitable irregularity in its form and in the
impulse which it had received, it had a motion of rotation;
and the axis of this rotation was of course the line which connected my pack with the volcanic projectile. About this line
my pack revolved, as I conceive, about twice in each of your
days; and as my expedition was made, as you terrestrials
would say, at the end of the moon's first quarter, that is when
the line of sun and moon was nearly perpendicular to the sun's
distance, the sun's rays shone into my eye-holes during a portion
of each revolution. In this way I had a sort of day and night
every six hours: for when the sun was at my back the stars on
the opposite side of the sky were visible, there being no atmosphere
to carry the sun's light among them and extinguish them, as
there is in the daytime of this your earth. And so I spun
on through the space which intervenes between yon moon and
this earth, and, I believe, reached the bounds of your atmosphere
about the twelfth day after I left the moon. I had found no
inconvenience for want of sleep during this time, for, as you
know, I had always been accustomed to a day of thirty of your
days, and habitually worked for a fortnight and slept for a
fortnight. When I arrived within the boundary of this your
terrestrial atmosphere, I was made aware of the fact by a sudden
and considerable increase of temperature. I suspect that the
outer parts of my pack of spongy asbestos became incandescent;
but the large porous mass which intervened between me and
the outside prevented my suffering from this, by intercepting
the communication of heat: I have reason to believe too that
the volcanic mass which carried me onwards became ignited
and fired on the outside. Such, as you know, is the condition

of most of the meteoric masses which have fallen to the earth. But this operation, if it took place in the case of my vehicle, I was from my position prevented from seeing.

Once arrived in your atmosphere, we rushed downwards rapidly and, as a spectator would have thought, madly, but yet the immense velocity with which we had travelled through blank space was soon materially diminished by the resistances. No doubt the velocity with which I fell through the air was very great; but, as you know, it was limited by the resistance of the air and was very little affected by the vast original velocity which had carried me away from the moon.

Through the air, then, I fell for some fifty of your miles, the air becoming denser and denser, so that I had soon no occasion for my vital-asbestos. I came to the earth on a vast plain in this Siberia, and found that the spongy mass of which I was the nucleus was quite effectual in preserving me from all inconvenience arising from the shock of the fall.

Having thus fallen from the sky to the earth, like Vulcan in the legend of your Greeks, I collected myself as well as I could. I crept out of my envelope and found that I was not even lame with my fall, as I have since read Vulcan was. I looked round me and gazed on the scenery; it appeared to me not very unlike our living side of the moon; though if I may venture to say so, not so beautiful as ours. I saw habitations near me, and hid myself behind trees that I might see some of the inhabitants before they saw me. When I saw them I was much surprised to find them so like our moon-men; although, to tell the plain truth, your earth-men appear to me, in a general way, to be more ferine and brutal, creatures of a lower order, than our moon-men. But still I saw that you were of a kind sufficiently like to us to make my intercourse with you not impossible. I had a hope of being able to learn some of your languages by living among you, listening to you, and imitating you. I had studied the nature of language in general, so as to be confident that in this I could succeed. I went to a city in China and did, in the course of a year, acquire the language which they there spoke. I have since acquired others. I was aware that such acquisition was necessary in order that I might take any further steps in the great object of my existence, the attempt to discover my brother on the surface of the earth. In the pursuit of that object I am now here. I do not know whether my narrative, so far as I have yet carried it, has removed your incredulity as to my really coming from the moon. If you have still any remaining scruples on that head, or if any difficulties occur to you with regard to the facts which I have very plainly and simply related to you, I think it will be better that you should mention your difficulties now, before we go to other matters; for I have other things to tell you in which I think you will be interested; and I have a great desire to discuss, with a person so intelligent as you are, the difference between your world and ours. I pause here, therefore, that I may listen to what you have to say.

CHAPTER XXI.

CONCLUSION.

I UNDERTOOK to trace the literary and scientific career of the late Master of Trinity College, and this task I have now accomplished. Others who knew him more intimately will describe his life as it was passed in the society of his friends and in the discharge of his official duties; and it is with some hesitation that I offer a few remarks which may seem to pass beyond the boundary traced out for my occupation.

My personal acquaintance with Dr Whewell was slight, though I met him occasionally in the course of University business. It has been handed down by tradition, and stated in print, that he was somewhat dictatorial and overbearing in conversation, though not unwilling to admit, at least implicitly, an error when decisively corrected. I cannot say for myself that I ever saw instances of the fault thus imputed to him, though he appeared to have less ease of manner than might have been expected in one who had been so much in the habit of associating with accomplished men and women. My own acquaintance with him however extended only to the later period of his life when advancing age and great sorrow had apparently subdued the vehement disputatiousness of character attributed to his early years. The slight recollections which I retain respecting some occasions in which I happened to be in his company are scarcely worth preserving, but I am glad to record that I sometimes enjoyed the benefit of his conversation.

I first spoke to him when undergoing examination for the Smith's Prizes; he was standing in Trinity Lodge near the portrait of himself painted by Samuel Lawrence, and though the occasion was little suited for the indulgence of any extraneous reflexion I could not help being struck with the skill of the artist in producing a work which, though slightly idealised, was still eminently faithful to the original. I have never seen any other likeness of Dr Whewell at all comparable to this, or indeed that gave me even moderate satisfaction. As an examiner for the Smith's Prizes Dr Whewell was not popular; he had almost ceased to take a personal interest in mathematics, and his questions were derived from methods and books that had gone out of date—for there are fluctuations of taste and fashion even in the serene realms of abstract science. Moreover he sometimes proposed problems which he himself had not fully worked out, and which were in fact too difficult for complete solution. Such exercises can only be good tests for students, provided they are informed how much they are expected to accomplish, and are not left to waste their time and courage in unavailing efforts.

Many years ago I was present at an official dinner which Dr Whewell gave when he was Vice-Chancellor of the University. One of the youngest persons of the party was speaking earnestly against the current system of mathematical examination, maintaining, and quite justly, that the papers set were too long and too difficult to be finished by the undergraduates in the time assigned to them, and he asserted that the University in allowing it to be supposed that such feats were really performed was guilty of *imposture*. The speaker was animated, and owing to this and to a sudden pause in the general conversation his last word was distinctly carried through the room. Dr Whewell who was at some distance turned towards the speaker and, addressing him by name, said slowly and gravely, "You are using very strong language with respect to the proceedings of the University;" then after a long pause he added to the relief of the audacious critic, "but I think not too strong." At the conclusion of the dinner Dr Whewell returned to the subject and discussed it very fully. He thoroughly acquiesced in the opinion that the mathematical examinations in the University were too severe, and complained of the unsatisfactory performance of the candidates in the Smith's Prizes Examinations, especially as he said that he studiously constructed his papers by an easier standard than that of the Senate House. He expressed especial displeasure at the conduct of those who left the examination room before the prescribed time had expired, without taking the trouble to look through their own papers for the purpose of correction and improvement; as he said "they will not condescend to revise what they have scribbled in tempestuous haste, but thrust it, barely legible, into the hands of the examiners."

Perhaps he was most interesting in company when the conversation turned on the history of science, for he had been personally acquainted with many of the most eminent philosophers of the present century and had watched the course of their studies. At one of the anniversary meetings of the Cambridge Philosophical Society he alluded to its commencement and progress, and to his own share in the proceedings; glancing from the Secretary's official seat at one end of the table to the President's at the other, he said that he had himself occupied every position in the philosophical firmament from the nadir to the zenith. After the dinner he delighted some of the junior members who were fortunate enough to be near him with reminiscences of heroes of the former generation, especially of Young and of Ampère, both famous for the marvellous variety of their attainments, and then not so well known as they have since become by the publication of able biographies.

In the year 1862 the British Association met for the third time in Cambridge. I do not know whether Dr Whewell protested, as he did on the occasion of the second visit, that the burden ought not to be again thrown on a place which had already endured it, but he certainly took much interest in the proceedings. He passed with Lady Affleck from one room to another of those occupied by the Sections, and was obviously gratified by the respect which was universally shewn to himself and to her. The late Professor Boole, so well known for his mathematical and philosophical attainments, attended the meeting, and took the opportunity of his stay in Cambridge to visit the grave of Robert Leslie Ellis at Trumpington; they were men of kindred genius and learning and had been drawn by mutual esteem into close friendship. Professor Grote who then resided at Trumpington brought to Mr Boole a request from Lady Affleck, the sister of Mr Ellis, that he would call at Trinity Lodge, and accordingly he saw her and Dr Whewell. He was much gratified with the interview, but somewhat surprised to find that she did not bear any great similarity in appearance or manner to her remarkable brother. In less than eight years not one of the four whom I have named as then living survived.

Late in his life I occasionally met Dr Whewell at the Board of Moral Sciences Studies established in the University, but the business was usually only of a formal character, and I recollect only one occasion on which he expressed an opinion on a matter of interest. The question was under consideration as to what amount of attention the students should be urged to give to Kant's *Kritik der Reinen Vernunft*, and a member of the board held it to be very important that the main principles of the work should be fully understood, but not advisable to exact a knowledge of all the details with which Kant had surrounded them; to this opinion Dr Whewell explicitly assented.

The last communication which I received from him was a correction of a mistake in my *History of Probability*. I had given to Professor J. D. Forbes the epithet *late*, confounding him with his cousin Professor Edward Forbes recently deceased. From Dr Whewell's note I learned for the first time how intimately he was acquainted with J. D. Forbes. He spoke to me about my book when last I saw him shortly before his death.

Dr Whewell was very fond of argument, as may be gathered from his letters and from the testimony of those who remember his conversation. His great attainments rendered him a formidable antagonist, as he was able to draw from all departments of knowledge the evidence which told in his own favour, not in any unfair spirit, but for the gratification of hearing the best arguments that could be urged against him.

His friendships were very numerous, and were maintained with a cordiality which gives strong evidence of the depth and sincerity of his feelings. With scientific men in the early part of his life he was widely acquainted; his travels on the continent, and his constant attendance at the meetings of the British Association, afforded him opportunities of personal intercourse of which he most willingly availed himself; but, as he acknowledges in his correspondence, he did not become familiar with the younger generation, the contemporaries of his later years. He was also inti-

mate with many of the most distinguished cultivators of other fields
of knowledge, as Bunsen, Empson, Guizot, Hallam, Hare, Humboldt, Jeffrey, Sir G. C. Lewis, Macaulay, Malthus and Milman.
The letters addressed to him by his correspondents, both literary
and scientific, shew not only respect and admiration for his great
attainments, but warm personal esteem and affection. I remember
the late Professor Brandis of Bonn spoke of him to me in terms
of strong commendation; the learned philosopher called him
Sir William Whewell, and when I suggested the omission of the
prefix he appealed to Shakespeare to convince me that it was
the proper term of respect for an English clergyman. Dr Whewell
was well acquainted with Mr Everett and Mr Bancroft, ambassadors from the United States to England, and this is the more
noteworthy because his opinions on secular and ecclesiastical
matters were in many respects contrary to theirs. He must have
been also in theory almost as distant from an eminent English
friend, Sydney Smith, who once in a letter asked him, "When are
you coming to thunder and lighten at the tables of the metropolis?" The well-known saying of the wit respecting him—
"science is his forte and omniscience is his foible"—was uttered as
he was leaving a small breakfast party: two others were present,
namely Samuel Rogers, and one who still survives with the honour
due to great achievements in science.

The subjects which Dr Whewell studied have been indicated
in the accounts given of his numerous publications; but a few
general observations may be added. His letters to Mr Rose imply
that in early life he had some intention of devoting considerable
time to the acquisition of languages, a project which is naturally
formed by students in the sanguine days of youth, and his note
books shew that he was extremely fond of etymological researches
and speculations. But the pressure on his leisure and his thoughts
produced by his scientific labours checked his linguistic zeal, and
compelled him to limit himself to moderate labours in that field.
His early Latin scholarship he always maintained. One of the
friends of his boyhood, having to return thanks for an honour
conferred on him by the French Academy of Sciences, appealed
to Dr Whewell for aid in expressing his sentiments in correct

Latinity, which he held to be the proper language for the inter-change of scientific courtesies. Dr Whewell wrote the inscriptions for the bust of Mr Sheepshanks and for the statue of Barrow, which are placed in the ante-chapel of Trinity College. I once heard him complain very much of the Latin prose composition which came under his notice officially in the University, not for faults of style, but for the absence of point and meaning; to this general censure he made one decided exception, namely, the speeches of the Public Orator of the period, the present Master of St John's College. Dr Whewell was an excellent German scholar, as appears from his published translations and from the evidence of well-qualified judges still living. Humboldt in a note to him thus explains the cause of having lost the pleasure of seeing him at Potsdam: "C'est votre admirable connoissance de la langue allemande qui m'a porté malheur. J'avois ordonné de laisser entrer tout gentleman *anglais* qui se présenteroit. Vous avez parlé allemand comme un habitant du pays!!"

Dr Whewell paid much attention to poetry, but I do not find in his publications or among his papers any very definite statement of his opinions with respect to the tests of poetical excellence, though judgments on particular works are delivered. In early life he was unfavourable to the poetry of what is called the *Lake School* generally, and to that of Wordsworth in particular, but his views considerably changed in subsequent years. He was always a great admirer of Southey's *Roderick*.

His sermons do not exhibit any special theological learning, and it is curious that with his power of study and his strong bent for system-making, of which he was himself conscious, he should have been so little attracted by divinity. Some of his correspond-ents in his early days recommended the subject to him, as well suited to his taste and ability, but he did not find it congenial. This has been attributed partly to his dislike of controversy on theological subjects, and partly to his earnest pursuit of science. Those who regret the want of precision which seems to belong to many modern English sermons, may speculate on the benefit which might have followed from the example and influence of so eminent a man as Dr Whewell. It is true that religion appeals to a more important part of human nature than even the intellectual, and enjoins principles and dispositions which will endure when knowledge shall have vanished away; but at the same time the minds of some men are so constituted that they long for order and coherence in their views of theology as well as of merely secular subjects. Such a disposition is justified by the fact that not a few of the most devout Christians have been remarkable for their endeavours to secure a logical theory as well as a vivid experience of religion, and it is stamped with the unquestionable authority derived from the argumentative form of apostolic teaching itself.

There seems to have been a peculiarity in Dr Whewell's mode of composing his works, if we may judge from various fragments relating to his scientific productions and to his early sermons. When he had secured a few hours of leisure, as for instance in the course of a foreign tour, he would write a brief essay on some part of the subject at which he was working; thus it might be on *Induction and Deduction,* or on the *Nature and Use of Hypothesis.* The only results which he proposed to himself apparently were to fix his ideas in his memory, and to gain facility in expressing them; for in all probability he did not recur explicitly to the papers thus written. When he finally sat down to prepare for the press I believe he never *copied* a former manuscript, revising as he went on, but wrote afresh. Mr Disraeli, in the preface to his last novel, strongly recommends the process of copying, and confirms his opinion by the example of Lord Byron; Dr Whewell might with advantage have adopted such a practice.

His style on the whole has been generally admired, though some critics have censured it as stiff and cumbrous. Lockhart complained of the review of Lyell's Geology as shewing *haste and slovenliness* when it came into his hands, and said that he had to rewrite it himself. Perhaps if Dr Whewell had copied it for the press he might have succeeded in satisfying the fastidious editor of the *Quarterly Review.* For my own part I think his writing is very good, vigorous enough to interest, and so clear as to prevent any misunderstanding; I except, however, the *Elements of Morality,* in which the author's power seems for a season to have deserted him, though it returned in all its vigour in the *Essay* on the Plurality of Worlds. His early letters are remarkable as exhibiting frequently three closely written pages with scarcely a correction or an erasure. Sir J. Herschel has well described the most obvious features of Dr Whewell's style in a passage towards the close of his review of the *History* and *Philosophy.* "Its chief characters are a remarkable occasional point and felicity of expression, and the almost systematic adoption, as a mode of illustration, of a great assemblage and variety of metaphorical allusion, much greater indeed than we should like to see adopted by an author less thoroughly imbued with his own meaning, and less capable of curbing the exuberance of a brilliant fancy into an entire subordination to his reason. We say systematic—for we have no doubt that it is intentional; and the object, moreover, is attained, the convergence of illustrations from so many different quarters rendering it perfectly impossible to mistake the point to which they are directed." Dr Whewell's publications and correspondence shew that he made a point, especially in his early years, of cultivating a Baconian habit of using simile and metaphor, and although this is not carried to excess in his own case, yet there are indications that he sometimes set too high a value on the productions of others on account of their exhibition of this peculiarity.

I touch but lightly on a subject which is, strictly speaking, beyond my province. In politics Dr Whewell in early life was not altogether averse from the Whig party, which included some of his firmest friends, as Sheepshanks and Peacock; but he became in the end a Conservative, though not in the strict sense of the term. It has been said that the decline into Conservatism is natural to Englishmen when they marry and settle in life, but marriage in this case can scarcely have conduced to such a result, since Mrs Whewell came from a family of decided Liberal principles. But he was never prominent in general politics. It was far otherwise in what may be called academical politics; in later life his great energy and influence were strongly exerted in resisting proposed reforms, with some exceptions which I have noticed on page 215. He himself regretted that he was thus compelled to oppose many friends whom he had warmly esteemed during his whole residence at Cambridge, and it seems to have escaped his suspicion that the change might be rather in himself than in them ; for in his early life he had been in favour of measures which subsequently he disregarded or opposed. In his own phraseology he finally regarded only the *Permanent* to the neglect of the *Progressive.* Such a change of opinion with him may, however, have been more apparent than real; he may still have felt that there were many faults in the academical system, but may have lost confidence in the methods proposed for correcting them, and it is only natural that hope and boldness should give place to doubt and caution as " progressive years strip from our life the illusions of its golden dawn." His historical tastes and studies would easily lead him to dwell more on the certain facts of the past than on the precarious promises of the future, and the annals of the University presented glories which were almost unrivalled among the academic institutions of Europe. From Cambridge had proceeded the reformers of religion, the chiefs of science, and the philanthropists who, in more recent times, had broken the fetters of the slave. Even the most ardent advocate for progress, though justly convinced that the best method of preserving our institutions is a wise adaptation of them to the circumstances of the age, warding off revolution by reform, may yet, when he looks back on such a noble history feel some misgiving, and hope rather than believe that the future will be not unworthy of the past.

Perhaps one of Dr Whewell's own philosophical doctrines might have been urged against his steadfast resistance to change in the University, namely, that ideas originally very imperfectly apprehended become clear in the course of time and study. It might have appeared obvious to an enquirer into the course of academic opinion that there was a sure, even if slow, approach to the recognition of important principles formerly doubted or denied—such as the inexpediency of compulsory celibacy, the impropriety of compulsory orders, and the injustice of excluding youths from the advantages of the University on account of the religious opinions of their parents. Thus the advocates of reform might have claimed some sympathy from Dr Whewell on the ground of his own published doctrines, for otherwise he would have fallen within the range of his own condemnation as recorded by Mr F. Newman: "The late celebrated Professor Whewell deliberately asserted that the great weakness of the English intellect was its inability to trust to broad principles." In some of the cases in which I should venture to differ from Dr Whewell's course of action, it is not improbable that he followed the advice of others who were not the best judges of the circumstances. Friends eminent in various branches of knowledge, to whom on their own subjects he had often referred with great advantage, were occasionally consulted, or offered their suggestions, on matters to which they had not given any special attention.

But although Dr Whewell was strongly opposed to the Royal Commission that visited the University, and to the reforms which were in consequence introduced, yet he loyally accepted the new constitution. He soon found that no evil results followed; nonconformists availed themselves of the partial removal of obstacles and entered the University, some of them selecting Trinity College itself, and they soon shewed, by exemplary conduct and by eminent ability, that their presence was nothing but a benefit to all. Dr Whewell himself could not help being struck with the fact, that in his own college those who advocated reform proved their sincerity and earnestness by regular attendance at the meetings held for the consideration of new statutes, and other legislative improvements, at great sacrifice of time and convenience; and he always appreciated the merit of a strict discharge of duty.

During the last years of his life he had quite overcome the unpopularity which once attached to him, he had lost much of the imperious combativeness which is said to have characterized his early years, and he was universally received with the respect justly due to his character and attainments. Such a man conferred honour on the whole University as well as on the great college of which he was the Head. He served the University faithfully and had promoted her interests, according to his best judgment, throughout his long career. Not the least of the many benefits which he conferred was the example he set of unceasing labour, for this was a permanent rebuke to that indolence which is the besetting failing of the place—not the grosser form of aimless waste of time, but the more seductive error which consists in the mere acquisition of knowledge which is never reproduced for the benefit of others.

The question whether his writings will live is frequently suggested with respect to an author who has occupied much attention during his own generation; and it must almost always be rash to predict with confidence that they will. For amongst the thousands of volumes which the press pours forth annually the chance of remote survival for single productions can be but small. Scientific research especially, from its very nature, though permanent in influence is transitory in form; the light which the philosopher displays guides to other investigations which far outshine his own, and leave his name in comparative obscurity. Much of Dr Whewell's own labour was of this kind; and he would himself have readily admitted that men must be content merely to hand on the torch of science to their successors. But whatever may be the fate of his works his memory cannot be forgotten in the University which he loved so well, and where there will always be a memorial of his munificence in the Professorship and Scholarships which bear his name. A strong testimony to his character and attainments is furnished by the opinion of those who knew him intimately. Few men now survive who can be strictly called his contemporaries—but some there are, eminent themselves, who though rather younger in years were associated with him at the time of his most vigorous power, and who speak with enthusiasm of his universal knowledge, his readiness to undergo any labour for the advancement of science, his punctual and zealous discharge of all that he thus undertook, and his generous sympathy with the pursuits and successes of his friends. One who was thoroughly acquainted with him, and whose own long career has been conspicuously bright and honourable, when lately recalling the incidents of that career, said that he had known many famous men but on the whole Dr Whewell was the foremost of them all.

THE END.

CAMBRIDGE : PRINTED BY C. J. CLAY, M.A. AT THE UNIVERSITY PRESS.